AF775003

POLYMER
OPTICAL FIBERS

POLYMER
OPTICAL FIBERS

Edited by

Hari Singh Nalwa, M.Sc., Ph.D
Formerly at
Hitachi Research Laboratory
Hitachi Ltd., Ibaraki, Japan

AMERICAN SCIENTIFIC PUBLISHERS
25650 North Lewis Way
Stevenson Ranch, California 91381-1439, USA

AMERICAN SCIENTIFIC PUBLISHERS
25650 North Lewis Way, Stevenson Ranch, California 91381-1439, USA
Tel.: (661) 254-0807, Fax: (661) 254-1207
E-mail: order@aspbs.com
URL: http://www.aspbs.com

Polymer Optical Fibers edited by Hari Singh Nalwa,

The image on the cover of this book is adapted from Chapter 5 Polymer-Based Integrated Optics Devices by A. Driessen, A. Leinse, S. Musa, R. M. de Ridder and Sulur with courtesy of all authors.

This book is printed on acid free paper. ♾

Copyright ©2004 by American Scientific Publishers.

All Rights Reserved.

No part of this publication may be reproduced, utilized, or tansmitted in any form or by any means, electronic or mechanical, including photocopy, recording, or by any information storage and retrieval system, without permission in writing from the publisher.

Authorization to photocopy for internal or personal use may be granted by American Scientific Publishers provided that $0.50 per page photocopied is paid directly to Copyright Clearance Center Inc., 222 Rosewood Drive, Danvers, MA 01923, USA. The fee is subject to change without any notice. American Scientific Publishers consent does not extend to copying for general distribution, promotion, creating new work or resale. Specific permission must be obtained from the publisher for such copying.

The information provided in this book is compiled from reliable sources but the contributing authors, editors, and the publisher cannot assume any responsibility whatsoever for the validity of all statements, illustrations, data, procedures, and other related materials contained herein or for the consequence of their use.

Library of Congress Catalog Number: 2002104601
International Standard Book Number: 1-58883-012-8

PRINTED IN THE UNITED STATES OF AMERICA
10 9 8 7 6 5 4 3 2 1

CONTENTS

Chapter 4. Polymer Optical Fiber Gratings

G. D. Peng and P. L. Chu

Chapter 5. Polymer-based Integrated Optics Devices

Alfred Driessen, Arne Leinse, Sami Musa, René M. de Ridder, and Sulur

Chapter 6. High Capacity Transmission Systems

Giok-Djan Khoe, Henrie van den Boom, and Idelfonso Tafur Monroy

Chapter 7. Plastic Optical Fiber Passive Devices

Liliana R. Kawase

Chapter 8. Second-Order Nonlinear Effects in Optical Fibers

I. C. S. Carvalho, W. Margulis, and B. Lesche

PREFACE

Polymer optical fibers (POFs) are in a great demand for the transmission and processing of optical communications compatible with the Internet, which is one of the fastest growing industries of modern times. Polymer optical fibers also have potential applications in telecommunication and high capacity transmission systems, WDM systems, gigabit Ethernet, power splitters and couplers, amplifiers, sensors, filters, lenses, scramblers, integrated optical devices, frequency up-conversion, etc. For practical applications in industry, many aspects of polymer optical fibers such as different classes of polymers, glass optical fibers, fiber fabrication techniques, laser sources used for optical fibers, mechanical properties, photosensitivity, poling and second harmonic generation, adhesion, donor–acceptor properties of surfaces, protective coatings, water resistance, stress corrosion, mechanical stability and deformation characteristics, temperature and pressure sensitivity, fiber elongation, optical fiber attenuation, planar optical waveguides, fiber gratings, refractive index distribution, etc., are discussed. This book summarizes such research activities on polymer optical fibers and their applications.

Chapter 1 by Chang-Won Park focuses on different fabrication techniques for step-index- and graded-index (GI) polymer optical fibers, convective diffusion in various coextrusion methods, and bandwidth estimation. Chapter 2 on gradient-index POFs by Chen et al., reviews multiplayer coextrusion, centrifugal field process, swollen-gel polymerization and theoretical analysis of refractive index distribution and bandwidth. Chapter 3 on gradient refractive index distribution of rare-earth doped POFs by Qijin Zhang outline qualitative description of graded index distribution forming process interfacial-gel polymerization, free-volume theory, and refractive index distribution in undoped and rare-earth doped POFs. In Chapter 4, G. D. Peng and P. L. Chu reviewed basic principles of optical fiber gratings. This chapter covers fabrication and tenability of POF Bragg gratings, photosensitivity of PMMA-based POFs, CYTOP POF gratings and thermal tuning and applications in add-drop multiplexer-demultiplexer, optical spectrum analyzer, WDM sensor arrays, etc. Chapter 5 by A. Driessen et al., provides an overview of polymer-based integrated optical devices for telecommunication as well as nonlinear optical polymers-based active devices such as optical waveguides, Mach-Zehnder interferometers, WDM, directional couplers, splitters and switches. G.-D Khoe et al. describes high capacity transmission systems including PMMA-based GI-POFs, perfluorinated GI-POFs, glass optical fibers and applications of POFs in home and office LAN and gigabit Ethernet. Chapter 7 by L. R. Kawase summarizes POF-based passive devices that includes optical couplers, filters, scramblers, tapers, lenses, Bragg Grating and WDM devices. Second-order nonlinear optical effects (NLO) in optical fibers are reviewed in Chapter 8 by Carvalho et al. The topics include optical and thermal poling process, second harmonic generation (SHG) in fibers, and poling of polymer waveguides.

This book contains eight state-of-the-art chapters written by research experts in the field. This book has about 800 bibliographic citations and hundreds of illustrations, figures, tables, and equations. This book should be a valuable reference source for scientists, researchers, upper-level undergraduate and graduate students working in the field of optical fibers, polymer fiber technology, telecommunication, nonlinear optics, polymer science and engineering, materials science, optical engineering, and device applications.

The editor and publisher thank all the authors for their valuable time and tremendous efforts they have put into writing these state-of-the-art review chapters.

Hari Singh Nalwa
Los Angeles, California, USA
January 2004

List of Contributors

Numbers in parenthesis indicates the page number on which the author's contribution begins.

Henrie van den Boom (103)
COBRA Institute, Eindhoven University of Technology, P.O. Box 513, Eindhoven, The Netherlands

Yung Chang (19)
Department of Chemical Engineering, National Taiwan University, Taipei, Taiwan 10617

I. C. S. Carvalho (127)
Departamento de Física, Pontifícia Universidade Católica do Rio de Janeiro, Gávea, P.O.Box 38071, 22452-970, Rio de Janeiro, Brazil

Wen-Chang Chen (19)
Department of Chemical Engineering, National Taiwan University, Taipei, Taiwan 10617

P. L. Chu (51)
Photonics and Optical Communications Group, School of Electrical Engineering and Telecommunications, University of New South Wales, Sydney, Australia

Alfred Driessen (73)
MESA$^+$ Institute, Lightwave Devices Group, University of Twente, Faculties of Electrical Engineering and Applied Physics P.O. Box 217, 7500 AE Enschede, The Netherlands

Liliana R. Kawase (121)
Fluminense Federal University, Av. Passos da Pátria 156, CP 1126025, CEP 24241-970, Niterói, RJ, Brazil

Giok-Djan Khoe (103)
COBRA Institute, Eindhoven University of Technology, P.O. Box 513, Eindhoven, The Netherlands

Arne Leinse (73)
MESA$^+$ Institute, Lightwave Devices Group, University of Twente, Faculties of Electrical Engineering and Applied Physics P.O. Box 217, 7500 AE Enschede, The Netherlands

B. Lesche (127)
Departamento de Física, UFJF, Campus Universitário, ICE, 36036-300, Juiz de Fora, MG, Brazil

Jui-Hsiang Liu (19)
Department of Chemical Engineering, National Cheng Kung University, Tainan, Taiwan 70101

W. Margulis (127)
Acreo, Electrum 236, Stockholm 16440, Sweden

Idelfonso Tafur Monroy (103)
COBRA Institute, Eindhoven University of Technology, P.O. Box 513, Eindhoven, The Netherlands

Sami Musa (73)
MESA+ Institute, Lightwave Devices Group, University of Twente, Faculties of Electrical Engineering and Applied Physics P.O. Box 217, 7500 AE Enschede, The Netherlands

Chang-Won Park (1)
Department of Chemical Engineering, University of Florida, Gainesville, Florida, USA

G. D. Peng (51)
Photonics and Optical Communications Group, School of Electrical Engineering and Telecommunications, University of New South Wales, Sydney, Australia

René M. de Ridder (73)
MESA+ Institute, Lightwave Devices Group, University of Twente, Faculties of Electrical Engineering and Applied Physics P.O. Box 217, 7500 AE Enschede, The Netherlands

Hong-Wen Su (19)
Department of Chemical Engineering, National Taiwan University, Taipei, Taiwan 10617

Sulur (73)
MESA+ Institute, Lightwave Devices Group, University of Twente, Faculties of Electrical Engineering and Applied Physics P.O. Box 217, 7500 AE Enschede, The Netherlands

Ming-Hsin Wei (19)
Department of Chemical Engineering, National Taiwan University, Taipei, Taiwan 10617

Qijin Zhang (31)
Department of Polymer Science and Engineering, University of Science and Technology of China, Hefei, Anhui 230026, People's Republic of China

About the Editor

Hari Singh Nalwa

Dr. Hari Singh Nalwa is the President and Chief-Executive Officer of American Scientific Publishers in Los Angeles, California. Previously, he was Head of Department and R&D Manager at the Ciba Specialty Chemicals Corporation in Los Angeles (1999–2000) and a staff scientist at the Hitachi Research Laboratory, Hitachi Ltd., Japan (1990–1999). While working at Hitachi Ltd., he was also concurrently an Honorary Visiting Professor at the Indian Institute of Technology in New Delhi (1996–1998). He has authored over 150 scientific articles in journals and books. He has 18 patents, either issued or applied for, on electronic and photonic materials and devices based on them.

Dr. Nalwa has published 73 books including *Ferroelectric Polymers* (Marcel Dekker, 1995), *Nonlinear Optics of Organic Molecules and Polymers* (CRC Press, 1997), *Organic Electroluminescent Materials and Devices* (Gordon & Breach, 1997), *Handbook of Organic Conductive Molecules and Polymers*, Vols 1–4 (John Wiley & Sons, 1997), *Handbook of Low and High Dielectric Constant Materials and Their Applications*, Vols 1–2 (Academic Press, 1999), *Handbook of Nanostructured Materials and Nanotechnology*, Vols 1–5 (Academic Press, 2000), *Handbook of Advanced Electronic and Photonic Materials and Devices*, Vols 1–10 (Academic Press, 2001), *Advanced Functional Molecules and Polymers*, Vols 1–4 (Gordon & Breach, 2001), *Photodetectors and Fiber Optics* (Academic Press, 2001), *Silicon-Based Materials and Devices*, Vols 1–2 (Academic Press, 2001), *Supramolecular Photosensitive and Electroactive Materials* (Academic Press, 2001), *Nanostructured Materials and Nanotechnology*—Condensed Edition (Academic Press, 2001), *Handbook of Thin Film Materials*, Vols 1–5 (Academic Press, 2001). *Handbook of Surfaces and Interfaces of Materials*, Vols. 4–5 (Academic Press, 2001), *Semiconductors and Semimetals*, Vol. 73 (Academic Press, 2001), *Advances in Surface Science*, Vol. 38 (Academic Press, 2001), *Handbook of Polyelectrolytes and Their Applications*, Vols 1–3 (American Scientific Publishers, 2002), *Magnetic Nanostructures* (American Scientific Publishers, 2002), *Nanoclusters and Nanocrystals* (American Scientific Publishers, 2003), *Handbook of Organic-Inorganic Hybrid Materials and Nanocomposites*, Vols 1–2 (American Scientific Publishers, 2003), *Handbook of Luminescence, Display Materials, and Devices*, Vols 1–3 (American Scientific Publishers, 2003), *Quantum Dots and Nanowires* (American Scientific Publishers, 2003). *Handbook of Photochemistry and Photobiology*, Vols 1–4 (American Scientific Publishers, 2003), *Polymer Optical Fibers* (American Scientific Publishers, 2004), *Encyclopedia of Nanoscience and Nanotechnology*, Vols 1–10 (American Scientific Publishers, 2004), and *Handbook of Nanostructured Biomaterials and Their Applications*, Vols 1-2 (American Scientific Publishers, 2004) . *The Handbook of Nanostructured Materials and Nanotechnology* edited by him received the 1999 Award of Excellence in Engineering Handbooks from the Association of American Publishers.

Dr. Nalwa is the founder and Editor-in-Chief of the *Journal of Nanoscience and Nanotechnology* (2001–). He also was the founder and Editor-in-Chief of the *Journal of Porphyrins and Phthalocyanines* published by John Wiley & Sons (1997–2000) and has served on the editorial boards of *Journal of Macromolecular Science-Physics* (1994–2000), *Applied Organometallic Chemistry* (1993–1999), *International Journal of Photoenergy* (1998–2001) and *Photonics Science News* (1995–). He has been a referee for many international journals including *Journal of American Chemical Society*, *Journal of Physical Chemistry*, *Applied Physics Letters*, *Journal of Applied Physics*, *Chemistry of Materials*, *Journal of Materials Science*, *Coordination Chemistry Reviews*, *Applied Organometallic Chemistry*, *Journal of*

Porphyrins and Phthalocyanines, *Journal of Macromolecular Science-Physics*, *Applied Physics*, *Materials Research Bulletin*, and *Optical Communications*.

Dr. Nalwa helped organize the First International Symposium on the Crystal Growth of Organic Materials (Tokyo, 1989) and the Second International Symposium on Phthalocyanines (Edinburgh, 1998) under the auspices of the Royal Society of Chemistry. He also proposed a conference on porphyrins and phthalocyanines to the scientific community that, in part, was intended to promote public awareness of the *Journal of Porphyrins and Phthalocyanines*, which he founded in 1996. As a member of the organizing committee, he helped effectuate the First International Conference on Porphyrins and Phthalocyanines, which was held in Dijon, France in 2000. He was on the organizing committee of the BioMEMS and Smart Nanostructures, (December 17–19, 2001, Adelaide, Australia).

Dr. Nalwa has been cited in the *2000 Outstanding Intellectuals of the 21st Century Order of Excellence*, *Dictionary of International Biography*, *Who's Who in Science and Engineering*, *Who's Who in America*, and *Who's Who in the World*. He is a member of the American Chemical Society (ACS), the American Physical Society (APS), the Materials Research Society (MRS), and the American Association for the Advancement of Science (AAAS). He has been awarded a number of presitgious fellowships including a National Merit Scholarship, an Indian Space Research Organization (ISRO) Fellowship, a Council of Scientific and Industrial Research (CSIR) Senior fellowship, a NEC fellowship, and Japanese Government Science & Technology Agency (STA) Fellowship.

Dr. Nalwa received a B.Sc. degree in Biosciences from the Meerut University in 1974, a M.Sc. degree in Organic Chemistry from the Indian Institute of Technology at Roorkee (formerly University of Roorkee) in 1977, and a Ph.D. degree in Polymer Science from the Indian Institute of Technology in New Delhi in 1983. His thesis research focused on the electrical properties of macromolecules. Since then, his research activities and professional career have been devoted to studies of electronic and photonic organic and polymeric materials. His endeavors include molecular design, chemical synthesis, spectroscopic characterization, structure-property relationships, and evaluation of novel high performance materials for electronic and photonic applications. He was a Guest Scientist at Hann-Meitner Institute in West Berlin, Germany (1983) and Research Associate at the University of Southern California in Los Angeles (1984 -1987) and State University of New York at Buffalo (1987 -1988). In 1988, he moved to the Tokyo University of Agriculture and Technology in Japan as a lecturer (1988 -1990), where he taught and conducted research on electronic and photonic materials. His research experience include ferroelectric polymers, electrically conducting polymers, polymer electrets, nonlinear optical materials for integrated optics, electroluminescent materials, low and high dielectric constant materials for microelectronics packaging, organic semiconductors, nanocrystalline and nanostructured materials, Langmuir-Blodgett films, high-temperature resistant polymer composites, epoxy-based photopolymers, rapid modeling and stereolithography.

CHAPTER 1

Fabrication Techniques for Plastic Optical Fibers

Chang-Won Park
Department of Chemical Engineering, University of Florida, Gainesville, Florida, USA

CONTENTS

1. INTRODUCTION

Plastic optical fiber (POF) was introduced by DuPont in the mid 1960s at approximately the same time glass optical fiber (GOF) was suggested as a transmission medium for optical communications. DuPont's product was a step-index fiber with a polymethyl methacrylate (PMMA) core and fluoropolymer cladding. Despite the fascinating ability to guide light, POF had a very large attenuation, limited applicability, and negligible commercial value. At that time, even GOFs had attenuation in excess of 1000 dB/km, which made optical communications use impractical [1]. Mitsubishi Rayon in Japan further developed DuPont's extrusion technology for POF, which led to the introduction of the Eska™ product in 1976 [2–4]. The attenuation of this product, which had a PMMA core and poly(fluoroalkyl methacrylate) cladding, was still as large as 300 dB/km and applicability was limited to low end uses like illumination.

In 1970, the attenuation of GOF had been reduced to about 20 dB/km in the wavelength region near 1 μm [5]; by the late 1970s, fiber loss below 1 dB/km could be achieved at a wavelength of 1.3 μm and the era of fiber-optic communications commenced. In the mean time, the attenuation of PMMA-based POF also had been reduced to about 160 dB/km at a wavelength of 650 nm by process improvements, and further reduced to about 20 dB/km at 680 nm

ISBN: 1-58883-012-8
Copyright © 2004 by American Scientific Publishers
All rights of reproduction in any form reserverd.

by the use of deuterated PMMA (PMMA-d8) [6]. All these POFs were step-index (SI) POFs. The first graded-index (GI) POF was fabricated by Ohtsuka and Hatanaka in 1976 by heat-drawing a graded-index plastic rod [7]. From the mid 1970s to the early 1980s, graded-index plastic rods were made by two different methods: two-step copolymerization [8–10] and photo-copolymerization [11–14]. GI plastic rods made by two-step copolymerization could not be heat-drawn due to their cross-linked network structure, whereas rods made by the photo-copolymerization method could be. However, the GI-POFs made by heat-drawing GI plastic rods had rather large attenuation (greater than 1000 dB/km at 670 nm) and, consequently, more interest at that time was shown in light-focusing plastic rod lens for which somewhat larger attenuation could be tolerated. Later, when the interfacial-gel polymerization method was introduced [15, 16], GI-POFs with much smaller attenuation could be fabricated [17–21].

The rapid advancement of fiber-optic communication technology during the past two decades led to the realization that GI-POFs could be used as a high-bandwidth data-transmission medium for local area networks or home networks despite their higher attenuation [22, 23]. The flexibility and durability of POFs make it possible to have a much larger core diameter, on the order of 1 mm compared to about 5–10 μm for single-mode GOFs [1, 24]. Because larger diameter makes splicing optical fibers much easier and allows the use of lower cost light sources and connectors, the economic advantage is very significant.

With the newly emerging areas of applications, extensive research has been conducted on both materials development and fabrication techniques for POFs. Because of the high commercial potential, most research was led by industrial concerns, especially Japanese companies, and tens, if not hundreds, of patents have been granted during the past decade. In this chapter, various techniques for the fabrication of POFs are described. There are numerous techniques available; obviously not all of them can be described here in detail. Although most reported methods have some unique features of their own, many appear to share common characteristics that make it possible to classify them into a few categories. Here, an attempt is made to cover as many different processes as possible by selecting a representative process in each category and providing a detailed description along with a discussion on common and different features between various processes in the same category. In doing so, major emphasis is given to describing the fabrication techniques, and little attention is given to the materials and specific conditions that are used. An extensive review on materials for POFs can be found elsewhere [17].

POFs can be classified into step-index and graded-index fibers in terms of their structural characteristics and utility, and these fibers are fabricated mostly by extrusion methods or preform methods. In Sections 2 and 3, these methods are described for the fabrication of SI-POFs and GI-POFs, respectively. Section 4.1 provides an analysis of convective diffusion, which is the underlying physical phenomenon for many GI-POF fabrication techniques. Molecular diffusion is a slow process that may limit the utility of various fabrication techniques for commercial applications. In Section 4.2, a simple approach to enhance the diffusion process is suggested that may be applicable to many existing processes through a simple modification. In Section 5, a brief discussion is given on bandwidth estimation using the ray-tracing method. Finally, in Section 6, a summary of the fabrication techniques for POFs is given. As pointed out previously, numerous patents can be found on the fabrication techniques for POFs; only a handful of them that appear to be most representative are listed in the reference section.

2. STEP-INDEX PLASTIC OPTICAL FIBERS

A step-index plastic optical fiber (SI-POF) has a simple structure of a concentric core and cladding. Consequently, rather simple methods can be applied for its fabrication. In this section, several representative techniques, which are either a continuous or a batch process, are described.

2.1. Continuous Extrusion Method

One of the most well-developed method for the production of SI-POF is the continuous extrusion method developed by Mitsubishi Rayon, which is described schematically in Figure 1 [2, 3]. In this method, a purified monomer (methyl methacrylate), an initiator, and a chain transfer agent are fed into a polymerization reactor, where polymerization reaction takes place. The polymer, which becomes the core of the SI-POF, is then fed into an extruder by a gear pump. The extruder is capable of devolatilization, which removes monomer residues and returns them to the reactor. The core material and a cladding material, which are fed by separate extruders, proceed into a coextrusion die (or a spinning block) where the concentric core–cladding structure of an SI-POF is formed. Although the description of this process may appear to be simple, numerous variables, including the purity of the monomer, the reaction temperature, the degree

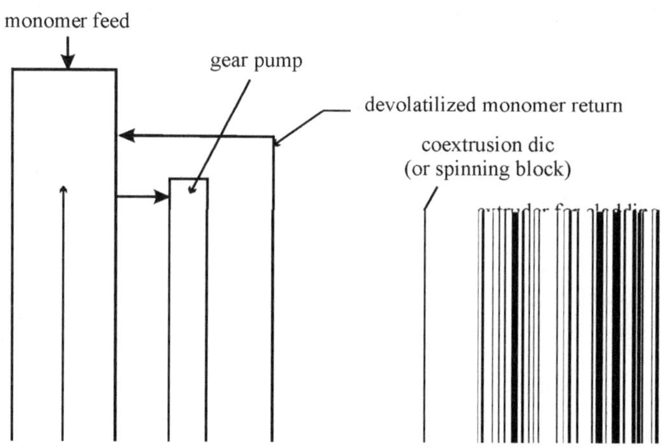

Figure 1. Schematic of a continuous coextrusion method.

of polymerization, the amount of initiator and chain transfer agent, and extrusion conditions, play an important role in determining the optical and physical properties of the fiber. Eska products from Mitsubishi Rayon, which currently dominate the SI-POF market, are made by this process. Similar processes also are used by Toray Industry and Asahi Chemical in Japan.

One important advantage of continuous extrusion is high production rate. This process, however, cannot be run for an indefinitely long period of time because the materials being processed thermally degrade over time, and the degraded material contaminates the fiber and deteriorates its optical property. Consequently, equipment cleaning is necessary at regular intervals and the fibers made at the beginning of an operation or before maintenance shut-down of the equipment tend to have poorer optical properties. To minimize the thermal degradation, special attention should be given to designing the spinning block so that the dead zone where thermal degradation occurs can be eliminated. The spinning block, where multiple strands of the concentric core–cladding structure are formed, is typically an assembly of several plates (Fig. 2). Each plate has both horizontal and vertical flow channels. The core and the cladding materials flow through the channels as they cascade down from the upper plate to the lower plate and form the concentric structure of the SI-POF. Dead zones, where the material residence time is long, causing thermal degradation, can be created by misalignment between the flow channels or by surface mismatch between the adjacent plates. Thus, the plane accuracy of the assembled surfaces between the plates is important in design and fabrication of the spinning block. It has been reported that by improving the plane accuracy of the assembled surfaces to less than 30 μm and by shielding the outer region of the assembled surfaces with heat resistant sealants, the extrusion process can be run continuously for a period of longer than two months while maintaining excellent optical properties of the POF [25].

The extrusion method can utilize ready-made amorphous polymers in pellet form, but contamination associated with materials handling is a serious problem that results in a high attenuation. Consequently, a closed system that integrates purification of monomers, polymerization, and extrusion is preferable.

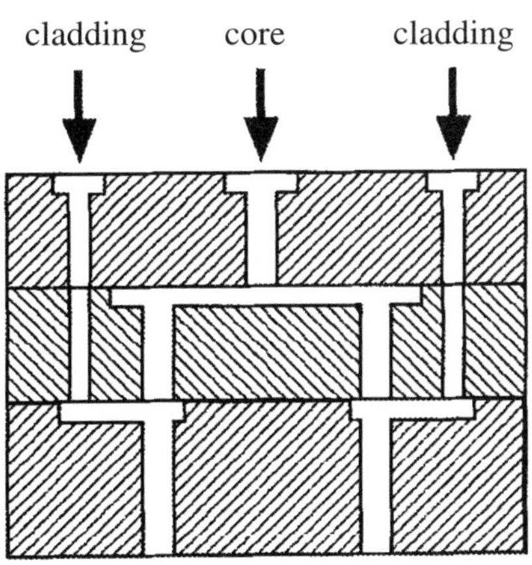

Figure 2. Schematic of a spinning block (Japanese Patent 07-270629).

2.2. Preform Method

Fabrication of POFs by the preform method involves two stages. In the first stage, a cylindrical preform 1–5 cm in diameter and up to 1 m in length is made. In the second stage, the preform is drawn into a fiber by the heat-drawing process. This is the conventional method for the manufacture of GOFs, and the same process, in principle, is applied for the manufacture of POFs.

2.2.1. Preform-Making Process

Figure 3 is a schematic of the modified chemical vapor deposition (MCVD) process, which is the most commonly used method to make a preform for GOF. In this method, a mixed vapor of $SiCl_4$ and O_2 is fed into a fused silica tube that is heated by a multiburner torch while rotating. The $SiCl_4$ reacts with O_2 inside the tube to form SiO_2, which is deposited onto the inside of the fused silica tube [26]. The reaction by-product, Cl_2, leaves the tube along with excess O_2. The refractive index is controlled by the addition of dopants. B_2O_3 or F is added to lower the refractive index for the cladding layer. Once the cladding layer is formed, the dopant is switched to GeO_2 or P_2O_5 to raise the refractive index for the core layer. Whereas a step change in dopant concentration results in a preform for SI-GOF, a preform for GI-GOF can be made by continuously changing the dopant concentration. In addition to MCVD, outside vapor deposition (OVD) and vapor axial deposition (VAD) methods also are used for the GOF preform in which the core and the cladding layers are deposited via flame hydrolysis on the outside of a rotating mandrel.

A similar process is applied to make a preform for POFs. First, cylindrical tube, which serves as the cladding layer, is made by polymerizing a cladding material inside a rotating cylindrical reactor. Materials that can be polymerized by the radical polymerization reaction typically are used and the reaction is induced thermally or by UV radiation using a photoinitiator. Due to the fast rotational speed of the reactor about its cylindrical axis, a tube of uniform thickness is formed once the reaction is complete. This tube is then removed from the reactor and filled with a core material mixed with an initiator and a chain transfer agent. The core

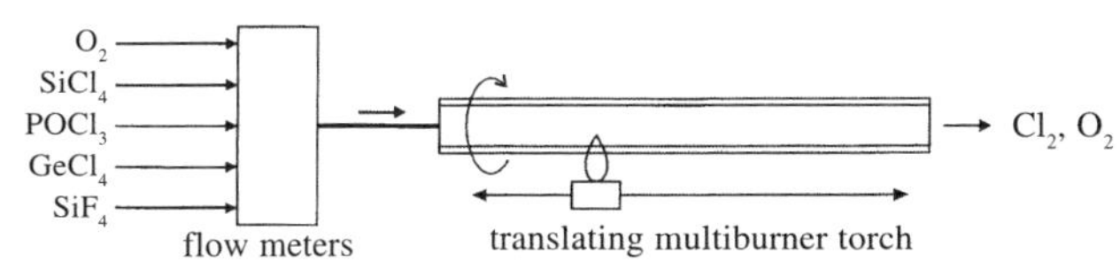

Figure 3. Schematic of the modified chemical vapor deposition process.

material is then polymerized to make a preform that is drawn into a fiber by the heat-drawing process. The tubular cladding layer also can be made by directly extruding a ready-made polymer. However, contamination can be serious, making this approach less preferable.

Advantages of the preform method are simplicity and versatility. Because thermal exposure is limited to the heat-drawing process, degradation occurs to a lesser degree compared to the continuous extrusion method. In addition, this process is easy to modify and the materials that are used can be readily changed. Despite these advantages, this batch process is not prevalent for the manufacture of SI-POFs, probably due to lower production rate than the continuous method. However, the versatility of the preform method makes it a preferred method for the manufacture of GI-POFs as is described in Section 3.

2.2.2. Heat-Drawing Process

The heat-drawing process is shown schematically in Figure 4. The preform is positioned vertically in the middle of the furnace (or oven) where its lower portion is heated locally to a drawing temperature. Both convective and radiative heat transfer mechanisms are important in heating the preform in the furnace [27–29]. When the lower part of the preform reaches a temperature beyond its softening point, it necks downward by its own weight due to gravity. Once this initiation of the drawing process is achieved, tension is applied to the fiber by drawing rollers and the fiber is drawn continuously while the preform is fed at a predetermined rate.

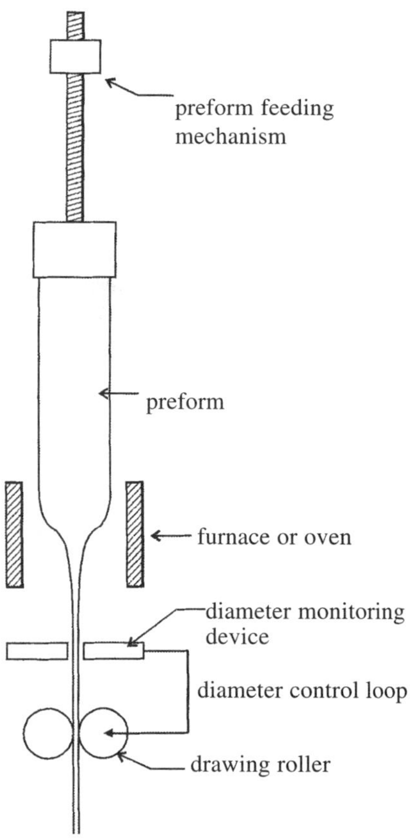

Figure 4. Schematic of a heat-drawing process.

The fiber diameter is continuously measured (typically by a laser micrometer) and the desired value is maintained by controlling the speed of the drawing roller [30]. The typical diameter of multimode GOFs is about 125–250 μm, whereas that of POFs is in the range of 0.5–1.0 mm; the drawing speed is typically 0.2–0.5 m/s. To minimize the transmission loss due to variation of the fiber diameter, the diameter must be controlled stringently [31]. Whereas the diameter of GOFs must be controlled to within 0.1%, less stringent tolerances of 2–3% are applied to POFs.

The steady drawing of fibers is a well-understood process for both glasses [32–34] and polymers [35–37]. Because fiber drawing is characterized by a free surface, it is subjected to a variety of instabilities [38]. The most prevalent instability modes are *tensile rupture* and *draw resonance* [38–41]. Tensile rupture is due to failure of the material in tension, whereas draw resonance is characterized by periodic oscillatioan of the fiber diameter that is sustained unless the response is driven to extreme amplitudes. Although the influence of convective heat transfer on draw resonance has been reasonably well documented, the influence of radiative heat transfer has not been rigorously investigated yet [42].

To minimize the transmission loss in POFs due to thermal damage, special attention also should be given to furnace (or oven) design. Because radiative heat transfer is dominant in the furnace, a shorter distance between the preform surface and the inner surface of the furnace is preferable. In addition, localizing the heating to a narrow zone where the preform is melted is preferable to minimize heat exposure. Thus, shorter length of the heating zone is desirable as long as sufficient heating is provided [43, 44].

2.3. Other Methods

There are a few other methods that are neither continuous extrusion nor preform. One of these methods is the *batch extrusion method* developed by Kaino et al. [45–47]. The apparatus for this method consists of three separate but connected vessels, a vacuum outlet, and a branch for nitrogen input. One vessel is to hold a monomer, another is for a chain transfer agent and/or an initiator, and the third is a reactor. The monomer (e.g., methyl metacrylate or styrene) in the monomer vessel is distilled under vacuum into the reactor. A chain transfer agent is then distilled into the reactor and mixed with the monomer. The reactor is then heated to polymerize the monomer under a sealed condition. When the reaction is complete, the temperature is raised to the extrusion temperature and fiber is drawn by pressurizing the reactor using nitrogen. The core material is drawn into a fiber and immediately clad with a cladding material. Although it is a batch process with a limited production rate, the overall setup is much simpler than the continuous extrusion method described in Section 2.1. Thus, thermal degradation is less likely to occur and the resultant fiber has been shown to have excellent optical properties.

A patent application by Mitsubishi Rayon in Japan described another method by which a mixture of monomers and a photoinitiator is poured into a heat-shrinkable tube and polymerized by photoradiation while the cladding tube shrinks [48]. It has been claimed that a high numerical aperture SI-POF was made by this method for outdoor display application.

3. GRADED-INDEX PLASTIC OPTICAL FIBERS

GI-POF has drawn considerable interest recently as a high bandwidth data transmission medium for short distance applications such as local area networks, office networks, and home networks. The bandwidth of a GI-POF is maximized by optimizing the distribution of the refractive index in the radial direction of the fiber. Theoretically, the refractive index profile is often described by the power-law model [49–51]

$$n(r) = \begin{cases} n_1\left[1 - 2\Delta\left(\dfrac{r}{R}\right)^{\alpha}\right]^{1/2} & \text{for } r \le R \\ n_2 & \text{for } r \ge R \end{cases} \tag{1}$$

Here r is the radial distance from the fiber axis, R is the radius of the fiber core, n_1 and n_2 are the refractive indices at $r = 0$ and $r = R$, respectively, $\Delta = (n_1^2 - n_2^2)/2n_1^2$, and α is the power-law exponent, which determines the refractive index profile as a function of radius. In particular, a parabolic index fiber corresponds to the case $\alpha = 2$.

The bandwidth of an optical fiber is limited by dispersion, which represents broadening of optical pulses as they propagate through the fiber. The dispersion of an optical fiber is divided into two parts: *intermodal* (or simply *modal*) and *intramodal.* Intramodal dispersion is also called *group-velocity dispersion* or *fiber dispersion.* In a geometrical optics description, the modal dispersion is attributed to different paths followed by different rays. For example, a ray propagating along the fiber axis has the shortest path, whereas the path is longer for oblique rays. Consequently, different rays with different paths arrive at the fiber output at different times unless their velocities are adjusted along their paths through spatial variation of the refractive index as in a GI fiber. For this reason, modal dispersion also can be called *multipath dispersion.* Intramodal dispersion has two contributions: *material* (or *chromatic*) *dispersion* and *waveguide dispersion.* Material dispersion is caused by a finite distribution of the wavelength of the incident light and the wavelength dependence of the refractive index of a material. Because nonideal incident light always has a wavelength distribution, material dispersion occurs unless the material is nondispersive. However, material dispersion is reportedly much smaller than modal dispersion unless the modal dispersion is extremely small, as is the case in single-mode fibers. Waveguide dispersion depends on fiber parameters such as core radius R and the fractional index difference Δ in Eq. (1).

When the refractive index profile is described by the power-law model, the bandwidth is maximized when

$$\alpha = 2 + \varepsilon - \Delta\frac{(4+\varepsilon)(3+\varepsilon)}{5+2\varepsilon} \tag{2}$$

where ε is the material dispersion parameter [50]. If the material dispersion is assumed to be negligible, the maximum bandwidth is achieved when $\alpha = (2 - 12\Delta/5)$. Because Δ is small, on the order of 0.01 for a typical GI-POF for fiber-optic communications, the optimum α is close to 2 and the refractive index profile is approximately parabolic. Studies indicate that material dispersion has a significant influence on the bandwidth of GI-POFs, and the optimum value of α for PMMA-based GI-POFs is about 2.3 when material dispersion is taken into account [22, 23].

It has been suggested that the bandwidth of POFs may be affected by mode coupling and differential mode attenuation (DMA) [52–54]. As mentioned previously, in a geometrical optics description, modal dispersion is attributed to different paths followed by different rays; hence the name multipath dispersion. In a modal description of wave propagation, modal dispersion is related to the different group velocities (or mode indices) associated with different modes. An optical mode refers to a specific solution of the wave equation that satisfies appropriate boundary conditions, and multiple guided modes (with different mode indices) propagate through a multimode optical fiber at different group velocities. Modes that belong to different groups interchange energy as they propagate, and this mode coupling may influence the bandwidth of an optical fiber [52]. Mode coupling, however, is negligible unless the length of an optical fiber is greater than 100 m [53–55]. Differential mode attenuation is related to the fact that loss (or attenuation) mechanisms, such as absorption or Rayleigh scattering, act differently on different modes. Consequently, attenuation characteristics vary depending on the mode. High order modes that have the fastest attenuation lose energy as they propagate through a fiber, and this DMA filtering effect is suggested to be the main reason for the bandwidth enhancement that has been observed experimentally with GI-POFs [53–55].

As previously described, the bandwidth of a GI-POF is quite sensitive to the refractive index profile. Thus, in manufacturing a GI-POF, the obtainable bandwidth of a GI-POF is directly related to the ability to control the profile. To date, several methods have been suggested for the manufacture of GI-POFs. Although continuous extrusion is the dominant method for the manufacture of SI-POF, the preform method appears to be the preferred approach for GI-POF because of its versatility. Preform methods that are more prevalent are described in the next section; a description of continuous methods follows in Section 3.2.

3.1. Preform Methods

3.1.1. Photo-copolymerization

The first GI-POF was fabricated by Ohtsuka and Hatanaka in 1976 by heat-drawing a graded-index plastic rod [7, 8]. At that time, graded-index plastic rods were made by two different methods: two-step copolymerization and photo-copolymerization [9–14]. In two-step copolymerization [9, 10], a polymerized rod is prepared and immersed into a comonomer that has a lower refractive index. The comonomer then diffuses into the rod to form a concentration gradient in the radial direction that is subsequently fixed by further polymerization. To prevent the immersed rod from dissolving too much while allowing the comonomer to diffuse inward, the prepolymerized rod has a cross-linked network structure. Consequently, such a GI rod cannot be drawn into POF. Conversely, a GI rod made by photo-copolymerization does not have a network structure, and can be heat-drawn into POF.

Figure 5 is a schematic of the photo-copolymerization apparatus used by Ohtsuka et al. [12]. A 2.9 mm diameter glass tube was filled with a monomer mixture [methyl methacrylate (MMA) and vinyl benzoate (VB)] and an initiator (benzoyl peroxide). The glass tube reactor was then positioned vertically in a constant temperature chamber and rotated about its vertical axis while being irradiated by UV. The UV radiation was applied locally by using a shade while the UV lamp traversed from the bottom of the tubular reactor to the top at a constant speed. The UV radiation quickly induced the polymerization reaction, forming a gel phase. Because the UV intensity was higher near the wall of the glass tube, the gel phase formed first on the inner wall of the glass tube and grew inward toward the center of the tube. The polymerization reaction also progressed from the bottom of the tube to the top as the UV lamp traversed upward. After irradiation, the glass tube reactor was heat-treated at a higher temperature (60 °C) to further react the remaining monomer.

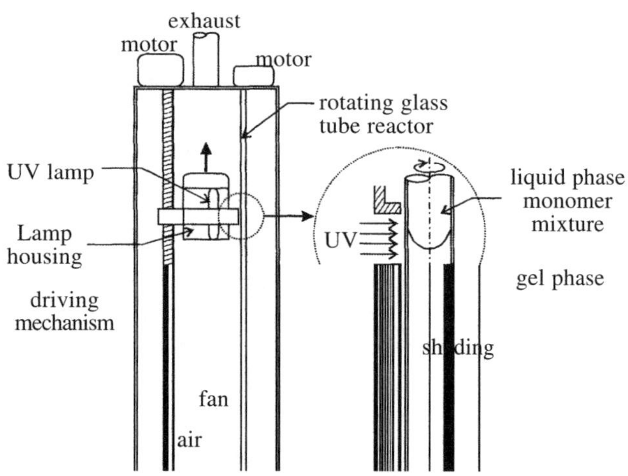

Figure 5. Schematic of the photo-copolymerization method

The criteria for the choice of monomers in the mixture prescribes one monomer (e.g., MMA) that has a lower refractive index than the other (e.g., VB) and has a higher reactivity ratio at the same time. Thus, the monomer with the lower refractive index polymerizes preferentially near the wall of the glass tube as the reaction propagates inward, thereby creating a gradient in the copolymer composition in the radial direction that yields a refractive index profile. Because the refractive index profile in this method is determined strictly by the reactivity ratios of the monomers, controlling the profile to a desired shape was difficult. Subsequently, a ternary system was adopted whereby a third monomer was added to achieve better control of the refractive index profile [13, 14].

The GI-POFs made by heat-drawing these GI rods had a large attenuation, greater than 1000 dB/km at 670 nm; hence, most interest at that time concentrated on the light-focusing plastic rod for various applications such as imaging systems, fiber coupling, switches, and multiplexers [56–58]. The large attenuation was ascribed to large-size heterogeneities on the order of 100 nm formed inside the material when the polymerization reaction took place at a temperature below the glass transition temperature [59, 60].

3.1.2. Interfacial-Gel Polymerization

The best known method for the fabrication of GI-POFs is the *interfacial-gel polymerization method*, which was pioneered by Koike and his co-workers [15, 16]. At first, this method was applied to the copolymerization of monomer mixtures with different reactivity ratios and different refractive indices as in the photo-copolymerization process described in the previous section [18, 61]. Later, it was applied to mixtures of a monomer and a nonreacting organic dopant [62]. In this method, a transparent polymeric tube (e.g., PMMA) is prepared and this tube is then filled with a mixture of two monomers (e.g., methyl methacrylate and vinyl benzoate) that is polymerized thermally while the tube is rotated. The inner wall of the tube is swollen by the monomer mixture as it forms a thin gel phase. Due to the "gel effect," the polymerization reaction is faster inside the gel phase than in the monomer bulk phase. Consequently, the reaction occurs preferentially on the inner surface of the tube and the copolymer phase grows inward toward the center of the tube as the reaction progresses. Due to the difference in the reactivity ratios, the composition of the copolymer changes gradually in the radial direction. If the monomer with the higher reactivity ratio has a smaller refractive index than the other monomer, the preform will have a gradually increasing refractive index toward the center. The refractive index profile obtained by this method depends on the relative ratios of the monomers in the mixture, and many different combinations of materials have been investigated [17].

When a mixture of a monomer and a nonreacting dopant is used (e.g., methyl methacrylate and benzyl benzoate), the inner wall of the tube is again swollen by the monomer as it forms a thin gel phase. The polymerization reaction that takes place by heating or by other means (such as UV irradiation) is faster in the gel phase; consequently, the thickness of the polymeric tube grows inward, forming a rod. Whereas the molecular volume of the dopant is typically larger than that of the monomer, the monomer diffuses into the gel phase faster than the dopant and reacts. Consequently, the dopant is rejected from the polymerization site and its concentration becomes gradually higher toward the center of the preform as the polymerization reaction progresses. Thus, a concentration gradient of the dopant is created in the radial direction. Because the dopant either increases or decreases the refractive index when dispersed in the polymer matrix, a positive or a negative refractive index profile is created. The preform is then heat-drawn into a fiber.

The interfacial-gel polymerization method and the photo-copolymerization method may appear to be very similar. In both methods, the refractive index profile is a consequence of the spatial variation of the copolymer composition that is created by the difference in the reactivity ratios of the monomers in the mixture. The essential difference between the two methods is that the polymerization reaction takes place preferentially on the inner wall of the tube due to the gel effect in the case of the interfacial gel polymerization method, whereas it is due to the radial gradient of the UV radiation intensity in the case of the photo-copolymerization method. In addition, the photo-copolymerization method uses a glass tube as the reactor, whereas a polymeric tube that serves as a cladding layer is used as the reactor in the interfacial-gel polymerization method.

Studies on light scattering loss in PMMA have suggested that large-size heterogeneities on the order of 100 nm form when the polymerization reaction is carried out below the glass transition temperature (T_g) and cause the large scattering loss, and that such heterogeneities can be reduced by heat treatment at a temperature sufficiently highter than T_g [59, 60]. It also has been revealed that copolymer-based GI preforms (e.g., methyl methacrylate–vinyl benzoate copolymer) have large-size isotropic heterogeneities that lead to increased scattering loss, whereas nonreacting organic dopant-based GI preforms (e.g., methyl methacrylate-benzyl benzoate) have no such heterogeneities [20]. In making copolymer-based GI preforms, when the difference in the monomer reactivity ratios is large, the monomer with the larger reactivity ratio polymerizes preferentially and is consumed almost completely at the initial stage of the reaction. Subsequently, the monomer with the smaller reactivity ratio polymerizes at the latter stage of the reaction. Consequently, the distribution of copolymer composition becomes nearly bimodal. This extreme distribution of copolymer composition appears to be responsible for excessive scattering loss [17, 20]. In dopant-based GI preforms, on the other hand, the dopant molecule is randomly located without any aggregation. Thus, large heterogeneous structures are not observed. Current research on interfacial-gel polymerization seems to put more emphasis on dopant-based GI preforms for this reason. Extensive efforts on this method have resulted in PMMA-based GI-POFs with attenuation lower than 200 dB/km at a wavelength of 650 nm and a bandwidth greater than 2 Gbit/s for 100 m [17–23].

Dopant-based GI preforms have a few critical drawbacks. Because the nonreacting dopant is distributed in a polymer matrix with a concentration gradient, molecular diffusion eventually diminishes the concentration gradient. Thus, the long-term stability of the refractive index profile is an area of concern. Furthermore, the dopant acts as a plasticizer, thereby decreasing the glass transition temperature of the polymer. To achieve a reasonably large numerical aperture (NA), a dopant concentration as high as 20% by weight is needed. At such a high concentration, the decrease in the glass transition temperature (T_g) is significant. For example, when a PMMA-based GI-POF was made using benzyl benzoate (BEN) at 20 wt.%, the refractive index at the center of the fiber was about 1.51, making the NA as large as 0.23 [63]. The T_g at the center, however, became lower than that at the outer surface (i.e., PMMA cladding) by more than 40°C. Whereas GI-POF needs to maintain its performance characteristics at service temperatures as high as 85 °C, thermal stability of the refractive index profile is another area of concern. Extensive studies have suggested that it is possible to achieve long-term stability at 85 °C with a careful choice of dopant [63, 64].

Although PMMA-based GI-POFs have a very high potential for short distance communication, perfluorinated polymer-based GI-POFs have drawn great interest [65–68]. Because perfluorinated polymers [e.g., poly(perfluorobutenyl vinyl ether (CYTOP™, Asahi Glass Co.) or poly (tetrafluoroethylene-*co*-dioxole) (Teflon AF™, of Du Pont)] do not have C-H bonds, they show very small intrinsic absorption loss, even in the wavelength range of 850–1300 nm. The capability to use a signal wavelength in this range is quite advantageous because transceivers that use the 850 or 1300 nm wavelength are readily available because they are also used for GOFs. In addition, the material dispersion of perfluorinated polymers is smaller than that of PMMA or GeO_2-doped silica [69]. Thus, perfluorinated polymer-based GI-POFs can have higher bandwidth than silica-based multimode fibers [70]. In fact, the very first commercial GI-POF is a perfluorinated polymer-based fiber called Lucina, which was introduced by Asahi Glass Company.

3.1.3. Centrifugal Method

In the centrifugal method, introduced by van Duijnhoven and Bastiaansen [71] in 1999, poly(tetrafluoropropyl methacrylate) (PTFPMA) solution dissolved in methyl methacrylate (MMA) monomer was put into a 5 mm diameter tube and subjected to a centrifugal field by rotating the tube about its cylindrical axis at speeds of up to 20,000 rpm at room temperature for 24 h. The monomer is then polymerized at an elevated temperature of 60~80 °C for 12 h while the centrifugal field is maintained. Subsequently, postpolymerization is carried out at 120 °C for 12 h to make a GI preform. Because the density of PTFPMA (1.496 g/cm^3) is greater than that of MMA (0.936 g/cm^3), PTFPMA migrates outward under the centrifugal field forming a concentration gradient. The concentration gradient is then fixed as the MMA polymerizes, thereby yielding a GI preform. The concentration profile created by this method is close to the ideal parabolic shape, and a quarter-pitch GI rod lens cut from this preform showed image inversion, but no other optical properties of the preform were reported. Although this laboratory scale experiment was conducted using a rather small radius tube, construction of a new apparatus with a larger diameter (5 cm) tube and a higher rotational speed (up to 60,000 rpm) was reported.

Although this method is capable of creating a graded-index profile, ultra high rotational speed may be problematic. Unless the tube dimension is very accurate, vibration of the apparatus at a very high rotational speed can be a serious problem. Furthermore, because the material shrinks due to increased density as the polymerization reaction proceeds, a cylindrical void that spans the tube axis will be formed under a very strong centrifugal field. Consequently, additional steps are required to fill the void and may alter the refractive index profile from the ideal parabolic shape.

3.1.4. Other Preform Methods

3.1.4.1. Chemical Vapor Deposition Figure 6 is a diagram of the chemical vapor deposition (CVD) method developed by Sumitomo Electric Industries [72]. A polymer and a refractive index modifier (or a dopant) are mixed in a vessel where the mixture is heated and vaporized. The vapor phase is then carried to a supply pipe by a carrier gas (e.g., N_2), and deposited on the inner surface of a rotating cylindrical tube. Gradually changing the mixing ratio between the polymer and the dopant in the supply vessel gradually varies the dopant concentration of the layer deposited on the inner surface of the tube. Thus, a preform with a gradient index profile can be made.

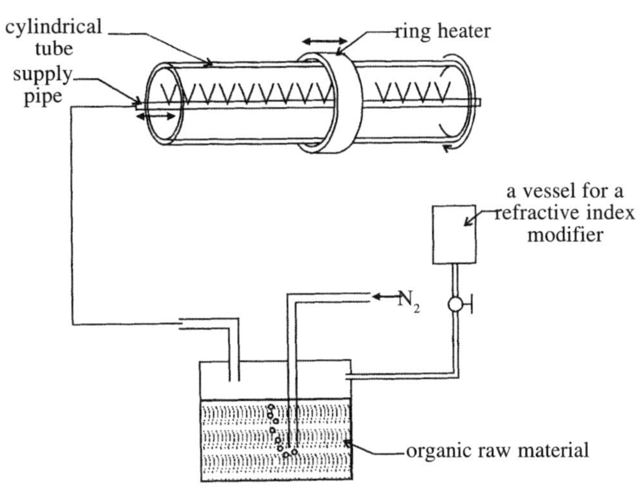

Figure 6. Schematic of the chemical vapor deposition method (U.S. Patent 5,851,666).

This method is similar to the MCVD process for GOF preforms, but a chemical reaction does not take place inside the rotating tube. Because the deposition (or coating) layer grows inward toward the center, a small hole remains in the middle of the preform after the deposition process is completed. Unless the hole is filled before the preform is drawn into a fiber, it creates a central dip in the refractive index profile.

Sumitomo also developed a similar method in which CVD is applied on the outer surface of a rotating rod [73]. This process is, in principle, the same as the OVD process for GOF preforms, although a chemical reaction is not involved. In this case, the deposition layer grows from the center to the outer periphery. Unlike the inner vapor deposition method, if the rod is a transparent polymeric material that can serve as a core, a central hole is not created.

3.1.4.2. Extraction of a Dopant Using a Nonsolvent In this dopant extraction method suggested by Tripathy, Chittibabu, and Kumar [74], a polymer is uniformly mixed with a refractive-index-modifying additive and is extruded into a rod. The rod is then immersed in a nonsolvent that permeates the rod and extracts a portion of the additive. Because the additive diffuses out of the polymer matrix due to the nonsolvent, a concentration gradient of the additive is created in the radial direction of the rod. This rod is then drawn into a GI-POF.

3.1.4.3. Inward Diffusion of a Dopant from Cladding to Core In this dopant diffusion method developed by Ilvashenko [75], a transparent cylindrical sheath (or cladding tube) is prepared by extrusion or other methods from a polymer mixed with a uniformly dispersed refractive-index-modifying dopant. The dopant dispersed in the sheath has a lower refractive index than the polymer of the tube. The tube is then filled with a polymerizable monomer that has a higher refractive index than the sheath. When the monomer is polymerized to form the core, a portion of the dopant in the sheath diffuses into the core material and creates a gradient index profile. If needed, a refractive-index-raising dopant can be mixed into the core monomer prior to filling the tube. A preform made by this process is drawn into a GI-POF.

3.1.4.4. Continuous or Stepwise Feeding of a Mixture Figure 7 is a schematic of the reaction chamber for the method suggested by Katoot [76]. First, polymeric tube that serves as the cladding is made using a cladding material (e.g., α, ω-dichloropropyl dimethyl siloxane). The tubular cladding is then inserted into a reaction chamber that is equipped with a heating jacket. Both ends of the cladding tube are fitted with end caps that are connected to a shaft with a material injection port in the middle. A mixture of monomers that has a refractive index higher than the cladding (e.g., α,ω- dichloro-propyldimethylsiloxane and bisphenol A polycarbonate) is added to the interior of the

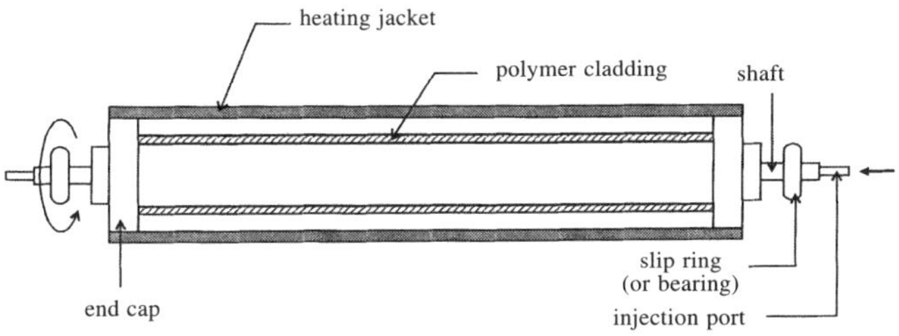

Figure 7. Schematic of the reaction chamber for the method developed by Katoot (U.S. Patent 5,861,129).

cladding either continuously or stepwise as the reaction chamber is heated and rotated. As the mixture polymerizes, the copolymer builds up on the inner surface of the cladding. The composition of the monomer mixture that is fed subsequently is varied so that the copolymer composition changes in a way that increases the refractive index gradually as the copolymer builds up. In this way, a gradient-index preform can be made when the tube is filled up and the reaction is completed. The preform is then heat-drawn to a GI-POF.

3.2. Continuous Methods

All the methods described in the previous section are batch processes in that a preform is made first, followed by heat-drawing to make GI-POFs. In this sections continuous methods are described where GI-POFs are made continuously and directly from the feed materials.

3.2.1. Internal Diffusion and Surface Evaporation

Figure 8 describes schematically the internal diffusion and surface evaporation (IDSE) process developed by Mitsubishi Rayon [77–79]. In this method, a polymer solution that contains additives is prepared using a monomer as a solvent. One example of the polymer solution is poly(2,2,3,3-tetrafluoropropyl methacrylate) dissolved in MMA along with 1-hydroxycyclohexyl phenyl ketone as an initiator and hydroquinone as an inhibitor [79]. Another example is vinylidene–tetrafluoroethylene copolymer and PMMA dissolved in MMA along with additives [77]. The polymer solution is then charged into a cylinder and pushed by a piston through a nozzle to form a strand fiber. The strand fiber proceed into a volatilization chamber where the monomer (e.g., MMA) is volatilized from the surface of the fiber by means of an inert gas (e.g., air, N_2, or Ar). Because the monomer evaporates from the fiber surface, a large concentration gradient of the monomer is created in the fiber near the outer periphery. Consequently, the monomer in the fiber diffuses in the radial direction and a concentration gradient that spans from the center of the fiber to the surface is created. This monomer concentration gradient is then fixed by a rather fast polymerization reaction induced by UV irradiation as the fiber passes through a curing zone. Thus, a GI-POF is made.

Tsai et al. [80] conducted a theoretical analysis of the convective diffusion process in the volatilization chamber that determines the monomer concentration profile. Their results

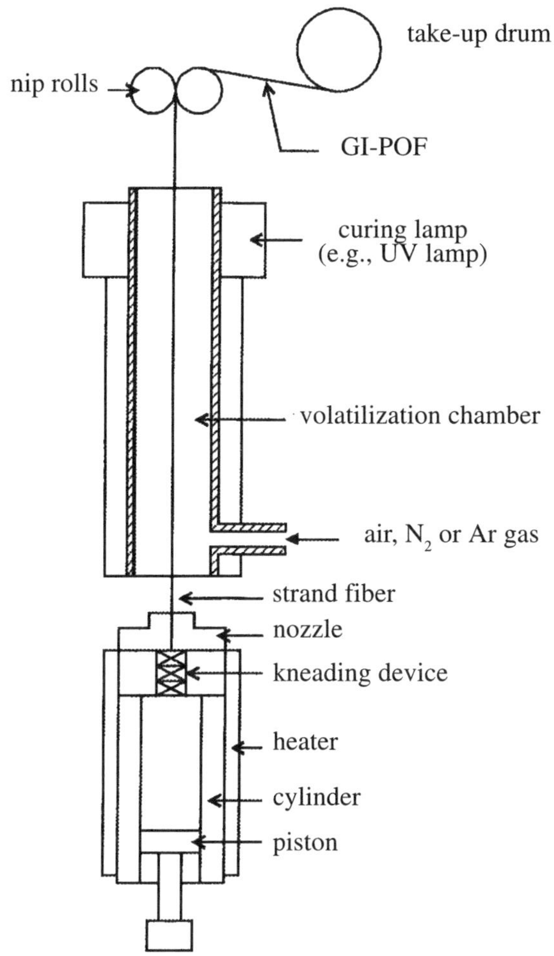

Figure 8. Schematic of the IDSE apparatus (U.S. Patent 4,822,122).

indicated that the balance between internal diffusion and surface evaporation is critical for obtaining a parabolic profile. If the evaporation is either too fast or too slow, the concentration profile deviates significantly from the parabolic shape. It has been suggested that the IDSE method could be improved by adopting a multilayer coextrusion process [81, 82].

3.2.2. Closed Coextrusion Method

The closed coextrusion method developed by Ho and coworkers [83–86] is shown schematically in Figure 9. A polymer solution for the core [e.g., PMMA dissolved in MMA and benzyl methacrylate (BzMA)] is placed in a material supply tank, and another polymer solution for the cladding (e.g., PMMA dissolved in MMA only) is put in another supply tank. These solutions are heated to a moderate temperature (60 °C) and fed into a concentric (or coextrusion) die by separate gear pumps. A bilayer concentric fiber is then extruded from the die and proceeds into an enclosed (or diffusion) zone that is maintained at a constant temperature. In the enclosed zone, the fiber surface is free (i.e., no contact with a solid surface) and mutual diffusion of the monomers

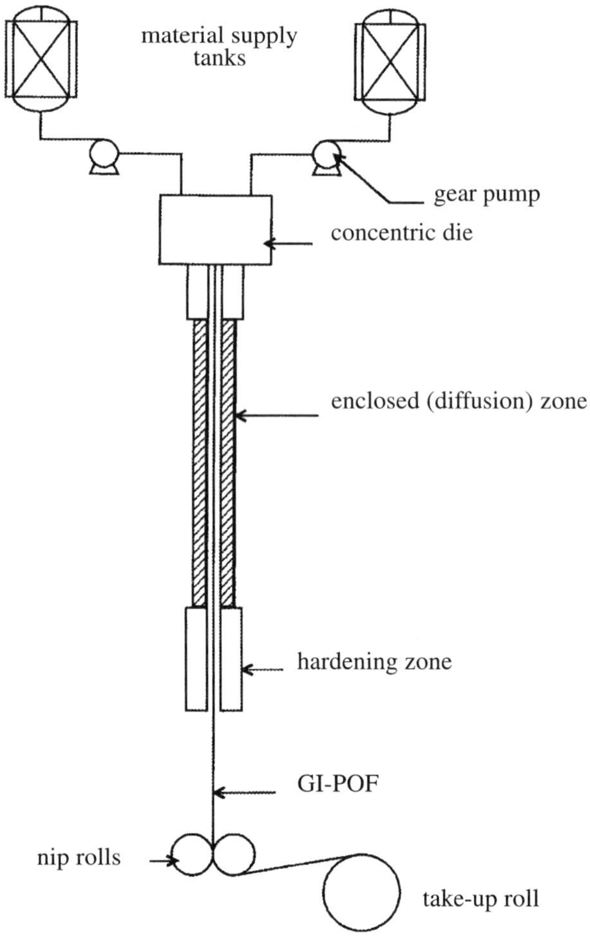

Figure 9. Schematic of the closed coextrusion process (U.S. Patent 5,555,525).

takes place at the core–cladding interface (i.e., BzMA diffuses outward to the cladding and MMA diffuses inward to the core.) Consequently, a concentration gradient of the monomers is created. Subsequently, the monomer concentration gradient is fixed by a rapid polymerization reaction induced by UV radiation in the hardening zone thereby forming a GI-POF.

In the case of the IDSE method described in Section 3.2.1, the mass flux of the monomer in the radial direction at the fiber surface is given by

$$\frac{\partial C}{\partial r} = -J \quad \text{at } r = R \tag{3}$$

where C is the monomer concentration, r is the radial coordinate, R is the radius of the strand fiber in the volatilization chamber, and J is the evaporation rate of the monomer. This surface condition describes the balance between the diffusional flux of the monomer at the fiber surface and its evaporation rate, and the negative sign is due to the fact that the monomer concentration decreases with r. In the case of the closed coextrusion method, on the other hand, the diffusion of the monomer occurs at the interface between the two layers and is driven by the concentration difference of the monomers in each layer. Evaporation of the monomer at the fiber surface is negligible because the diffusion zone is simply a constant temperature chamber without any forced flow of an inert gas. Thus, the surface condition in this case is given as

$$\frac{\partial C}{\partial r} = 0 \quad \text{at } r = R \tag{4}$$

which indicates that the fiber surface is impenetrable to the monomers; hence the name *closed* coextrusion. This no-flux condition requires the slope of the concentration profile (or the refractive index profile) at the fiber surface (i.e., at $r = R$) to be zero as depicted in Figure 10, whereas the slope of the refractive index profile at $r = R$ has a finite value for the fiber made by the IDSE method. Thus, a truly parabolic profile cannot be created by the closed coextrusion method due to the inevitable inflection to a positive curvature toward the periphery of the fiber. However, as will be described in Section 5, a GI-POF that has a much higher bandwidth than SI-POFs still can be produced by this method.

3.2.3. Coextrusion Method of Koike and Nihei

Figure 11 describes another coextrusion method which was introduced by Koike and Nihei [87]. A polymer (material B in Fig. 11) is placed in a cylinder and heated to a molten state. The polymer is then pressurized and extruded in the form of a fiber through nozzle B. When a diffusible material (material A) is injected through nozzle A from the inner cylinder, it diffuses radially into the polymer from the central axis as it flows toward nozzle B. When the polymer is drawn into a fiber through nozzle B and cooled, a GI-POF is obtained because the diffused material forms a concentration gradient in the radial direction. The concentration gradient may vary

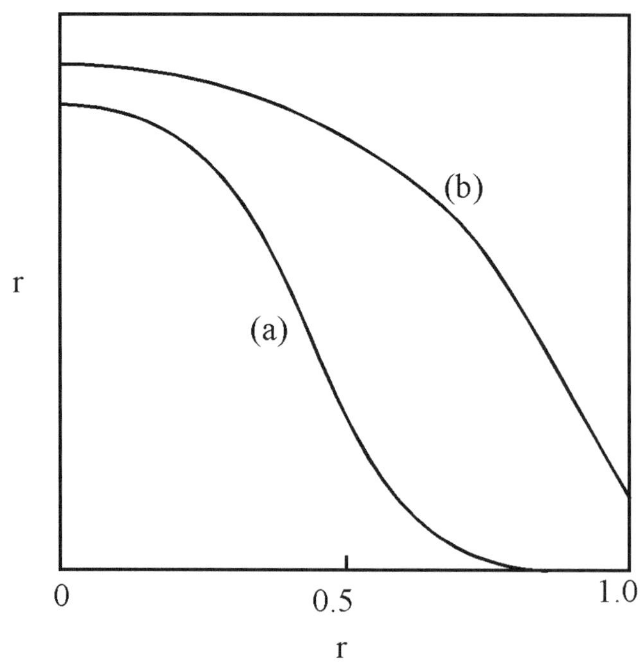

Figure 10. Refractive index profiles obtained by (a) the closed coextrusion method and (b) the IDSE method.

depending on the relative flow rate of the two materials and the length of the diffusion zone (d in Fig. 11). Because diffusion is a very slow process, the penetration depth in the radial direction is very small compared to the length of the diffusion zone. Consequently, the concentration profile of material A created by this method is similar to that created by the closed coextrusion method [i.e., (a) in Fig. 10].

3.2.4. Coextrusion Method of Sohn and Park

The coextrusion method introduced by Sohn and Park is called a *diffusion-assisted coextrusion process* [88]. In this process, core and cladding materials that contain refractive-index-modifying additives are fed into a coextrusion die by two separate feeders, which can be of any type such as extruders or gear pumps (Fig. 12). After a concentric core–cladding structure is formed in the coextrusion die, the materials proceed into a diffusion zone, where the additives diffuse from the core to the cladding or vice versa, depending on which layer contains the additives. Whereas the material is solidified in a nonequilibrium state, a continuously varying additive concentration profile (or refractive index profile) is created in the radial direction. The diffusion zone in this method is an extension of the coextrusion die, where the outer surface of the cladding material is still in contact with the solid wall, the dimension of which can be varied to control the diffusional characteristics of the additives.

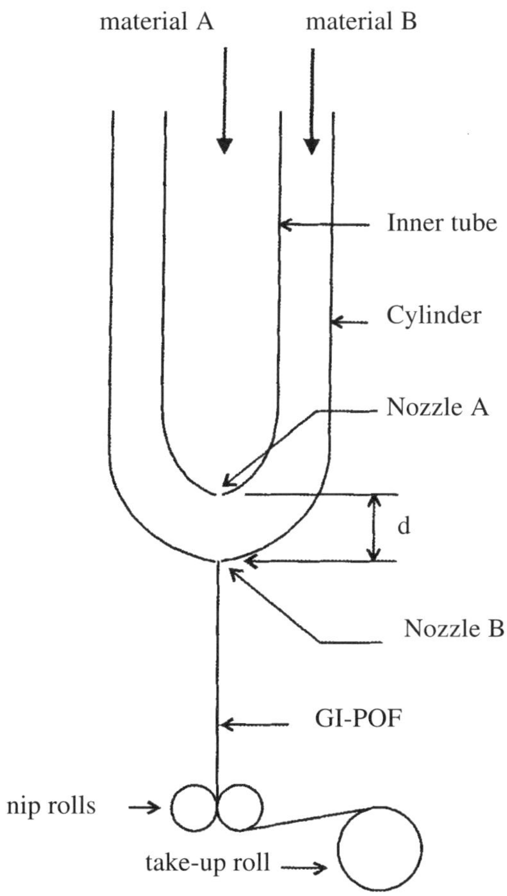

Figure 11. Schematic of the coextrusion method developed by Koike and Nihei (U.S. Patent 5,593,621).

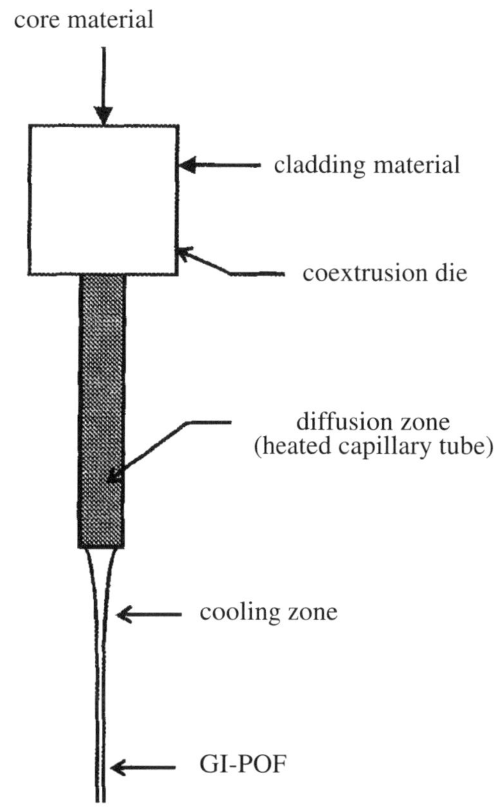

Figure 12. Schematic of the coextrusion method developed by Sohn and Park.

This method may appear to be similar to the closed coextrusion process of Ho et al. [83–86]. Unlike the diffusion zone of the diffusion-assisted coextrusion process, the enclosed zone of Ho's process is a heating chamber where the fiber surface is stress-free (i.e., contacts with a gas phase). Due to this difference, the processing characteristics as well as the materials that can be processed by each method are different.

The coextrusion methods described in Sections 3.2.1–3.2.4 involve diffusion of either monomers or additives. Although they are continuous processes, diffusion is a very slow process and it requires a substantial residence time (or contact time) to obtain a desired concentration profile. Consequently, the material throughput rate must be low, which is one of the major drawbacks of these methods.

3.2.5. Coextrusion Method of Park and Walker

Unlike all the other coextrusion methods described herein, the method developed by Park and Walker [89, 90] does not rely on diffusion of a material to create a concentration profile. Their method utilizes a specially designed coextrusion die called a GI die block (GDB), and the refractive index profile is created solely by mechanically mixing two polymers that have different refractive indices inside the GDB.

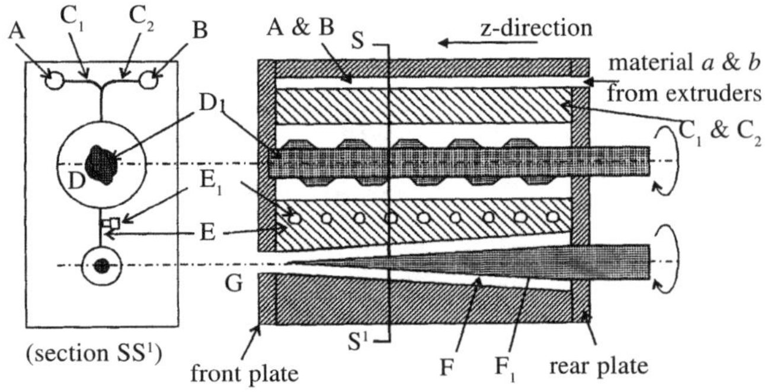

Figure 13. Schematic of the GI die block (U.S. Patent 6,267,915).

Figure 13 schematically describes the GDB, which consists of three major sections: manifolds (A and B), a mixing chamber (D), and a conical feed section (F). A three-dimensional view of the GDB that shows these three sections in the absence of the front and back plates, and the two rotating elements (D_1 and F_1) is shown in Figure 14. The slit channel C_1 that connects manifold A to the mixing chamber has a decreasing gap along the *z* direction, whereas slit channel C_2 has an increasing gap. Thus, the flow rate of material *a* is higher at a smaller value of *z*, whereas the converse situation holds for material *b*. Consequently, the right-hand side of the mixing chamber D (i.e., at a smaller value of *z*) is richer in material *a* and its relative concentration decreases along the *z* direction. The material in the mixing chamber that has a varying composition of *a* and *b* in the *z* direction is mixed uniformly by a rotating mixer blade D_1 while maintaining the varying blend composition in the *z* direction. If the refractive index of material *a* is higher than that of material *b*, the refractive index of the blended material in the mixing chamber D will decrease continuously in the *z* direction. Although the flow in the GDB is driven by pressure gradient, *z*-directional axial mixing in the mixing chamber is minimal because the pressure gradient in the *z* direction in the mixing chamber is insignificant. Therefore, a polymer blend that has a continuously varying refractive index along the *z* direction can be prepared in the mixing chamber D.

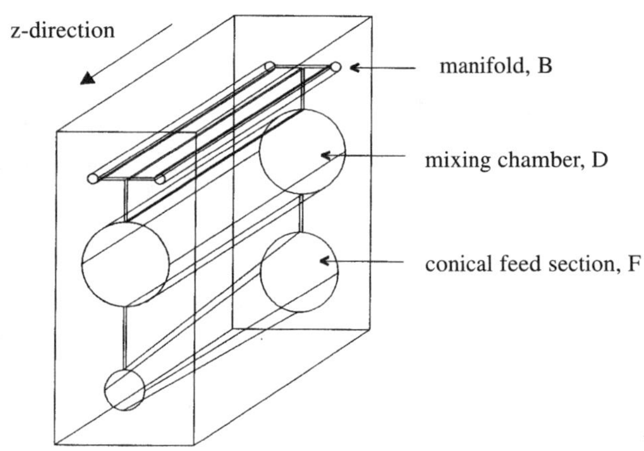

Figure 14. Three-dimensional view of the GI die block.

The homogeneous blend prepared in the mixing chamber D then flows through the slit channel E to the conical feed section F, where a rotating cone F_1 is placed. The material flowing to the conical feed section F at a location with a smaller value of *z*, which is richer in material *a*, is positioned closer to the surface of the rotating cone F_1. As that material flows along the *z* direction in the conical feed section F, it is coated by the incoming lower refractive index material at larger *z*. While the material flows downstream in the axial direction, the rotating cone F_1 distributes it uniformly in the circumferential direction. Due to the high viscosity of polymers, a laminar flow is maintained and the circumferential distribution of the material by the rotating cone converts the axial-directional refractive index profile achieved in the mixing chamber to a radial-directional profile. Consequently, the material that leaves the circular exit hole G has a refractive index profile that decreases in the radial direction. When this material is stretched as in a fiber spinning process, a GI-POF finally is obtained.

The refractive index profile of the GI-POFs fabricated by this method is determined mainly by the axial profile (i.e., *z*-directional profile) of the blend composition in the mixing chamber D, which is determined by the dimensions of the slit channels C_1 and C_2. Once the rheological properties of the optical polymers *a* and *b* are known, the dimensions of the slit channels C_1 and C_2 can be determined accordingly to achieve a predetermined refractive index profile in the mixing chamber. In addition, minor adjustment of the refractive index profile can be achieved by adjusting the *z*-directional flow rate profile in the slit channel E. This minor tailoring of the refractive index profile is achieved via flow restricters E_1, which control the gap of the slit channel E.

This method is capable of controlling the refractive index profile by mechanical means. Although the GDB is a new and novel concept, its major elements are reminiscent of existing extrusion technologies. The flow in the manifolds and the connecting slit channels is similar to that in a slit die, which is often used in a flat film casting process. In addition, the mixing chamber resembles the static kneading mixers used for polymer melt blending. Thus, the well-established design concepts for slit dies or static mixer can be applied directly to the design of GDB. Although the description given here involved two polymers with different refractive indices, the polymers can be copolymers with different comonomer compositions or polymers that contain refractive-index-modifying additives at different concentrations. Thus, the materials used in most other processes also can be utilized for the GDB method.

3.3. Multilayer POFs

The refractive index profile of a GI-POF should be close to a parabolic shape to achieve maximum bandwidth, as discussed in the beginning of Section 3. However, multilayer

POFs, which consist of core, cladding, and more than one intermediate layers, also offer much higher bandwidth than SI-POFs. Several methods have been suggested for the manufacture of multilayer POFs. Most methods used to make SI-POFs easily can be modified to make multilayer POFs. In addition, some methods described previously for GI-POFs (e.g., centrifugal method, CVD method, Katoot's method) also can be used to fabricate preforms that have a concentric multilayer structure.

In the method introduced by Perry and Witcher [91], the conventional coextrusion process is used to make a fiber of three or more concentric layers. The multilayer fiber is then passed through a cylindrical furnace where the fiber is exposed to a temperature higher than its melting point. It is claimed that the stepwise index profile can be smoothed out by the diffusion of polymer molecules between adjacent layers, forming a smooth profile GI-POF. However, considering the fact that the diffusivity of polymer molecules is extremely low, it is unlikely that a smooth index profile can be achieved. Nevertheless, as will be described in Section 5, a small change in the refractive index profile can result in a substantial increase in the bandwidth.

Although the fabrication method has not been published, a high bandwidth multilayer POF was introduced in 1999 by Mitsubishi Rayon under the product name Eska-Miu™. It is a four-layer POF with a numerical aperture of 0.25 and a bandwidth of 250 MHz for 100 m. It meets the specification of IEEE1394 category S400, and is to be applied for the backbone of home networks.

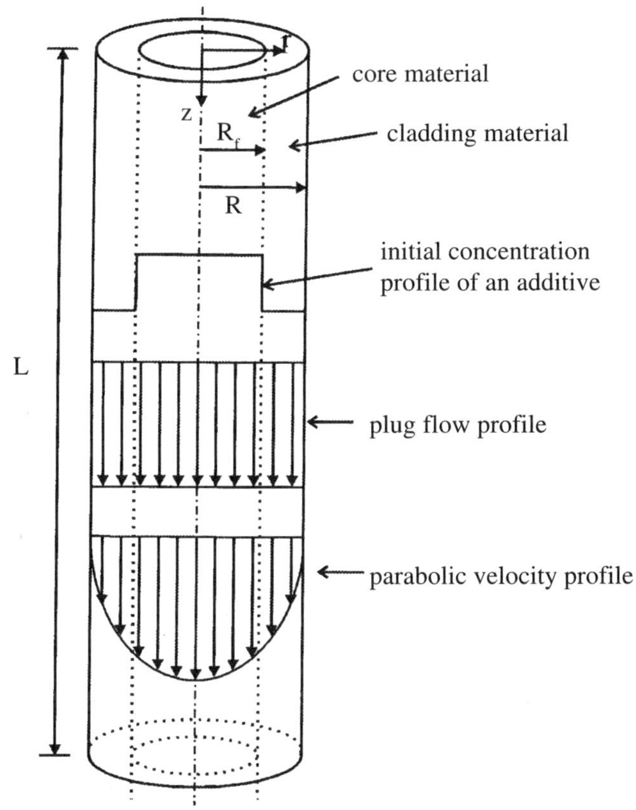

Figure 15. Initial concentration and velocity profiles in the diffusion zone.

4. CONVECTIVE DIFFUSION IN VARIOUS COEXTRUSION METHODS

4.1. Diffusion of an Additive at the Core-Cladding Interface

This section provides a brief analysis of the diffusion of an additive at a polymer–polymer interface. Although this analysis specifically relates to the diffusion-assisted coextrusion process described in Section 3.2.4, it is also relevant to other processes such as the IDSE method (Sect. 3.2.1) and the closed coextrusion method (Sect. 3.2.2).

The additive concentration profile is determined mainly in the diffusion zone. Thus, only the convection diffusion in the diffusion zone is considered, and the flow situation there is described schematically in Figure 15. The additive concentration profile (hence the refractive index profile) created in the diffusion zone can be described by the convective diffusion equation

$$\mathbf{u} \cdot \nabla x_A - D_A \nabla^2 x_A = 0 \tag{5}$$

were $\mathbf{u}$ is the velocity field, x_A is the additive concentration profile, and D_A is the diffusivity of the additive. For simplicity, we consider a case in which the core and the cladding materials are the same, and only the core contains an additive (e.g., PMMA with BEN for the core and pure PMMA for the cladding). The concentration of an additive may be as high as 20% by weight, which is large enough to influence the rheological properties of the core material significantly. For simplicity, however, the change in the rheological properties is neglected.

Under the prescribed assumptions, Eq. (5) gives the following dimensionless equation in a cylindrical coordinate system:

$$u\frac{\partial x}{\partial z} = k^2\left(\frac{\partial^2 x}{\partial r^2} + \frac{1}{r}\frac{\partial x}{\partial r}\right) \tag{6}$$

Here k^2 is a dimensionless parameter defined as $1/(\varepsilon^2 Pe)$, where ε is the ratio of the two characteristic length scales, R/L, and Pe is the Péclet number defined as UL/D_A. R and L are the radius and the length of the diffusion zone, and U is the average velocity of the material in the diffusion zone. The scalings for the radial and the axial coordinate r and z are obviously R and L, respectively. The axial velocity u is scaled by U and the additive concentration x is scaled by the initial concentration in the core, x_0. Because the diffusion zone has a large aspect ratio [i.e., $\varepsilon < O(10^{-2})$], the diffusional flux in the axial direction is negligibly small compared to that in the radial direction. Thus, Eq. (6) represents the balance between axial convection and radial diffusion.

The initial and boundary conditions are

$$x(r,0) = 1 \quad \text{when } r \leq R_f$$
$$x(r,0) = 0 \quad \text{when } r > R_f \tag{7}$$
$$\frac{\partial x}{\partial r} = 0 \quad \text{at } r = 0 \tag{8a}$$

and

$$\frac{\partial x}{\partial r} = 0 \quad \text{at } r = 1 \tag{8b}$$

The initial conditions (7) represent the step concentration profile at the inlet to the diffusion zone (i.e., at $z = 0$). R_f is the location of the interface between the core and the cladding, which can be varied by adjusting the flow rate ratio of the core and the cladding materials. Boundary conditions (8a) and (8b) represent the symmetry at the center and the no-flux condition at the wall, respectively.

In the IDSE method or the closed coextrusion method, the material that is diffusing is not an additive but a monomer. In addition, the fiber surface in the diffusion zone (or volatilization zone) is stress-free. Thus, the axial velocity u (scaled by the average velocity U) for both cases is simply 1. In the IDSE method, there is no cladding; thus, $R_f = 1$. Furthermore, because evaporation of the monomer at the fiber surface is induced by the flow of an inert gas, the boundary condition at the fiber surface should be [i.e., Eq. (8b) should be replaced by]

$$\frac{\partial x}{\partial r} = -h \cdot x \quad \text{at } r = 1 \tag{9}$$

Here h is the local mass transfer coefficient, which is a function of the flow characteristics of the inert gas. Equation (9) is, in fact, the same as Eq. (3) in Section 3.2.2.

In the diffusion zone of the diffusion-assisted coextrusion method (Sect. 3.2.4), the no slip condition applies at $r = 1$. Thus, the velocity field is more complicated than a plug flow. To proceed with the analysis, it is further assumed that the velocity profile is parabolic. This simplification is equivalent to assuming that the core and the cladding materials are Newtonian and have the same viscosity. This assumption may be an oversimplification because the materials in the actual process are not Newtonian. However, as will be discussed later, the results obtained for a plug flow (i.e., closed coextrusion method) are not very much different from those for a parabolic flow, which indicates the small influence of the velocity field. Therefore, the Newtonian assumption may not alter the conclusions drawn from this analysis.

The solution to Eqs. (6), (7), and (8) can be obtained by following a similar procedure for a classical heat transfer problem tackled by Graetz [92] and Siegel et al. [93]. Because details of the solution can be found elsewhere [82, 88], a mathematical description of the solution is omitted here and the results are given only in a graphical form. In Figure 16, the additive concentration profile obtained by the diffusion-assisted coextrusion method is plotted for three different cases. The additive concentration profiles are directly related to the refractive index profiles through the Lorentz–Lorenz equation [94]. Unless the density difference between the polymer and the additive is very large, a linear relationship holds between the additive concentration and the refractive index. Thus, the concentration profiles given in Figure 16 should match the refractive index profiles closely.

The value of the diffusivity for Figure 16 was chosen because it represents an additive that has a molecular weight of about 500 diffusing into PMMA at 200 ~ 230 °C [95, 96]. The three concentration profiles that correspond to various values of the core–cladding interface position are similar to each other; only minor differences are evident. The diffusion of the additive occurs to a slightly greater extent when the interface position is farther away from the center axis (i.e., for a larger value of R_f), due to the larger interface area through which the diffusion takes place. In addition, the lower velocity near the outer radius of the die, which makes the material residence time longer, also may contribute to this slightly enhanced diffusion.

Figure 16 indicates that the variation of additive concentration (or refractive index) in the radial direction is restricted to the region near the core–cladding interface unless the material residence time is very long. In addition, the no-flux condition at $r = 1$ makes the slope of the concentration profile at the fiber surface zero, as was discussed previously. Consequently, the profile is quite different from the ideal parabolic shape. Thus, the very large bandwidth that can be obtained with an optimum parabolic profile is not expected. Nevertheless, as will be described in Section 5, a small change in the concentration profile can result in a substantial increase in the bandwidth compared to that of the step-index profile. In addition, smoothening of the refractive index profile is expected to improve the attenuation of the

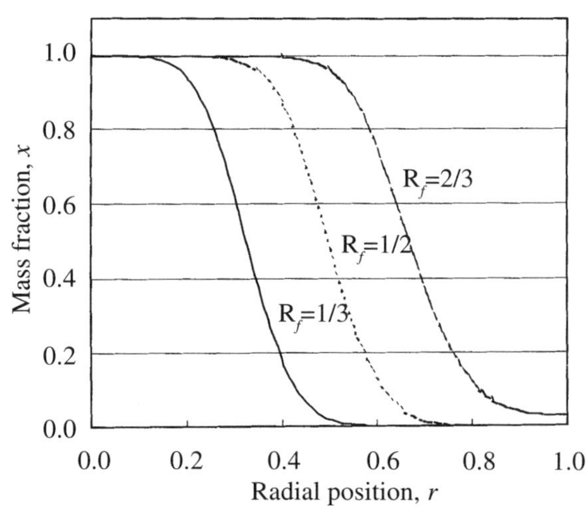

Figure 16. Additive concentration profiles at various values of R_f ($k^2 = 6.73$, $R = 0.384$ cm, $L = 50$ cm, $Q = 100$ g/h, $D_A = 10^6$ cm^2/s).

optical fiber, because the sharp interface between the core and the cladding of SI-POFs is often a source of significant scattering loss.

The concentration profiles obtained by the closed coextrusion process under the same processing conditions as Figure 16 are shown to be very similar to those in Figure 16, although the diffusion takes place to a slightly greater extent in the plug flow situation than in the parabolic flow situation [88], due to the shorter contact time between the core and the cladding materials in the case of parabolic flow. Although the average velocity (hence, the average residence time) is the same for both processes, the local velocity of the core material is higher than that of the cladding in parabolic flow. Consequently, the net contact time between the neighboring fluid elements across the core–cladding interface is shorter and the extent of diffusion is decreased.

Theoretical analysis can be extended easily to the case of diffusion in multilayer coextrusion. Obviously, the concentration profiles for multilayer fibers show larger variation than two-layer fibers under the same processing conditions. As will be shown, the bandwidth also increases with increasing number of layers. In principle, the concentration profile can be manipulated to match the ideal parabolic profile closely by using multilayer processes.

4.2. Enhanced Diffusion by a Concentric Annuli Structure

Molecular diffusion is a very slow process that may limit the utility of various fabrication techniques for commercial applications. A simple modification, however, can enhance the diffusion process substantially [97]. Because the rate of mass transfer is proportional to the area through which diffusion occurs, increasing the area of contact will enhance the diffusion process. In Figures 17 and 18, a simple modification of a coextrusion die is shown for the diffusion-assisted coextrusion method and the closed coextrusion method, respectively. The presence of the inner mandrel increases the interfacial area substantially, thereby increasing the rate of mass transfer. For example, if the diameter of the inner mandrel is 1 cm, the interfacial area increases by a factor of more than 10 compared to the case of a 1 mm core without the mandrel when the lengths of the diffusion zones are the same. Consequently, the mass transfer rate is expected to increase by the same factor. Recent analyses and experiments have, in fact, shown the expected increase of the mass transfer rate due to this modification [97].

5. BANDWIDTH ESTIMATION BY RAY ANALYSIS

In this section, the bandwidth characteristics of the fibers that have refractive index profiles that are significantly different from a parabolic shape are estimated by ray analysis. The bandwidth of GI fiber that has a refractive index profile that follows the power-law model [Eq. (1)] can be estimated by

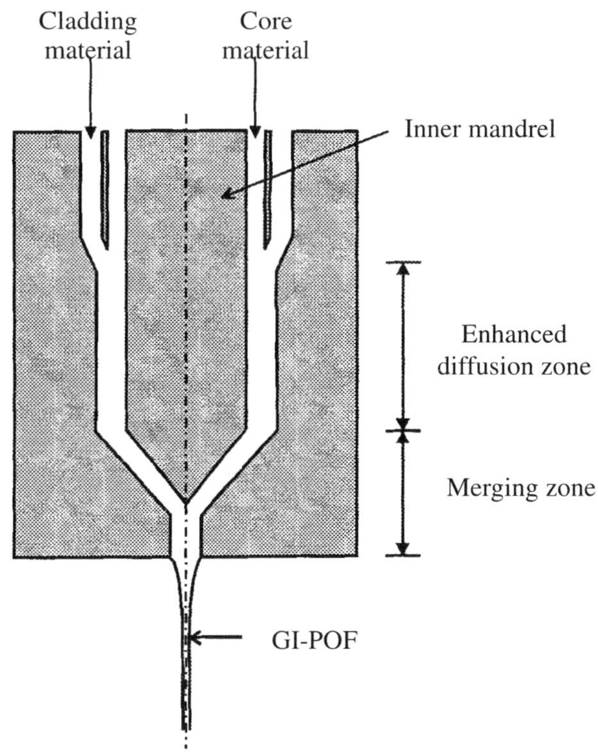

Figure 17. Modification of a coextrusion die for the diffusion-assisted coextrusion method.

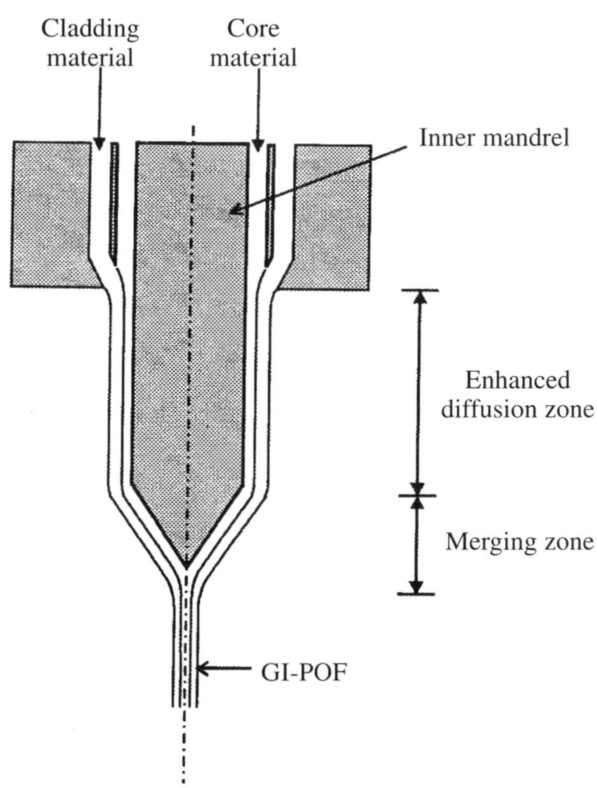

Figure 18. Modification of a coextrusin die for the closed coextrusion method.

the Wentzel–Kramers–Brillouin (WKB) approximation [51]. However, when the refractive index profile is given by a rather complicated form other than the power-law model, a simple analytic expression for the bandwidth may not be available. Although several different approaches can be applied for estimation of the bandwidth, ray analysis is adopted here because of its simplicity. In geometrical optics, which can be applied when the dimension of optical elements is larger than the wavelength of light by orders of magnitude, light is considered to propagate in a straight line in a homogeneous medium. In an inhomogeneous medium, the trajectory of the light ray is described by [98]

$$\frac{d}{ds}\left(n\frac{d\mathbf{r}}{ds}\right) = \nabla n \tag{10}$$

where $\mathbf{r}$ is the position vector of a point on the ray path and s is the arc length of the propagating ray measured from a fixed point along the trajectory. In Figure 19, a ray path in an optical fiber with core radius R and refractive index $n(r)$ is shown.

The direction of a propagating ray at a radial position r can be defined by two angles θ and ϕ: θ is the angle between the tangent line to the ray and the fiber axis (i.e., z axis); ϕ is the angle that the projection of the tangent line onto the fiber cross section makes with the azimuthal direction. It can be shown from Eq. (10) that two parameters κ and l defined by

$$\kappa = n(r)\cos\theta(r) \tag{11}$$

$$l = \frac{r}{R}n(r)\sin\theta(r)\cos\phi(r) \tag{12}$$

remain constant along a given ray path. Thus, all rays propagating in a fiber can be completely defined by the values of these two invariants [99]. In the preceding equations, κ accounts for the translational invariance of the fiber along its axis, whereas l is related to the skewness of a ray. Although some of the rays are meridional ($l = 0$) that is, their travel paths sit on the plane that possesses the fiber axis (i.e., z axis in Fig. 19), many of them are skew rays, which do not pass through the fiber axis, but propagate along helical paths. It can be shown that all the rays that satisfy the inequality

$$n_2 \leq \kappa \leq n_1 \tag{13}$$

are guided by the optical fiber. Here n_1 and n_2 denote the refractive indices at the center and at the outer edge of the optical fiber.

An input light impulse can be considered as a packet of rays that have various values of κ and l. Thus, the incident rays travel through various paths inside the optical fiber and their net travel distances vary as they exit the fiber. Consequently, their travel time through the optical fiber has a spread (or distribution) that is called the modal dispersion. In a GI optical fiber, the local speed of light is higher at a position further away from the center of the fiber, because the refractive index is smaller. Thus, a ray propagating through a longer helical path will travel with a higher average velocity, thereby reducing the modal dispersion.

In ray optics, estimation of the modal dispersion can be achieved by calculating the net travel time and transmitted power of each ray in the packet following its propagation path from inlet to outlet. The bounded rays follow a periodic path within the fiber core as shown in Figure 19, and the ray transit time over an axial distance z for a given value of κ and l is given as [100]

$$\tau = \frac{2z}{cz_p}\int_{r_{ic}}^{r_{tp}} \frac{n^2(r)}{\sqrt{g(r)}}dr \tag{14}$$

where

$$g(r) = n^2(r) - \kappa^2 - l^2(r/R)^2$$

where c is the speed of light in vacuum and z_p is the axial half period, which is given as

$$z_p = 2\kappa\int_{r_{ic}}^{r_{tp}} \sqrt{g(r)}dr \tag{15}$$

The integration limits r_{ic} and r_{tp} are the radii of the so-called *inner caustic* and *turning point caustic,* respectively, between which the ray path is bounded (Fig. 19). These caustics are zeros of the function $g(r)$ [i.e., $g(r_{ic}) = 0$ and $g(r_{tp}) = 0$].

For a uniform Lambertian source, which is the most typical source in practice and approximates the output of a light emitting diode, the intensity distribution of input light is given as [99]

$$I = I_0\cos\theta \tag{16}$$

and the power transmitted by the rays, which have the values of the two invariants in the range of κ to $(\kappa + d\kappa)$ and l to $(l + dl)$, is

$$dP = P_0 z_p(\kappa, l)d\kappa\, dl \text{ where } P_0 = 4\pi R I_0/n_0^2 \tag{17}$$

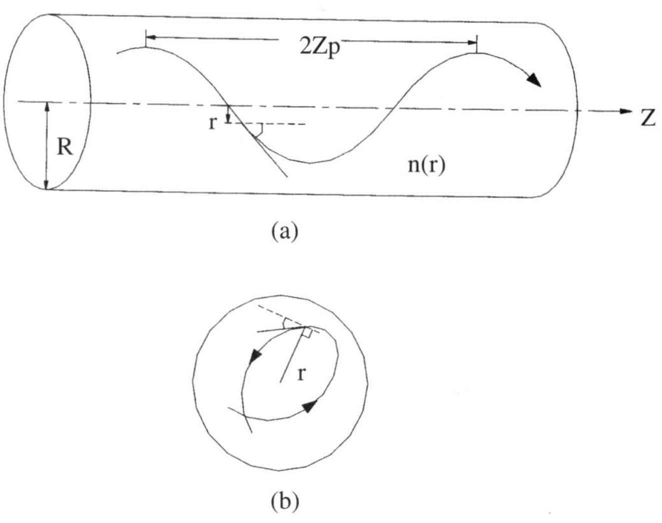

Figure 19. Schematic of a skew ray path in an optical fiber: (a) r-z plane; (b) cross-sectional view.

The transmitted power also can be expressed in terms of transit time to give impulse response function $Q(\tau)$ as

$$dP = Q(\tau)d\tau \tag{18}$$

Following the approach of Barrell and Pask [100], the $\kappa - l$ domain is discretized into a finite number of subdomains, and by calculating the transit time and the transmitted power in each subdomain, the impulse response curves for various refractive index profiles are obtained. The bandwidth is then calculated as the inverse of four times the standard deviation of the impulse response curve. In Figure 20, various refractive index profiles are given for which the bandwidth has been estimated by the prescribed ray analysis. In calculating the bandwidth, the $\kappa - l$ domain was divided into 1000×500 subdomains and the case for which $n_1 = 1.515$ and $n_2 = 1.498$ was considered.

As discussed in the beginning of Section 3, the impulse response or the bandwidth of POFs also is affected by other factors such as material dispersion, mode coupling, and differential mode attenuation [52–55]. These effects are either negligible or act in such a way as to increase the bandwidth of the POFs. Thus, the current analysis, in which the material dispersion, DMA, and mode coupling have not been considered, may provide the lower limit of the bandwidth.

The bandwidth estimates for a 100 m fiber that has the refractive index profiles given in Figure 20 are shown in Table 1. The bandwidth of the two-layer SI fiber is calculated to be 154 Mbps (megabits per second), which is consistent with the result of the WKB approximation. As the number of layers in a multilayer POF increases, the bandwidth increases almost linearly. It is interesting to note that the bandwidth of a four-layer POF, estimated to be 242 Mbps for 100 m, is close to the reported value for the Eska-Miu of Mitsubishi Rayon. When the number of layers is increased to nine, the bandwidth reaches about 400 Mbps for 100 m.

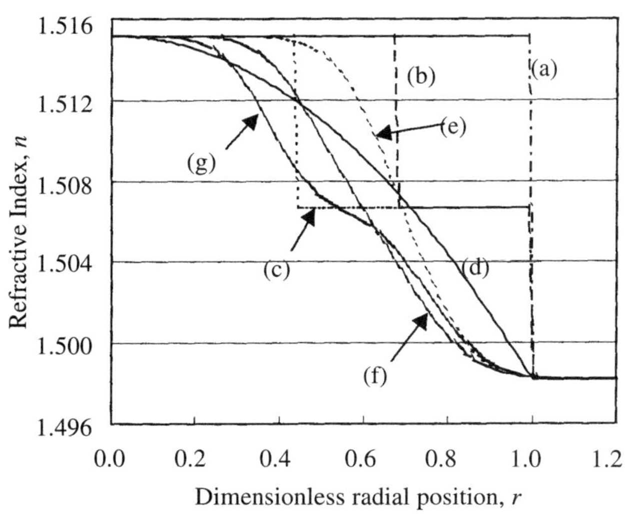

Figure 20. Various refractive index profiles used for bandwidth estimates: (a) two-layer step index; (b) three-layer step index ($R_f = 0.4, 0.6$); (c) three-layer step index ($R_f = 0.3, 0.7$); (d) ideal graded index ($a = 2.0$); (e) diffusion-assisted two-layer fiber; (f) diffusion-assisted three-layer fiber ($R_f = 0.4, 0.6$); (g) diffusion-assisted three-layer fiber ($R_f = 0.3, 0.7$).

Table 1. Bandwidths obtained by ray analysis for various refractive index profiles given in Figure 20 ($n_1 = 1.515$, $n_2 = 1.498$; NA = 0.226)

Optical fiber	Bandwidth (Mbps for 100 m)
Two-layer SI fiber (Fig. 20a)	154
Three-layer SI fiber (sharp interfaces; Fig. 20b)	200
Three-layer SI fiber (sharp interfaces; Fig. 20c)	180
Four-layer SI fiber (sharp interfaces)	242
Six-layer SI fiber (sharp interfaces)	327
Nine-layer SI fiber (sharp interfaces)	415
Diffusion-assisted two-layer fiber (Fig. 20e)	319
Diffusion-assisted three-layer fiber (Fig. 20f)	369
Diffusion-assisted three-layer fiber (Fig. 20g)	413

If the sharp interfaces of a multilayer POF are smoothened by diffusion, the increase in bandwidth is quite significant. The bandwidth of a SI-POF is more than doubled to 319 Mbps for 100 m when the sharp interface of the SI fiber is smoothened by additive diffusion. Considering a rather small change in the refractive index profile, this large increase in bandwidth may have significant implications. As Table 1 indicates, bandwidth higher than 400 Mbps for 100 m is obtainable only with three layers if the shape interfaces are smoothened by diffusion, whereas nine layers are required for the same bandwidth if the interfaces are sharp.

6. SUMMARY

Various techniques for the fabrication of POFs have been discussed. Many of the methods, especially those for the fabrication of GI-POFs, have some unique features and are capable of producing GI-POFs with high bandwidth. The bandwidth estimated by ray analysis indicates that even a small variation of the refractive index profile from a step-index shape can result in a significant increase in bandwidth.

REFERENCES

1. G. P. Agrawal, "Fiber-Optic Communication Systems," 2nd ed. Wiley-Interscience, New York, 1997.
2. Mitsubishi Rayon Co., U.K. Patent 1,431,157, 1974.
3. Mitsubishi Rayon Co., U.K. Patent 1,449,950, 1974.
4. C. Emslie, *J. Mater. Sci.* 23, 2281 (1988).
5. F. P. Kapron, D. B. Keck, and R. D. Maurer, *Appl. Phys. Lett.* 17, 423 (1970).
6. T. Kaino, K. Jinguji, and S. Nara, *Appl. Phys. Lett.* 42, 567 (1983).
7. Y. Ohtsuka and Y. Hatanaka, *Appl. Phys. Lett.* 29, 735 (1976).
8. Nippon Selfoc Co., U.S. Patent 3,955,015, 1976.
9. Y. Ohtsuka, T. Senga, and H. Yasuda, *Appl. Phys. Lett.* 25, 659 (1974).
10. Y. Ohtsuka, T. Sugano, and Y. Terao, *Appl. Opt.* 20, 2319 (1981).
11. Y. Ohtsuka and I. Nakamoto, *Appl. Phys. Lett.* 29, 559 (1976).
12. Y. Ohtsuka, Y. Koike, and H. Yamazaki, *Appl. Opt.* 20, 280 (1981).
13. Y. Koike, H. Hatanaka, and Y. Ohtsuka, *Appl. Opt.* 23, 1779 (1984).
14. Y. Ohtsuka and Y. Koike, *Appl. Opt.* 24, 4316 (1985).
15. Y. Koike, Y. Takezawa, and Y. Ohtsuka, *Appl. Opt.* 27, 486 (1988).

16. Y. Koike, N. Tanio, E. Nihei, and Y. Ohtsuka, *Polym. Eng. Sci.* 29, 1200 (1989).
17. Y. Koike, *Polymer* 32, 1737 (1991).
18. Y. Koike and E. Nihei (Kawasaki), U.S. Patent 5,253,323, 1993.
19. T. Ishigure, E. Nihei, and Y. Koike, *Appl. Opt.* 33, 4261 (1994).
20. Y. Koike, T. Ishigure, and E. Nihei, *J. Lightwave Technol.* 13, 1475 (1995).
21. T. Ishigure, A. Horibe, E. Nihei, and Y. Koike, *J. Lightwave Technol.* 13, 1686 (1995).
22. T. Ishigure, E. Nihei, and Y. Koike, *Appl. Opt.* 35, 2048 (1996).
23. T. Ishigure, M. Satoh, O. Takanashi, E. Nihei, T. Nyu, S. Yamazaki, and Y. Koike, *J. Lightwave Technol.* 15, 2095 (1997).
24. J. Senior "Optical Fiber Communications," Prentice-Hall, London, 1985.
25. Toray Industry, Japanese Patent Publication 07-270629, 1995.
26. S. R. Nagel, J. B. MacChesney, and K. L. Walker, in "Optical Fiber Communications" (T. Li, Ed.), Vol. 1, Chap. 1. Academic Press, San Diego, 1985.
27. S. H.-K. Lee and Y. Jaluria, *J. Mat. Proc. Manufacturing Sci.* 3, 317 (1995).
28. S. H.-K. Lee and Y. Jaluria, *Int. J. Heat Mass Transfer* 40, 843 (1997).
29. A. M. Mescher, H. M. Reeve, and J. B. Dixon, "Proceedings of the International Polymer Optical Fibers Technical Conference," Cambridge, MA, 2000, p. 26.
30. T. Izawa and S. Sudo, "Optical Fibers: Materials and Fabrication," Kluwer, Boston, 1987.
31. D. Markuse, *Appl. Opt.* 14, 3021 (1975).
32. L. R. Glicksman, *Glass Technol.* 9, 131 (1968).
33. G. Manfre, *Glass Technol.* 10, 99 (1969).
34. J. A. Burgman, *Glass Technol.* 11, 110 (1970).
35. S. Kase and T. Matsuo, *J. Appl. Polym. Sci.* 11, 251 (1967).
36. M. M. Denn, *Ann. Rev. Fluid Mech.* 12, 365 (1980).
37. C.-W. Park, *AIChE J.* 30, 197 (1990).
38. A. Ziabicki, in "Man-made Fibers" (M. F. Mark, S. M. Atlas, and E. Cernia, Eds.), p. 13. Interscience, New York, 1967.
39. J. R. A. Pearson and M. A. Matovich, *Ind. Eng. Chem. Fund.* 8, 605 (1969).
40. S. Kase, *J. Appl. Polym. Sci.* 18, 3279 (1974).
41. W.-S. Lee, C.-W. Park, *J. Appl. Mech.* 62, 511 (1995).
42. F. T. Geyling and G. M. Homsy, *Glass Technol.* 21, 95 (1980).
43. Sumitomo Electric Industries, Japanese Patent Publication 07-234323, 1995.
44. Sumitomo Electric Industries, Japanese Patent Publication 07-234325, 1995.
45. T. Kaino, M. Fujiki, and S. Nara, *J. Appl. Phys.* 52, 7061 (1981).
46. T. Kaino, M. Fujiki, S. Nara, and S. Oikawa, *Appl. Opt.* 20, 2886 (1981).
47. Nippon Telegraph & Telephone, U.K. Patent 2,089,352B, 1981.
48. Mitsubishi Rayon Co., Japanese Patent Publication 07-104133, 1995.
49. P. Halley, "Fiber Optic Systems," Wiley, New York, 1987.
50. D. Gloge and E. A. J. Marcatili, *Bell System Tech. J.* 52, 1563 (1973).
51. R. Olshansky and D. B. Keck, *Appl. Opt.* 15, 483 (1976).
52. A. F. Garito, J. Wang, and R. Gao, *Science* 281, 962 (1998).
53. T. Ishigure, M. Kano, and Y. Koike, "Proceedings of the 8th International Polymer Optical Fiber Conference," Chiba, Japan, 1999, p.106.
54. T. Ishigure, M. Kano, and Y. Koike, *J. Lightwave Technologies* 18, 959 (2000).
55. G. Yabre, *J. Lightwave Technol.* 18, 869 (2000).
56. K. Matsushita and M. Toyama, *Appl. Opt.* 19, 1070 (1980).
57. J. C. Palais, *Appl. Opt.* 19, 2011 (1980).
58. W. J. Tomlinson, *Appl. Opt.* 19, 1127 (1980).
59. Y. Koike, N. Tanio, and Y. Ohtsuka, *Macromolecules* 22, 1367 (1989).
60. Y. Kioke, S. Matsuoka, and H. E. Bair, *Macromolecules* 25, 4807 (1992).
61. Y. Koike and E. Nihei, U.S. Patent 5,382,488, 1995.
62. Y. Koike, U.S. Patent 5,541,247, 1996.
63. M. Sato, T. Ishigure, and Y. Koike, *J. Lightwave Technol.* 18, 952 (2000).
64. N. Tanio, H. Kato, Y. Koike, H. E. Bair, S. Matuoka, and L. L. Blyler, Jr., *Polymer J.* 30, 56 (1998).
65. Y. Koike, M. Naritomi, H. Murofushi, and N. Sugiyama, U.S. Patent 5,760,139, 1998.
66. Y. Koike and M. Naritomi, U.S. Patent 5,783,636, 1998.
67. Y. Koike, H. Murofushi, and N. Sugiyama, U.S. Patent 5,916,971, 1999.
68. Y. Koike, "Proceedings of the International Polymer Optical Fiber Technical Conference," Cambridge, MA, 2000, p. 1.
69. T. Ishigure, Y. Koike, and J. W. Fleming, *J. Lightwave Technol.* 18, 178 (2000).
70. T. Ishigure and Y. Koike, "Proceedings of the International Polymer Optical Fiber Technical Conference," Cambridge, MA, 2000, p. 14.
71. F. G. H. van Duijnhoven and C. W. M. Bastiaansen, *Appl. Opt.* 38, 1008 (1999).
72. T. Nonake, T. Hosoya, Y. Kobayashi, and Y. Matsuda, U.S. Patent 5,851,666, 1998.
73. T. Nonake, T. Hosoya, Y. Kobayashi, and Y. Matsuda, U.S. Patent 5,891,570, 1999.
74. S. Tripathy, K. G. Chittibabu, and J. Kumar, U.S. Patent 5,935,491, 1999.
75. V. M. Ilvashenko, U.S. Patent 6,086,999, 2000.
76. M. W. Katoot, U.S. Patent 5,861,129, 1999.
77. Mitsubishi Rayon Co., U.S. Patent 4,822,122, 1989.
78. Mitsubishi Rayon Co., U.S. Patent 4,852,982, 1989.
79. Mitsubishi Rayon Co., Japanese Patent 2-16505, 1990.
80. C. Tsai, T. Liu, Y. Chang, and W. T. W. Tseng, *Chem. Eng. Sci.* 52, 221 (1997).
81. N. Toyoda, Y. Mishina, R. Murata, Y. Uozu, M. Oda, and T. Ishimaru, U.S. Patent 5,390,274, 1990.
82. W. Chen, Y. Chang, and J. Hsu, *J. Phys. Chem. B* 103, 7584 (1999).
83. B. C. Ho, J. H. Chen, W. C. Chen, Y. H. Chang, S. Y. Yang, J. J. Chen, and T. W. Tseng, *Polym. J.* 27, 310 (1995).
84. W. C. Chen, J. H. Chen, S. Y. Yang, and J. Y. Cheng, Y. H. Chang, and B. C. Ho, *J. Appl. Polym. Sci.* 60, 1379 (1996).
85. B. C. Ho, J. H. Chen, S. Y. Yang, and Y. H. Chang, U.S. Patent 5,555,525, 1996.
86. B. T. Liu, W. C. Chen, and J. P. Hsu, *Polymer* 40, 1451 (1999).
87. Y. Koike and R. Nihei, U.S. Patent 5,593,621, 1997.
88. I.-S. Sohn and C.-W. Park, *Ind. Eng. Chem. Res.,* 88: 40, 3740 (2001).
89. C.-W. Park, B. S. Lee, J. K. Walker, and W. Y. Choi, *Ind. Eng. Chem. Res.* 39, 79 (2000).
90. C.-W. Park and J. K. Walker, U.S. Patent 6,267,915, 2001.
91. G. A. Perry and C. E. Witcher, U.S. Patent 5,235,660, 1993.
92. M. Jakob, "Heat Transfer," Wiley, New York, 1962.
93. R. Siegel, E. M. Sparrow, and T. M. Hallman, *Appl. Sci. Res. A* 7, 386 (1958).
94. L. A. Hornak, "Polymers for Lightwave and Integrated Optics," Dekker, New York, 1992.
95. D. Ehlich and H. Sillescu, *Macromolecules* 23, 1600 (1990).
96. G. Heuberger and H. Sillescu, *J. Phys. Chem.* 100, 1525 (1996).
97. K. S. Yoon and T. H. Kang, personal communication.
98. M. Born and E. Wolf, "Principles of Optics," 6th ed. Pergamon, New York, 1999.
99. A. W. Snyder and J. D. Love, "Optical Wavequide Theory," Chapman and Hall, New York, 1983.
100. K. F. Barrell and C. Pask, *Appl. Opt.* 19, 1298 (1980).

CHAPTER 2

Gradient-Index Polymer Optical Fibers: Analysis of Fabrication Techniques

Wen-Chang Chen
Department of Chemical Engineering, National Taiwan University, Taipei, Taiwan 10617

Jui-Hsiang Liu
Department of Chemical Engineering, National Cheng Kung University, Tainan, Taiwan 70101

Yung Chang, Ming-Hsin Wei, Hong-Wen Su
Department of Chemical Engineering, National Taiwan University, Taipei, Taiwan 10617

CONTENTS

1. INTRODUCTION

Gradient-index polymer optical fibers (GI POFs) have attracted extensive interest in light of their versatile applications in optical communication, imaging, sensing, and collimating [1–4]. The originally developed GI POFs were based on the poly(methyl methacrylate) (PMMA) and had an optical loss of ~100–400 dB/km in the visible wavelength, which is only suitable for short range data communication. During the last few years, Professor Koike of Keio University has collaborated with Asahi Chemical Company to develop perfluorinated GI POF for optical communication using the

ISBN: 1-58883-012-8
Copyright © 2004 by American Scientific Publishers
All rights of reproduction in any form reserverd.

near infrared wavelength [1–3]. The theoretical attenuation limit of perfluorinated (PF) GI POF was estimated to be as low as 0.3 dB/km, which is comparable to that of a silica fiber in the near infrared region. GI POFs based on PF polymer that have a transmission speed greater than 10 Gbit/s at the near infrared wavelength have been reported [5]. GI POFs used as optical fiber amplifiers by doping with organic dyes also have been reported [6]. By cutting the prepared GI POFs into short lenses and assembling the lenses into an array, they could be used as essential components of the Selfoc lens array in fax machines and scanners [7].

Basically, three essential properties are considered for GI POFs: refractive index distribution (RID), optical loss, and bandwidth. The RID inside a GI POF can be described by

$$n(r) = n_1\left[1 - 2\delta\left(\frac{r^g}{R}\right)\right]^{1/2}, \quad 0 \leq r \leq R \qquad (1)$$

$$\delta = \frac{n_1 - n_2}{n_1} \qquad (2)$$

where n_1 and n_2 are the refractive indices at the center axis and the periphery of an optical fiber, R is the radius, and g is a constant that characterizes the RID. The bandwidth of a GI POF was found to be maximized at the g value of 2.0 [8]. Therefore, the optimal RID for a GI POF is a parabolic curve. However, Ishigure et al. [9] demonstrated the important influence of material dispersions on the bandwidth of GI POF. They concluded that the g values should be in the range of 1.7–3 to achieve high transmission speed (>1 Gbit/s) with a light source spectral width greater than 2 nm. Controlling both $\Delta n > 0.01$ and $g \cong 2$ is very important for applications of GI lens to imaging applications [7].

Consideration of both material compositions and essential process parameters may enable the RID of a GI POFs to be controlled. Basically two kinds of material systems have been studied extensively in the literature [10]. The first is the polymer/monomer mixture [e.g., (PMMA)/methyl methacrylate (MMA)/benzyl methacrylate (BzMA) [11]. The mutual diffusion of monomers inside the fiber forms a composition distribution and then polymerizes to become a GI POF by thermal or photopolymerization. In this case, the diffusion rates of the monomers inside the polymer are very important for controlling the RID. Another materials system is the polymer/unreactive dopant mixture [12, 13] [e.g., PMMA/benzyl benzoate (BB)]. In this case, the dopant diffuses into the polymer matrix to form a composition distribution and then a refractive index profile is produced. In the former case, the long-term stability of the RID can be maintained. However, the difference in the reactivities of various monomers causes phase separation and large scattering loss; therefore, the optical losses of the prepared GI POFs are basically larger than 100 dB/km. In the latter case, there is a lower optical loss than in the previous case because excess scattering loss from the difference of polymerization speed of monomers does not exist. However, dopant diffusion might occur with time or environment, and thus high temperature or long-term stability is poor for the prepared GI POFs [13]. Many methods for preparing GI POFs, including two-stage copolymerization [14], photocopolymerization [15], centrifugal process [16], interfacial-gel copolymerization [11], and coextrusion [7, 17–19], have been reported in the literature. Among these methods, interfacial-gel copolymerization, the centrifugal process, and coextrusion are the most promising approaches for practical applications. In the first two processes for preparing GI POFs, a preform with a large core diameter is prepared first. Then it is melt and drawn into fiber to create silica optical fiber. In the third process, a GI POF is drawn directly from the reaction mixture. Our laboratory has done extensive experimental [17–19] and theoretical work [20–23] on interfacial-gel copolymerization, centrifugation, and multilayer coextrusion for preparing GI POFs. It was found that the RID of the GI POFs can be tuned by material formulations and system parameters.

For the properties of optical loss and bandwidth of GI POFs, the molecular structure, morphology, and the fiber structure need to be considered. The dispersion in an optical fiber is divided into three parts: modal dispersion, material dispersion, and waveguide dispersion. Consideration of both modal and material dispersions is required to optimize the RID of the high bandwidth GI POFs.

The aim of this chapter is to provide a comprehensive description of the GI POFs prepared by the multilayer coextrusion process, the centrifugal field process, and swollen-gel polymerization. With respect to these fabrication techniques, this chapter illustrates how to control the RID and the bandwidth of GI POFs via the essential process parameters and material formulas.

2. GI POFs PREPARED BY MULTILAYER COEXTRUSION

2.1. The Process

Multilayer coextrusion is a versatile technology for preparing GI POFs. The details of the process have been described in the literatures [7, 17–19]. Here, a brief summary is given. Figure 1 shows a schematic representation of an N-layer internal diffusion and surface evaporation (IDSE) coextrusion process. Material tanks **1, 2,**$\ldots$, ***N*** that contain solutions of a polymer and monomers are heated at a sufficiently high temperature. The contents of each tank are fed by the gear pumps $\boldsymbol{G_1, G_2, \ldots, G_N}$ to the concentric die ***D***. The tanks are arranged so that the RIDs of the contents inside the concentric die are in a decreasing order from the center to the periphery of the die. The volume ratios of the polymer solutions from different tanks can be adjusted to vary the RID. An N-layer composite monofilament is then extruded out of the orifice of the die and fed into a diffusion zone ***Z*** of length of L. As the monofilament goes through the diffusion zone, the monomers contained in each layer diffuse across the monofilament, leading to a certain RID. A hot nitrogen gas is purged through the diffusion zone, and

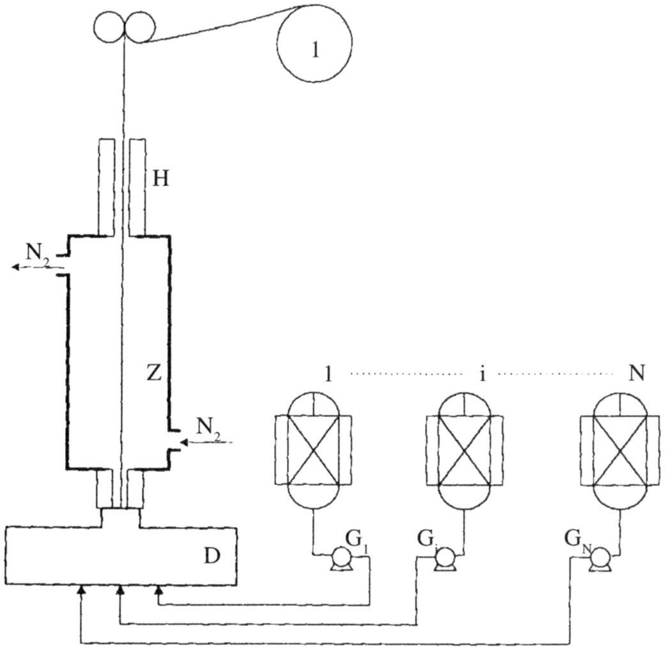

Figure 1. Schematic representation of N-layer IDSE coextrusion process, $G_1, G_2, \ldots, G_n$ represent gear pumps, D is a concentric die, Z and H are the diffusion zone and the hardening zone, respectively, and I is the take-up roll. Adapted with permission from W. C. Chen et al., *J. Phys. Chem. B* 103, 7584. Copyright © 1999, American Chemical Society.

mass transfer between the gas phase and the monofilament is allowed to occur, which affects the RID of the latter. The monofilament is then fed through a hardening zone ***H***, where it is hardened by ultraviolet (UV) lamps. The polymer fiber, which has the prescribed RID, is taken up through rollers by a take-up roll ***I***. The monomer diffusion inside the polymer mixture, and the mass transfer between the fiber periphery and the purged gas lead to an approximately quadratic RID.

A two-layer closed coextrusion process was developed in our laboratory to produce GI POF [17–19]. In those studies, the inner layer was composed of PMMA and the monomers BzMA and MMA, and the outer layer was composed of PMMA and MMA. In addition to the effect of monomer diffusion, the nitrogen purging along the fiber preform is very important for controlling RID. The experimental results revealed that only about 60% of a GI fiber had an approximate parabolic RID in the absence of nitrogen purging [17]. This situation exists mainly because the outer boundary of a closed coextrusion process is impermeable to BzMA and, therefore, the RID became flat near the peripheral region of a fiber. To adjust the RID of the prepared GI POFs toward a parabolic distribution, an open, two-layer coextrusion process with IDSE was developed that successfully enlarged the portion of the parabolic RID [18, 19].

The RID of GI POFs can be further controlled by increasing the number of layers in the coextrusion process. For example, Toyoda et al. [7] developed a multilayer coextrusion process for preparing GI POFs. They found that a parabolic RID could be achieved by gradually increasing the number of layers from three to five. The conclusion of Toyoda et al. [7] is of practical significance and stimulates our interest in the theoretical understanding of the process.

2.2. Theoretical Analysis of RID

The theoretical work on a multilayer coextrusion process for preparing GI POF has been done by our laboratories in last few years [20–23]. Here, a brief summary is provided. Figure 2 shows the coordinates adopted for the mathematical modeling. Here R_{fi} denotes the radius of ith layer and R is the radius of the diffusion zone. Let r and z be the radial distance and the distance from the entrance of the diffusion zone, respectively. Suppose that the process is operated under an isothermal condition and the diffusivities of monomers are functions of the mass fraction of the host polymer only. Whereas the density of the filament remains approximately the same in the diffusion zone, it is assumed that the change in the bulk density of the filament is negligible. Suppose that the diffusion of monomers during the hardening period is negligible. In this case, the variation of the mass fraction of monomer M, x, at a steady state operation can be described by

$$u\frac{\partial x}{\partial z} = D_r\left(\frac{\partial^2 x}{\partial r^2} + \frac{1}{r}\frac{\partial x}{\partial r}\right) + D_z\frac{\partial^2 x}{\partial z^2} \tag{3}$$

where u denotes the extrusion velocity, and D_r and D_z are the effective diffusivities of monomers in the r and z directions,

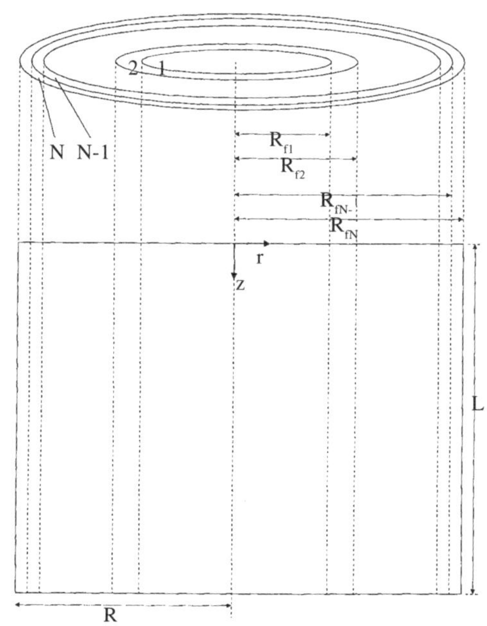

Figure 2. Coordinates adopted in the mathematical modeling, r and z are the radial and axial distances, respectively, L is the length of the diffusion zone, $R_{f1}, R_{f2}, \ldots, R_{fN}$ represents the radius of each layer. Adapted with permission from W. C. Chen et al., *J. Phys. Chem. B* 103, 7584. Copyright © 1999, American Chemical Society.

respectively. The boundary conditions associated with Eq. (3) are summed as

$$x \text{ is finite at } r = 0 \tag{4a}$$

$$-D_r \frac{\partial x}{\partial r} = kx \quad \text{at} \quad r = R \tag{4b}$$

Here, k represents the mass transfer coefficient of monomer M between the fiber periphery and the purging gas. The dimensionless mass fractions of monomers at the outlet of the diffusion zone ($z = L$) can be obtained in the form

$$x^*_{\mathrm{M}} = 2\sum_{m=1}^{\infty}\left(k^* x^*_{\mathrm{M},N} J_0(\lambda_m) + \sum_{i=2}^{N}(x^*_{\mathrm{M},i-1} - x^*_{\mathrm{M},i}) \times \lambda_m R^*_{f_{i-1}} J_1(\lambda_m R^*_{f_{i-1}}) / J_0^2(\lambda_m)(k^{*2} + \lambda_m^2)\right) \times \exp(-\lambda_m^2 z^*_M) J_0(\lambda_m r^*) \tag{5}$$

where x^*_{M} is the mass fraction of the monomer M. In the expressions, J_0 and J_1 are the Bessel functions of the first kind of orders 0 and 1, respectively, and λ_m is the positive root of $k^* J_0(\lambda_m) = \lambda_m J_1(\lambda_m)$.

The transformation of the fiber composition to RID can be performed as follows. Lorentz and Lorenz [24, 25] suggested that the optical property of a nonabsorbing medium and its chemical structure correlate through an additive rule. The RI of a composite material, for example, can be represented by [26]

$$n_d = \sqrt{\frac{1 + 2\phi}{1 - \phi}} \tag{6a}$$

with

$$\phi = \frac{\left(\sum_{\mathrm{M}} \frac{n^2_{d,\mathrm{M}} - 1 x_{\mathrm{M}}}{n^2_{d,\mathrm{M}} + 2\rho_{\mathrm{M}}}\right)}{\left(\sum_{\mathrm{M}} \frac{x_{\mathrm{M}}}{\rho_{\mathrm{M}}}\right)} \tag{6b}$$

where $n_{d,\mathrm{M}}$ and ρ_{M} are the refractive index (RI) and density of the component M. Therefore, the RI of a GI POF produced by the N-layer extrusion process can be determined by the combination of Eqs. 5 and (6a). The index exponent, g, of the RID can be estimated by fitting the predicted data with Eq. (1).

The applicability of the present model was justified by fitting it to the experimental data of Toyoda et al. [7] Their experimental conditions are summarized in Table 1.

Table 1. Experimental conditions [7] used in the multilayer coextrusion process.[a]

Item	Two layer	Three layer Mixture 1 (inner layer)	Four layer	Five layer
PMMA (%)	58	52	52	52
Bz MA (%)	28	35	35	35
MMA (%)	14	13	13	13
		Mixture 2 (second layer)		
PMMA (%)	60	50	50	50
BzMA (%)	0	15	15	15
MMA (%)	40	35	35	35
		Mixture 3 (third layer)		
PMMA (%)	—	50	50	50
BzMA (%)	—	—	—	—
MMA (%)	—	50	50	50
		Mixture 4 (fourth layer)		
PMMA (%)	—	—	47	47
BzMA (%)	—	—	—	—
MMA (%)	—	—	40	40
FPMA (%)	—	—	13	13
		Mixture 5 (outer layer)		
PMMA (%)	—	—	—	40
BzMA (%)	—	—	—	—
MMA (%)	—	—	—	18
FPMA (%)	—	—	—	42
R (mm)	0.5	0.59	0.6	0.6
u (cm/min)	254	40	40	40
T (°C)	80	60	60	60
L (cm)	45	45	45	45
Radius ratio				
Two layer	0.5 (R^*_{f1}):1(R^*_{f2})			
Three layer	0.76 (R^*_{f1}):0.96(R^*_{f2}):1(R^*_{f3})			
Four layer	0.75 (R^*_{f1}):0.94(R^*_{f2}):0.98(R^*_{f3}):1(R^*_{f4})			
Five layer	0.73 (R^*_{f1}):0.92(R^*_{f2}):0.96(R^*_{f3}):0.98(R^*_{f4}):1(R^*_{f5})			

Source: Adapted from with permission from W. C. Chen et al., *J. Phys. Chem. B* 103, 7584. Copyright © 1999, American Chemical Society.

[a] The RI values of PMMA, poly(BzMA), and poly(FPMA) used in the present study are 1.490, 1.568, and 1.422, respectively. The densities of PMMA, poly(BzMA), and poly(FPMA) used in the present study are 1.188, 1.179, and 1.496, respectively.

A theoretical model was successfully developed to predict the multilayer IDSE coextrusion process for preparing GI POFs via the three layer process, as shown in Figure 3.

The RID of a GI POF created by the multilayer process was affected by the following essential parameters: the mass transfer coefficient of monomer across the solid–gas interface, k, the diffusivities of monomers, the radius ratio of each layer, and the number of layers [20–22]. An example of the effect of the essential parameter is given in Figure 4, which shows the simulated RID of a fiber at various dimensionless mass transfer coefficients k^* for three-layer IDSE coextrusion. The index exponents, g, of k^* equal to 0, 5, and ∞ are 1.01, 2.13, and 1.58, respectively; that is, as k^* approaches a very small or a very large value, the RID deviates significantly from the parabolic RID. The Δn value is enhanced as k^* approaches infinity, but it decreases to a small value as k^* approaches zero. The effects of other essential parameters on the RID of the GI POFs are summarized as follows [22, 23]. The diffusion rate of the monomer BzMA shows a more significant effect on the RID than does that of MMA, because BzMA controlled the RI difference in each layer. The order of the degree of the RID close to a parabolic distribution is five layer > four layer > three layer > two layer. The radius ratio and the composition of each layer both must be considered to obtain the parabolic RID.

2.3. Theoretical Analysis of Bandwidth

The bandwidth is an important property for a GI POF applied to communication systems. Several theoretical analyses have been developed to analyze the bandwidth of GI POFs [9, 27–30]. The bandwidth analysis by Wentzel–Kramers–Brillouin (WKB) [7, 31] is relatively easy. Here, the WKB method was applied to analyze the bandwidth of the GI POFs prepared by the multilayer coextrusion process. In the WKB method, the basic assumptions are that nonlinear terms such as distribution loss and mode coupling are negligible. According to the basic assumptions, the modal delay per unit length, $\tau(x)$, is derived as [31]

$$\tau(x) = \frac{N_1}{c}\left[1 - \frac{\delta(4+\varepsilon)}{g+2}x^{2g/g+2}\right][1 - 2\delta x^{2g/g+2}]^{-1/2} \quad (7)$$

where g is the group index, c is the velocity of light, ε is the parameter type and N_1 is the profile dispersion parameter

$$N_1 = n_1 - \lambda\frac{dn_1}{d\lambda} \quad (8)$$

$$= \frac{-2n_1}{N_1}\frac{\lambda}{\delta}\frac{d\delta}{d\lambda} \quad (9)$$

where the derivative terms such as $\frac{d\lambda}{dn_1}$ and $\frac{d\lambda}{d\delta}$ can be estimated from a three-term Sellmeier equation [32],

$$n^2 - 1 = \sum_i \frac{A_i\lambda^2}{\lambda^2 - \lambda_i^2} \quad (10)$$

where A_i and λ_i are the oscillator strength and the oscillator wavelength, respectively. If we approximate Eq. (1) by a Taylor series in power of $x^{2g/2g+2}$ and estimate the first three moment of a full impulse response, then the intermodal and intramodal pulse broadening can be derived as

$$\sigma_{\text{intermodal}} = \frac{LN_1\delta}{2c}\frac{g}{g+1}\left(\frac{g+2}{3g+2}\right)^{1/2} \times\left[C_1^2 + \frac{4C_1C_2\delta(g+1)}{2g+1} + \frac{4\delta^2C_2^2(2g+2)^2}{(5g+2)(3g+2)}\right]^{1/2} \quad (11)$$

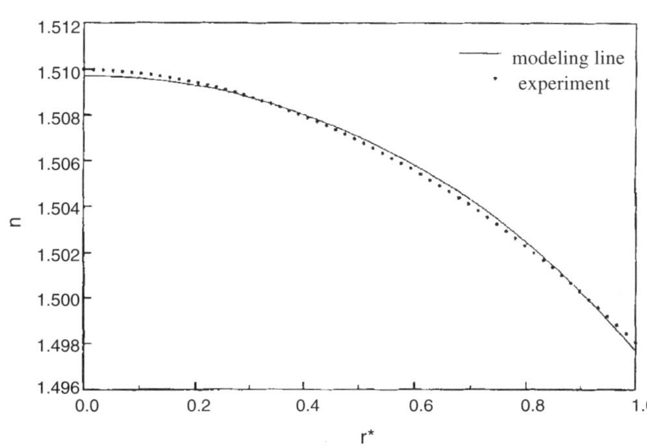

Figure 3. Variation of the RI of a fiber n as a function of its dimensionless radius, r^* in the three-layer IDSE process. The experimental conditions are given in the second column of Table 1. Adapted with permission from W. C. Chen et al., *J. Phys. Chem. B* 103, 7584. Copyright © 1999, American Chemical Society.

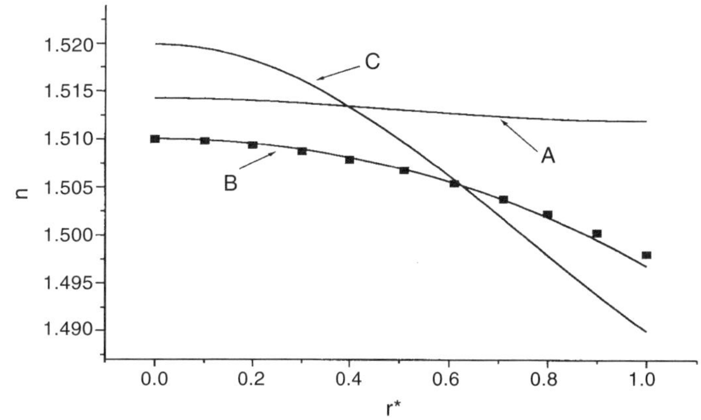

Figure 4. Variation of n of a GI POF as a function of its dimensionless radius r^* at various k^*; curve A, k^*® 0; B, $k^* = 5$; C, k^*®. Parameters used are $z^*_{\text{BzMA}} = 0.165$, $z^*_{\text{MMA}} = 0.165$, $R^*_{f1} = 0.76$, $R^*_{f2} = 0.96$. Adapted with permission from W. C. Chen et al., *J. Phys. Chem. B*, 103, 7584. Copyright © 1999, American Chemical Society.

$$\sigma_{\text{intramodal}} = \frac{\sigma_s L}{\lambda c}\left[\left(-\lambda^2\frac{d^2 n_1}{d\lambda}\right)^2 - 2\lambda^2\frac{d^2 n_1}{d\lambda}(N_1\delta) \times C_1\left(\frac{2g}{2g+2}\right) + (N_1\delta)^2 C_1^2 \times \frac{(2g)^2}{(g+2)(3g+2)}\right]^{1/2} \quad (12)$$

where

$$C_1 = \frac{g-2-\varepsilon}{g+2} \quad (13)$$

$$C_2 = \frac{3g-2-2}{2(g+2)} \quad (14)$$

In Eqs. (11) and (12), L is the length of the fiber and σ_s is the root-mean-square width of the light source. The total root-mean-square width of the impulse response, σ_{total}, and the maximum 3 dB bandwidth (BW) can be obtained by the equations [30]

$$\sigma_{\text{total}} = (\sigma_{\text{intermodal}} + \sigma_{\text{intramodal}})^{1/2} \quad (15)$$

and

$$BW = \frac{0.1874}{\sigma_{\text{total}}} \quad (16)$$

The WKB method has been adapted here to estimate the bandwidth of GI POFs prepared by the multilayer coextrusion process. The three-layer coextrusion process shown in the third column of the Table 1 was used as an example. Figure 5 shows the dispersion of the refractive index of the fiber core and the periphery for three-layer coextrusion at $k^* = 5$. The experimental results shown in this figure were obtained from analysis of n and k in the wavelength range of 190–900 nm. The refractive index dispersion was then fitted by the Sellmeier equation, and the parameters of A_1–A_3 and λ_1–λ_3 in Eq. (10) were obtained. The fitting parameters were used subsequently to determine the values of n, $dn/d\lambda$, and $d^2n/d\lambda^2$ at different wavelengths. Thus, the dispersion of the bandwidths with respect to wavelength was obtained by Eqs. (11)–(16). Figure 6 shows the BW variation with respect to g and optical wavelength of the GI POF fabricated by three-layer coextrusion. As can be observed from the figure, the maximum value of the estimated BW increases with increasing operating wavelength. This is due to the decreased materials dispersion caused by increased operating wavelength. Another characteristic of Figure 6 is the position of the maximum bandwidth shifts from 1.55 to 2.33 as the wavelength increases from 650 to 1550 nm. This shift is due to both the materials dispersion and Δn. Figure 7 shows the variation of the pulse width with the g value at 650 nm. At the fixed wavelength of 650 nm, the materials dispersion is fixed and thus the intramodal dispersion shows a constant value. However, the intermodal dispersion is varies significantly at different g value. At $g = 1.55$, the intermodal dispersion reaches the minimum and thus maximum BW is achieved at this g value, because $\sigma_{\text{intermodal}}$ is determined by g, δ, and materials

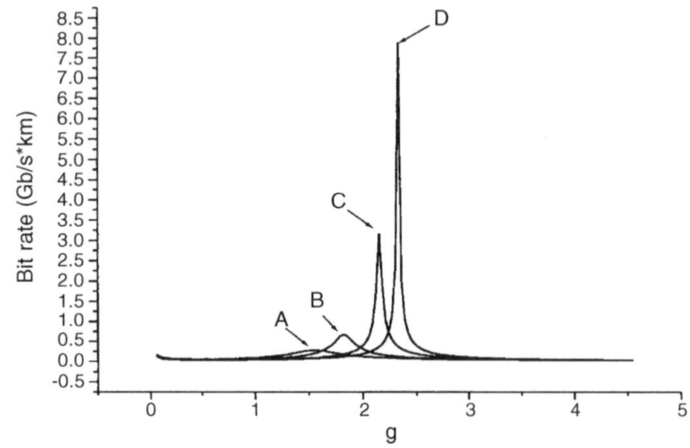

Figure 6. Simulated BW of a GI POF prepared by the three-layer coextrusion process at various operating wavelengths with $k^* = 5$; curve A, 650 nm; B, 850 nm; C, 1300 nm; D, 1550 nm.

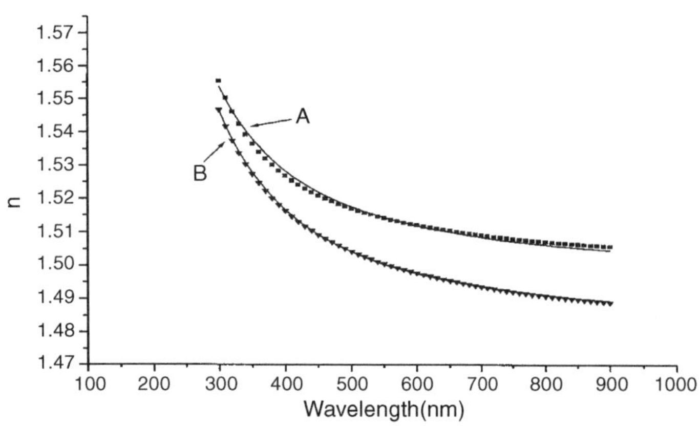

Figure 5. Variation of n as a function of wavelength of the fiber core and the periphery prepared by three-layer coextrusion. The experimental RID data as a function of wavelength at the core center (■) and at the periphery (▼) of a fiber. Curves A and B are line fits from the Sellmeier equation, Eq. (10).

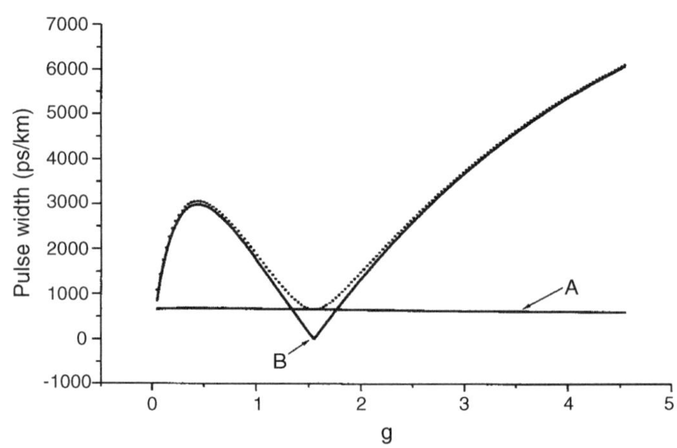

Figure 7. Simulated pulse width of a GI POF prepared by the three-layer coextrusion process with various exponent indices at 650 nm; curve A, Intramodal pulse width; B, Intermodal pulse width; dotted line, total pulse width.

dispersion from Eq. (11) and thus the optimal g value for maximum BW is not at $g = 2$. Based on the predictive results shown in Figures 6 and 7, the following conclusions can be made: (1) the maximum value of the estimated BW depends on the operating wavelength; (2) the maximum bandwidth of a GI POF may not be in the RID with a parabolic distribution ($g = 2$). This explains why the optimum properties for the GI POF depend on different systems and operating processes. Previous studies [22, 23] have shown that the RID can be controlled by the greater number of layers created in the multilayer coextrusion process. This process offers a convenient way to approach the maximum BW value of a GI POF.

3. GI POFs PREPARED BY A CENTRIFUGAL FIELD PROCESS

3.1. The Process

The centrifugal process is regarded as a potential approach for the preparation of GI polymer rods because of the ability to obtain large diameter GI rods. Duijhoven and Bastiaansen successfully prepared GI polymeric rods that had diameters as large as 0.5 cm by this method. Figure 8 shows a schematic representation of the preparation of GI polymeric rods by the centrifugal process. Duijhoven and Bastiaansen discussed the requirements of rod diameter and rotating speed for the preparation of GI rods on an experimental basis. They found that a parabolic RID is obtained when the rotating speed is greater than 15,000 rpm. However, the effects of the other key parameters on the RID of the GI polymeric rods under a centrifugal field were not explored fully in their study. Our laboratory has successfully developed a theoretical method to analyze the effects of material formula and system parameters on RID. In this section, the theoretical analysis is extended to both the RID and optical bandwidth of GI polymeric rods fabricated by the centrifugal field process.

3.2. Theoretical Analysis of RID

The centrifugal field process generates GI polymeric rods from a mixture of polymer and monomer the mass transfer phenomena. The variation of the composition by the centrifugal process was derived previously [33]. Here, a short description of the theoretical analysis is given. For a binary mixture in a centrifugal process that comprises A (polymer) and B (monomer or dopant), the variation of the concentration of component A is

$$\frac{\partial x_A}{\partial r^*} = k x_A (1 - x_A) r^* \tag{17a}$$

where r^* is r/R and

$$k = \frac{\omega^2}{R_g T}(\rho_A - \rho_B)\frac{M_A}{\rho_A}R^2 \tag{17b}$$

Integrating Eq. (17a) with the initial condition yields the composition x_A,

$$x_A = \frac{1}{1 + C\exp\left(-\frac{1}{2}k(r^*)^2\right)} \tag{18a}$$

$$C = -\frac{\exp\left(\frac{1}{2}k(1 - x_{A0})\right) - 1}{\exp\left(-\frac{1}{2}k x_{A0}\right) - 1} \tag{18b}$$

where x_{A0} is the polymer composition in the mixture initially. Variation of the concentration of component B also can be obtained in similar form. The transformation of the rod composition to RID can be performed from the Lorentz and Lorenz equation, Eq. (6) Therefore, the RID of the GI rod created via the centrifugal process can be determined by combination of Eqs. (6a) and (18a).

The applicability of the present model was justified by fitting it to the experimental data [16]. Figure 9 shows the variation of the RID of a GI polymeric rod, Δn, as a function

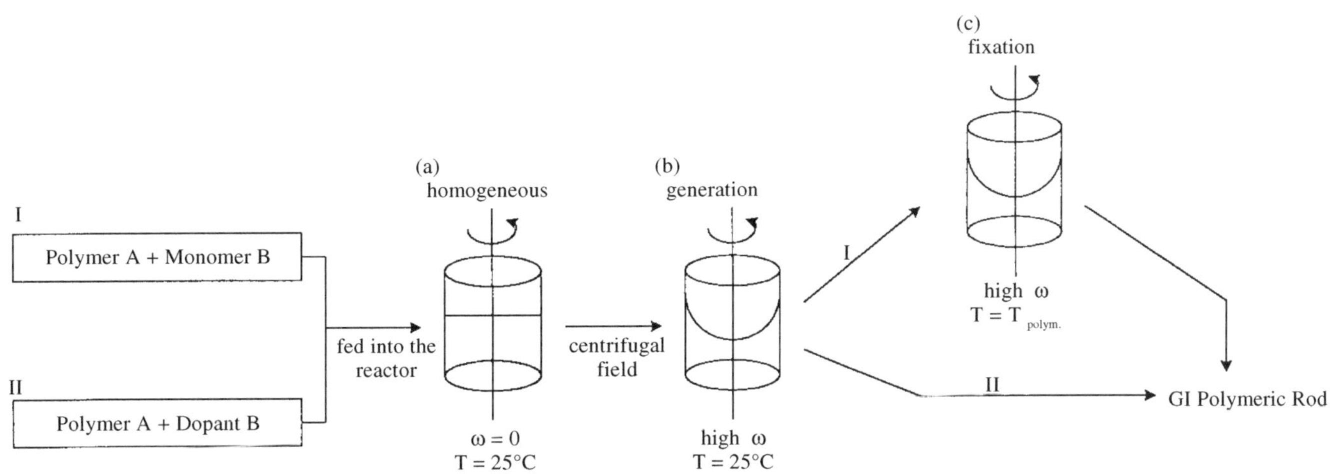

Figure 8. Schematic representation of the preparation of GI polymeric rods by the centrifugal field process. Adapted with permission from W. C. Chen et al., *J. Polym. Sci. Part B. Polym. Phys.* 38, 1764. Copyright © 2000, John Wiley & Sons.

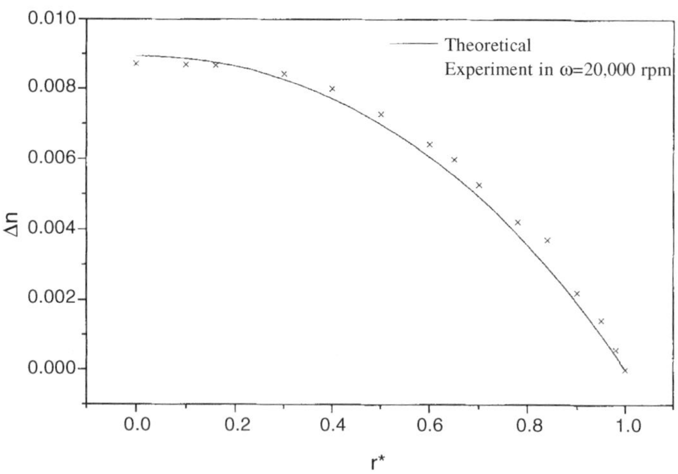

Figure 9. Variation in Dn of the rod as a function of the dimensionless radius r^*. The simulated conditions were 30/70 wt% of PTFPMA/MMA, a rotating speed of 20,000 rpm, a PTFPMA molecular weight of 335,000, a radius of 0.25 cm, and a fixed temperature of 25 °C. +, experimental data [16];–, values based on the model. Adapted with permission from W. C. Chen et al., *J. Polymer. Sci.: Part B. Polymer. Phys.* 38, 1764. Copyright © 2000, John Wiley & Sons.

of its dimensionless radius r^*. The results predicted by both the present model and the experimental data [16] are presented. As can be observed from Figure 9, the performance of the present model is satisfactory.

The effects of the essential parameters, including the molecular weight of the polymer, the composition and component properties (refractive index and density) of the mixture, the rotating speed, and the diameter of the GI rod, have significant effects on the RID [33]. Here, the effect of the dopant on RID in the materials system is used as an example. Figure 10 shows the variation of the refractive index (n) of the rod as a function of the dimensionless radius, r^*, for different dopants: DBME, BB, BIE, and DOP. The physical

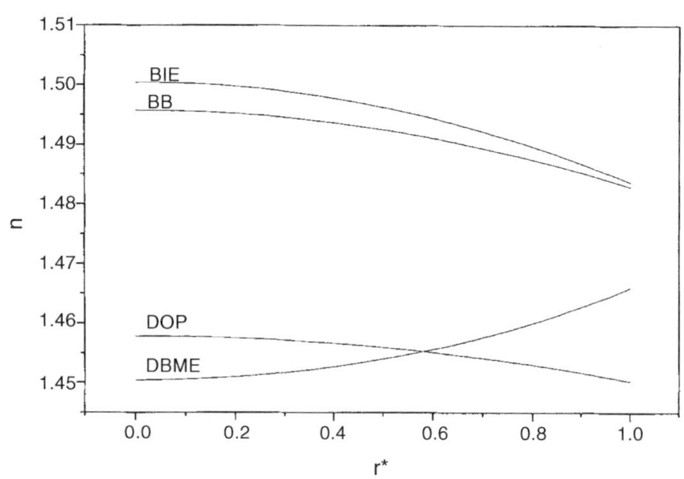

Figure 10. Variations in the refractive index (n) of the rod as a function of the dimensionless radius r^* for different dopants. The simulated conditions were 60/40 wt% of PTFPMA/dopant (DBME, BB, BIE, and DOP), a rotating speed of 20,000 rpm, a radius of 0.25 cm, and a fixed temperature of 25 °C. Adapted with permission from W. C. Chen et al., *J. Polym. Sci.: Part B. Polym. Phys.* 38, 1764. Copyright © 2000, John Wiley & Sons.

Table 2. Refractive indices and densities of the studied materials for the centrifugal process.

Property/material	N	ρ (kg/m^3)	Ref.
PTFPMA	1.422	1496	9
PHFIP 2-FA	1.349	1795.5	26
MMA	1.490	936	33
DBME	1.538	2180	33
BB	1.568	1120	33
BIE	1.564	982	33
DOP	1.486	981	33
DBP	1.490	1045	26

Source: Adapted with permission W. C. Chen et al., *J. Polym. Sci.: Part B. Polym. Phys.* 38, 1764. Copyright © 2000, John Wiley & Sons.

properties of the polymers and dopants are shown in Table 2. The RID for DBME exhibits a reverse trend compared to the RIDs for the other dopants; that is, the refractive index increases from the rod center to the periphery. The refractive indices of the four dopants are greater than the index of PTFPMA. However, the density of DBME (2180 kg/m^3) is greater than the density of PTFPMA (1496 kg/m^3), whereas the other three dopants have lower densities than PTFPMA, as shown in Table 2. These results for the RID due to different dopants can be proven by Eq. (17a). If n_A is less than n_B, $\frac{\partial x_A}{\partial r^*}$ must be greater than 0 to obtain a decreasing refractive index from the center to the periphery (i.e. $\frac{\partial n}{\partial r^*} < 0$). Hence, k from Eq. (17b) must be larger than 0, and this leads to $\rho_A > \rho_B$. On the contrary, if n_A is greater than n_B, ρ_A must be less than ρ_B to obtain an increasing refractive index from the center to the periphery. The order of the component refractive index in the mixture must be the reverse of the component density; that is, if the polymer refractive index is less than the monomer (or dopant), a polymer density greater than that of the monomer (or dopant) is required to obtain the parabolic RID. The dimensionless group k is one of the most important parameters for the centrifugal process because it is equal to the dimensionless driving force that separates polymer and dopant. Hence, the RIDs of GI POFs prepared by the centrifugal process are controlled mainly by k.

Variation of other process parameters significantly modifies the RID of the prepared GI rod. The enhancement of polymer molecular weight, polymer composition, rotating speed, and rod radius increases the Δn value and thus modifies the RID, because these factors enlarge the difference between the refractive index contributions of the polymer and the monomer.

3.3. Theoretical Analysis of Bandwidth

The bandwidth of GI POFs prepared by the centrifugal field method also was analyzed by the WKB method. The methodology for estimating the bandwidth was described in Section 2.3. The materials system of a perfluorinated polymer/dopant system (PHFIP 2-FA/DBP) was investigated in this study. The physical properties of the materials system are listed in the Table 2. Figure 11 shows the variation of BW and g with the k of Eq. (17 b) for the PHFIP 2-FA/DBP

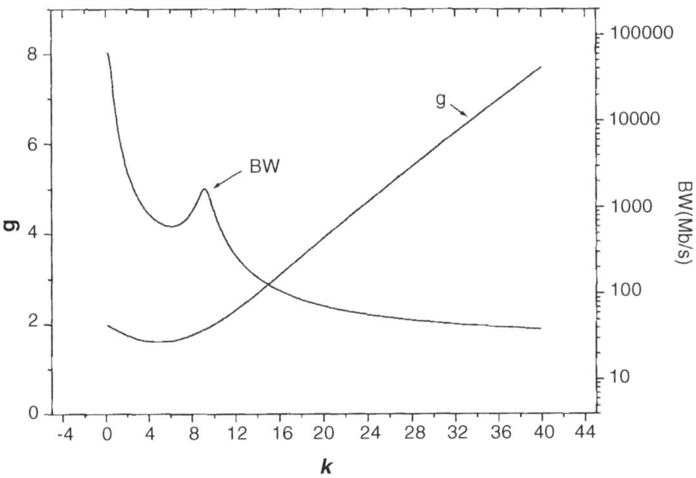

Figure 11. Variation of bandwidth and g versus k for the materials system of PHFIP 2-FA/DBP prepared by the centrifugal process.

materials system created by the centrifugal field process. The maximum BW around 1 Gbit/s is observed at $g = 1.879$ for the assumed fiber length of 100 m. The fluorinated polymer system can minimize the materials dispersion and, thus, the intramodal dispersion can be regarded as a constant in this case. Thus, the bandwidth is controlled by the intermodal dispersion, which is significantly affected by the RID (or g). The maximum BW observed at $g = 1.879$ is shown in Figure 11. The deviation of maximum BW from $g = 2.0$ is because $\sigma_{\text{intermodal}}$ is determined by g, δ, and materials dispersion from Eq. (11). It will be interesting to study the effect of g, k, δ, and λ on the bandwidth of the GI POFs fabricated by the centrifugal field process.

4. GI POFs PREPARED BY SWOLLEN-GEL POLYMERIZATION

4.1. The Process

Interfacial-gel polymerization is a common process for preparing GI polymeric rods [34–39]. In this process, a monomer mixture is filled into a polymer to be first, (e.g., PMMA tube). The monomer mixture is generally a combination of monomers with various refractive indices (e.g., MMA/BzMA). The solubility of the monomer is similar to that of the polymer tube, which is easily swollen by the monomer mixture to form a gel phase near the wall of the polymer tube. In the case of a monomer mixture of BzMA/MMA, for example, the MMA penetrates into the gel phase and polymerizes easier than BzMA due to the smaller molecular size of MMA. Therefore, a GI copolymer gradually forms in the direction from the periphery to the center of the rod at the polymerization temperature. In this example case, the BzMA content increases from the periphery to the center. Thus, a GI polymeric rod is formed because the BzMA has a higher refractive index than the MMA. A GI POF can then be drawn from the prepared GI polymeric rod. However, the GI polymeric rod produced from interfacial-gel copolymerization generally contains large voids and bubble defects [37–39], due to the difference in the polymerization speeds of the monomers in the polymer tube. Furthermore, because the monomer mixture is not refilled into the polymer tube, void spaces are produced by shrinkage during monomer polymerization.

A two-stage interfacial-gel copolymerization process, "swollen-gel polymerization", has been developed to solve the void defects [40]. Figure 12 shows the swollen-gel polymerization process for preparing a GI polymeric rod. The first stage, shown in Figure 12a and b, maintains the monomer mixture in the polymer tube at a low temperature of T_1 for t_1 hours. The purpose of the first stage is to achieve a swelling gel state of the polymer tube without significant polymerization. Hence T_1 was below the polymerization temperature of the monomer mixture (e.g., 40 °C). Then the temperature of the polymer tube was raised to polymerization temperature T_2 (80 °C in this case) for t_2 hours and the void-free GI polymeric rod was obtained. The monomer system used in the previous study was MMA/BzMA, MMA/bromonaphthalene (BN), MMA/benzyl phenyl acetate (BPAc), MMA/benzyl salicylate (BSA), MMA/triphenyl phosphate (TPP), and MMA/dibromobenzene (DBB). BzMA was copolymerized with MMA. BN, BPAs, BSA, TPP, and DBB were nonreactive additives with MMA that have refractive indices of 1.660, 1.555, 1.580, 1.550, and 1.574, respectively. All the preceding systems obtained void-free GI polymeric rods and GI POFs could then be drawn from these rods. However, the combination of MMA and nonreactive additive achieved better optical characteristics than the MMA/BzMA system in the previous study [40]. Hence, the BW analysis will focus on the MMA/nonreactive additive system in the following section.

4.2. Theoretical Analysis of Bandwidth

The bandwidth of GI POFs prepared by swollen-gel polymerization also was analyzed by the WKB method described in Section 2.3. Figure 13 shows the RI dispersion of the materials systems MMA/BN, MMA/BPAc, and MMA/BSA for preparing GI polymeric rods. The RI dispersion was then fitted by the Sellmeier equation, Eq. (10), and used for the BW analysis. Figure 14 shows a simulated BW of a GI polymeric rod prepared by various compositions of PMMA/BSA at 650 nm. The maximum bit rate is similar for the different compositions of BSA in PMMA. However, the optimum

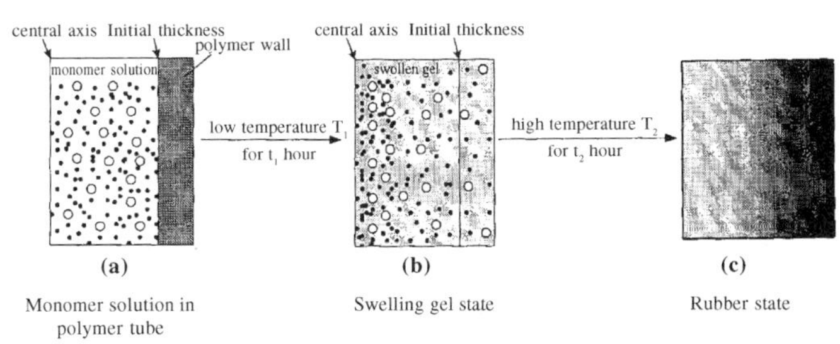

Figure 12. Preparation of GI polymeric rod by swollen-gel polymerization: O, M_1; •, M_2.

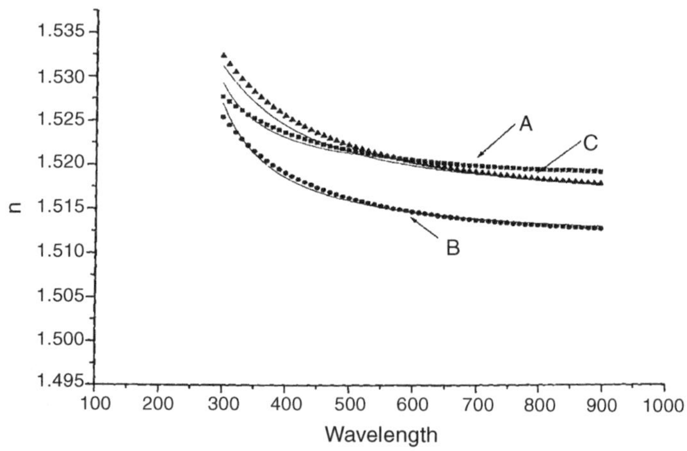

Figure 13. Variation of n with wavelength of the GI rod core prepared by swollen-gel polymerization. Experimental RID data: ■, BN; ●, BPAc; ▲, BSA. The mixing ratio of PMMA/nonreactive additive = 3/1. Curves A, B, and C are the line fits from the Sellmeier equation, Eq. (10).

g value for the maximum bit rate increases with increasing BSA composition, probably due to the smaller dispersion of BSA in comparison with PMMA, and thus the optimum g value gradually red shifted to 2.0. Figure 15 illustrates the materials dispersion of PMMA, PMMA/BN, PMMA/BSA, and PMMA/BPAc with wavelength. Observe from this figure that the dispersion of the nonreactive additive is smaller than PMMA. Hence, the effect of the materials dispersion on BW decreases with increasing additive composition. Hence, the optimum g value for maximum BW shifts toward 2.0 with increasing additive composition in PMMA, as shown in Figure 14. Figure 16 shows the variation of pulse width with the g value. Observe that intramodal dispersion remains constant at various g value, but intermodal dispersion varies significantly. This is because the nonreactive additive has a smaller dispersion than PMMA and thus materials dispersion is not important in this case.

The effect of materials dispersion on BW can be demonstrated further by the relationship between bit rate and

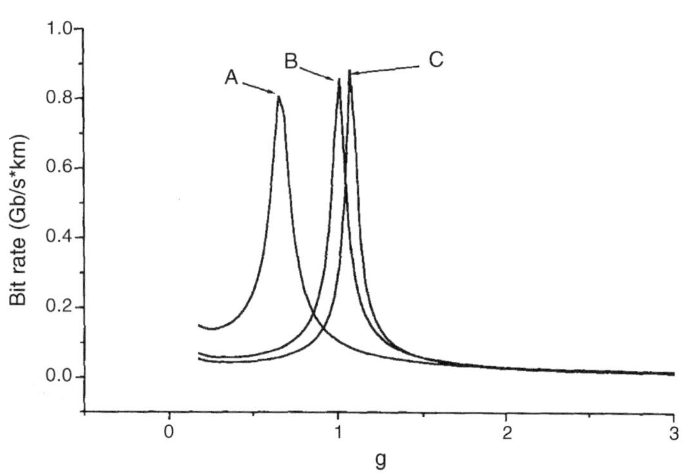

Figure 14. Simulated BW of a GI rod prepared by swollen-gel polymerization at various g values. Composition ratios of PMMA/BSA: curve A, 4/1; B, 3/1; C, 2/1.

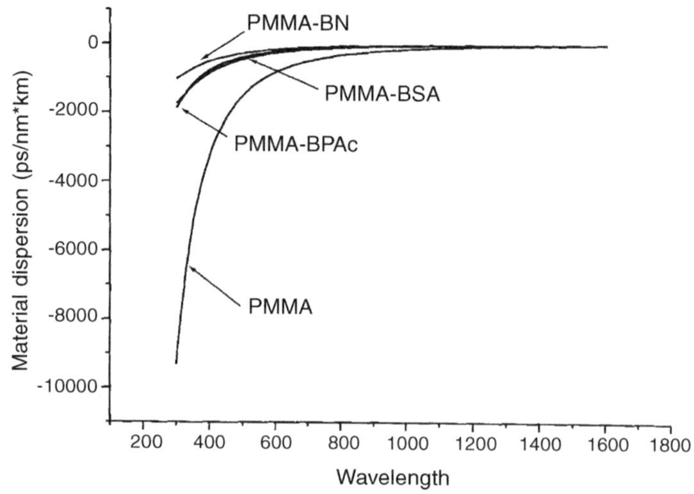

Figure 15. Simulated materials dispersion versus wavelength of various materials system.

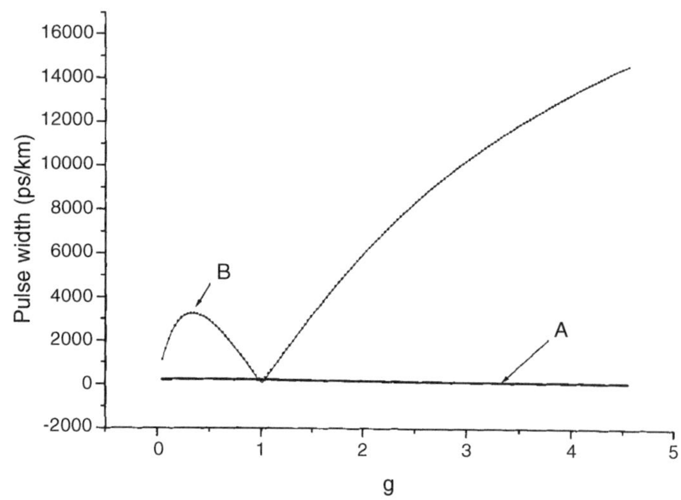

Figure 16. Simulated pulse width of the PMMA/BSA GI rod prepared by swollen-gel polymerization versus g at 650 nm; curve A, intramodal pulse width; B, intermodal pulse width; dotted line, total pulse width.

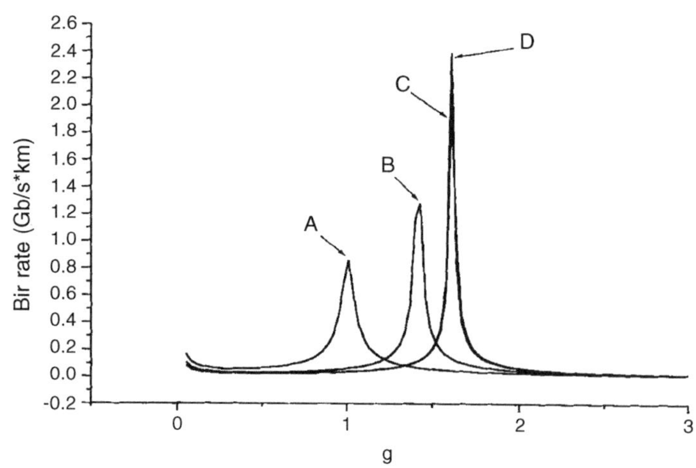

Figure 17. Variation of the simulated BW of a PMMA/BSA (3/1) GI rod prepared by swollen-gel polymerization versus g at various operating wavelengths; curve A, 650 nm; B, 850 nm; C, 1300 nm; D, 1550 nm.

g value at various wavelengths, as shown in Figure 17. The gradual shift toward 2.0 of the optimum g value for maximum BW suggests that the materials dispersion becomes less significant at smaller wavelengths. Hence, also observe that the maximum BW is enhanced at increasing wavelengths, because only the modal dispersion is considered at long wavelengths.

5. CONCLUSION

In this chapter, theoretical analysis was applied to analyze the RID and bandwidth of the GI POFs prepared by multilayer coextrusion, the centrifugal field process, and swollen-gel polymerization. Optimum essential parameters were required to obtain GI POFs that had the suitable RID. For the multilayer IDSE coextrusion process, the effects of the RID in a fiber were analyzed via the essential parameters, including the mass transfer coefficient of monomer across the solid–gas interface, k, the diffusivities of monomers, and the radius ratio of each layer on RID. For the centrifugal field process, component properties (refractive index and density), mixture composition, rotating speed, and rod radius all showed significant effects on the RID of a polymeric rod. The bandwidth of the GI POFs was analyzed by the WKB method. The BW was affected by g, δ, and materials dispersion. The materials dispersion showed a significant effect on the optimum g value for maximum BW.

ACKNOWLEDGMENTS

We thank the National Science Council of Taiwan, Republic of China, for financial support of this work. Permission of using figures and table from American Chemical Society and John Wiley & Sons, Inc., to use figures and tables also is appreciated greatly.

REFERENCES

1. Y. Koike and W. Daum, Eds., "Proceeding of the 7th International Plastic Optical Fibers Conference," Berlin, October 5–8, 1998.
2. Y. Koike, Ed., "Proceeding of the 8th International POF Conference," Chiba, Japan, July 14–16, 1999.
3. Y. Koike, Ed., "Proceeding of the International POF Technical Conference," Cambridge, MA, September 5–8, 2000.
4. L. A. Hornak, Ed., "Polymers for Lightwave and Integrated Optics," Dekker, New York, 1992.
5. X. Quan, *Polymer Prepr.* 39, 1279 (1999).
6. A. Tayaga, T. Kobayashi, S. Nakatsuka, E. Nihei, K. Sasaki, and Y. Koike, *Jpn. J. Appl. Phys.* 36, 2705 (1997).
7. N. Toyoda, Y. Mishina, R. Murata, R. Y. Uozu, M. Oda, and T. Ishimaru, U.S. Patent 5,390,274, 1995.
8. R. Olshansky and D. B. Keck, *Appl. Opt.* 15, 483 (1976).
9. T. Ishigure, E. Nihei, and Y. Koike, *Polymer J.* 28, 272 (1996).
10. A. Baran, in "Proceeding of the 7th International Plastic Optical Fibers Conference", Berlin, October 5–8, 1998, p. 43.
11. Y. Koike, Y. Takezawa, and Y. Ohtsuka, *Appl. Opt.* 29, 2686 (1990).
12. Y. Koike and R. Nihei, U. S. Patent, 5,593,621, 1997.
13. L. L. Blyler, Jr., "Proceeding of the 7th International Plastic Optical Fibers Conference, Berlin, October 5–8, 1998.
14. Y. Ohtsuka, *Appl. Phys. Lett.* 23, 2347 (1973).
15. Y. Ohtsuka and I. Nakamoto, *Appl. Phys.Lett.* 29, 559 (1976).
16. F. G. H. Duijhoven and C. W. M. Bastiaansen, Adv. Mater. 11, 567 (1999).
17. B. C. Ho, J. H. Chen, W. C. Chen, S. Y. Yang, J. J. Chen, and T. W. Tseng, *Polymer J.* 27, 310 (1995).
18. W. C. Chen, J. H. Chen, S. Y. Yang, J. Y. Cherng, Y. H. Chang, and B. C. Ho, *J. Appl. Polym. Sci.* 60, 1379 (1996).
19. W. C. Chen, J. H. Chen, S. Y. Yang, J. J. Chen, Y. H. Chang, B. C. Ho, and T. W. Tseng, *ACS Symp. Ser.* 672, 71 (1997).
20. B. T. Liu, W. C. Chen, and J. P. Hsu, *Polymer* 40, 1451 (1999).
21. B. T. Liu, M. Y. Hsieh, W. C. Chen, and J. P. Hsu, *Polym. J.* 31, 233 (1999).
22. W. C. Chen, Y. Chang, and J-P. Hsu, *J. Phys. Chem. B* 103, 7584 (1999).
23. Y. Chang, W. C. Chen, M. S. Wei, and W. C. Wu, in "Optical Polymers: Advances in Optical Fibers and Waveguides", (J. P. Harmon and G. K. Noren, Eds.). American Chemical Society, Washington, DC, 2001.
24. H. A. Lorentz, *Wied. Ann. Phys.* 9, 641 (1880).
25. L. V. Lorenz, *Wied. Ann. Phys.* 11, 70 (1880).
26. D. W. Van Krevelen, "Properties of Polymers", 3rd ed., Chap. 10. Elsevier, Amsterdam, 1990.
27. A. F. Garito, J. Wang, and R. Gao, *Science* 281, 962 (1998).
28. T. Ishigure, Y. Koike, and J. W. Fleming, *J. Lightwave Technol.* 18, 178 (2000).
29. G. Yabre, *J. Lightwave Technol.* 18, 869 (2000).
30. T. Ishigure, M. Kano, and Y. Koike, *J. Lightwave Technol.* 18, 959 (2000).
31. R. G. Hunsperger, "Integrated Optics: Theory and Technology", 3rd ed., pp. 3–6. Springer-Verlag, Berlin, 1991.
32. G. R. Fowles, "Introduction to Modern Optics", 2nd ed. Holt, Rinehart and Winston, New York, 1975.
33. W. C. Chen, Y. Chang, and M. S. Wei, *J. Polym. Sci. Polym. Phys.* 38, 1764 (2000).
34. Y. Koike, Y. Takezawa, and Y. Ohtsuka, *Appl. Opt.* 27, 486 (1988).
35. Y. Koike, N. Tanio, E. Nihei, and Y. Ohtsuka, *Polym. Eng. Sci.* 29, 1200 (1989).
36. Y. Koike, *Polymer* 32, 1737 (1991).
37. S. Y. Yang, Y. H. Chang, B.C. Ho, W. C. Chen, and T. W. Tseng, *J. Appl. Polym. Sci.* 56, 1179 (1995).
38. S. Y. Yang, B. C. Ho, Y. H. Chang, J. J. Chen, T. W. Tseng, and M. S. Lee, Proc. *SPIE* 2000, 396 (1993).
39. S. Y. Yang, Y. H. Chang, B. C. Ho, W. C. Chen, and T. W. Tseng, *Polym. Bull.* 34, 87 (1991).
40. J. H. Liu and H.-T. Liu, *J. Appl. Polym. Sci.* 64, 849 (1997).

CHAPTER 3

Gradient Refractive Index Distribution of Rare-Earth-Doped Polymer Optical Fibers

Qijin Zhang
Department of Polymer Science and Engineering, University of Science and Technology of China, Hefei, Anhui 230026, People's Republic of China

CONTENTS

1. INTRODUCTION

Polymer or plastic optical fiber (POF) has received lots of interest in the past decades because it can be fabricated to have a larger core diameter, which results in a POF that is a promising candidates for fiber to the desktop [1]. However, the increased fiber core diameter is accompanied by an increase in the number of optical signal modes and consequent optical pulse broadening. The unwanted pulse broadening is due mainly to modal dispersion, which makes components of the signals in different modes need different travel times and ultimately, at the far end of the fiber under transmission, results in pulse overlap and a garbled communication signal [2]. To overcome and compensate for modal

ISBN: 1-58883-012-8
Copyright © 2004 by American Scientific Publishers
All rights of reproduction in any form reserved.

dispersion, the refractive index of the fiber core can be made into a gradient distribution from a high index in the fiber core center to a low index in the outer core region. An optical fiber with such a refractive index distribution is often called a graded (or gradient) index (GI) optical fiber [3].

It is known that the bandwidth can be maximized by optimizing the shape of the graded index distribution within the GI fiber core. From the point of theory, a versatile refractive index profile is the power law function [4]

$$n(r) = n_1[1 - 2(r/R)^g\delta]^{1/2} \qquad r \leq R \tag{1.1}$$

in which

$$\delta = \frac{n_1^2 - n_2^2}{2n_1^2} \approx \frac{n_1 - n_2}{n_1}, \qquad n_1 > n_2 \tag{1.2}$$

Equations (1.1) and (1.2) show that the core refractive index is a function $[n(r)]$ of the radial position (r), and the cladding refractive index is a constant (n_2). The highest value of $n(r)$ is $n(0) = n_1$ and the lowest value occurs at the core radius $r = R$, $n(R) = n_2$, as illustrated in Figure 1. The value of g, called the grade profile parameter, determines the steepness of the profile. The bandwidth is maximized when

$$g = g_{\text{opt}} = 2 + \varepsilon - \delta\frac{(4+\varepsilon)(3+\varepsilon)}{5+2\varepsilon} \tag{1.3}$$

in which ε is the parameter of the material dispersion, the value of which is dependent on wavelengths and kinds of materials [5]. For a multimode POF, much more attention is paid to the dispersion caused by the differences in the group velocities of the modes, than to the idea that the bandwidth of the GI POF may be dominated by the material dispersion [6]. So, Equation (1.3) can be simplified as

$$g_{\text{opt}} \approx 2 - \frac{12}{5}\delta \tag{1.4}$$

There are, in general, two techniques for fabricating POF. One is the continuous extrusion process in which the monomer, which contains traces of initiator and chain transfer agent, is fed continuously into the reactor and unclad fiber continuously is drawn from the die. The other technique involves the separate fabrication of a preform and then drawing this preform into fiber in a manner similar to standard glass fiber manufacture. The second technique is preferred for the fabrication of graded index polymer optical fibers (GI POFs) and also of rare-earth-doped GI POFs, because the dopant can be controlled accurately when it is introduced into the fiber core at the fabrication stage of the preforms. Furthermore, the preform can be characterized before it is drawn into fibers and, in this way, the fiber properties can be controlled better.

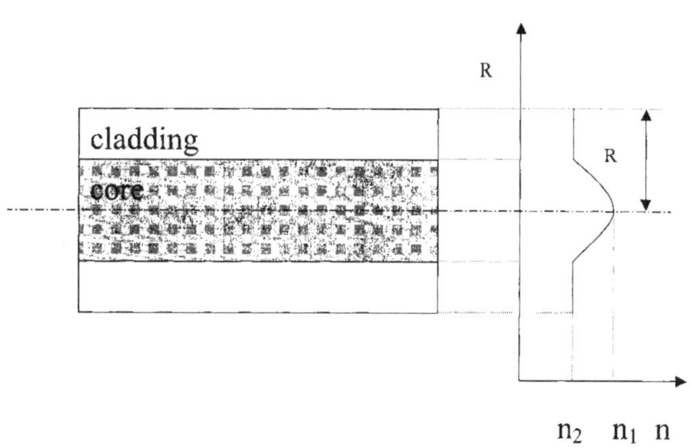

Figure 1. Schematic side elevation of the refractive index distribution in a cross section of graded index POF.

Many idiographic methods have been developed in the past years to obtain GI POFs. Descriptions of some of them can be found in the reported works. For example, a closed extrusion process based on the continuous extrusion process was developed by Chen and Co-workers [7]. The mechanism of this method can be explained by mutual diffusion of monomers that have different refractive indices in a host polymer. Other methods used for preform fabrication include two-stage copolymerization, photo copolymerization, centrifugal molding, interfacial-gel polymerization and copolymerization, and swollen-gel polymerization [8]. Among these methods, interfacial-gel polymerization was the first to be developed (by Koike et al. in Japan), but there is no quantitative analysis about its mechanism, because the polymerization process is affected by too many factors; some qualitative explanation can be obtained in terms of the gel effect in bulk polymerization. In the past several years, based on the experimental work on interfacial-gel polymerization, a detailed analysis, both quantitative and qualitative, has been made in terms of the free-volume theory. On the other hand, the preform method also has been used to make rare-earth-doped GI POFs, in which rare-earth-doped GI preforms first are made by the interfacial-gel polymerization technique and then are drawn into GI POFs.

In the first half of this review, a qualitative description of the gradient-index-forming mechanism is given. Then, after a short introduction about the theory used in the analysis, a theoretical model of the interfacial-gel polymerization is put forward to analyze the mechanism semiquantitatively. Subsequently, applications of the model in our experimental work are discussed. In the second part of the review, the work on rare-earth-doped GI POF is introduced and the effect of doped rare earth ions on the gradient index distribution is analyzed semiquantitatively.

2. QUALITATIVE DESCRIPTION OF GI DISTRIBUTION-FORMING PROCESS DURING INTERFACIAL-GEL POLYMERIZATION

2.1. Historical Background

As mentioned in Section 1, there are many methods for making graded index polymer rods. The earliest work concerned with interfacial-gel polymerization can be traced back to the 1970s. Ohtsuka [9] made plastic rods that had a cross-sectional quadratic gradient refractive index distribution by two-step copolymerization of diallyl esters. This procedure involves the immersion of a partially polymerized rod into a comonomer of lower refractive index. The comonomer diffuses into the rod and sets up the refractive index gradient distribution, which is "fixed" by further polymerization. The main monomers used are diethylene glycol bis(ally-carbonate) for the prepolymer rod and fluorinated methacrylates. The polymer produced from diallyl esters is predominantly cross linked. This cross linking is advantageous in that it allows the comonomer to penetrate the prepolymer rod without dissolving it too extensively. However, the intermolecular bonds within a cross-linked material prevent it from flowing under the action of heating. Thus, the obtained preforms cannot be melt-processed into POF. To overcome this deficiency, a new preparation method—photocopolymerization—was developed by the same group [10]. In this method, methacrylic esters, such as methyl methacrylate, are photocopolymerized with vinyl benzoates. Two monomers that have different refractive indices and reactivity ratios are sealed in a glass tube that contains an ultraviolet-sensitive initiator. The tube is then rotated about its vertical axis and illuminated externally by ultraviolet light. The monomers are chosen such that their reactivity ratios cause the lower refractive index material to polymerize preferentially near the outside of the tube during the early part of the polymerization. Varying degrees of copolymerization then take place until the polymerization is dominated by the second monomer during the latter stage of the polymerization, toward the center of the polymer rod formed in the tube. The obtained rod, which has a cross-sectional GI distribution, can be converted into plastic fibers by a heat-drawing process, because the rod is fusible because it consists of linear copolymers.

In 1988, a new interfacial-gel copolymerization method was first proposed to prepare steric gradient refractive index polymer optical waveguides [11], and subsequently, to prepare preforms of graded index POF [12]. With the interfacial-gel copolymerization technique, copolymerization first takes place within the thin gel phase formed on the wall of polymer substrate, because the rate of copolymerization of the monomers inside the gel is much higher than in the monomer liquid due to the gel effect. At the same time, because the reactivity ratio of each monomer is different, one monomer is preferentially polymerized within the gel phase. Under the condition wherè the monomer is of lower refractive index than another monomer (two-component system, e.g.), a gradient refractive index profile is formed perpendicular to the original surface of the polymer substrate, because the composition of the copolymer being formed changes gradually in this direction. By 1994, interfacial-gel polymerization was further developed by the same group make graded index POF [13]. The difference between interfacial-gel polymerization and copolymerization lies in that only one monomer and an unreactive component are included in the interfacial-gel polymerization. The unreactive component can be chosen so that it is completely miscible with monomer, such as the two-component system of methyl methacrylate (MMA) and bromobenzene (BB), which is difficult for the interfacial-gel copolymerization technique because the requirements of comonomer include not only compatibility with another monomer, but also a suitable reactivity ratio to obtain graded index POF with no scattering loss. For this same reason, interfacial-gel polymerization has received more attention and since then, much work has been concentrated all selecting suitable unreactive components or dopants to optimize the properties of GI POFs [14]. Because interfacial-gel polymerization is a relatively simple system, analysis of its mechanism has been carried out concurrently.

2.2. Qualitative Description of Interfacial-Gel Polymerization

Interfacial-gel polymerization (IGP) usually begins by making a polymer tube that can be dissolved by monomer and form a gel layer. Because poly(methyl methacrylate) (PMMA) is the polymer most often used for POF, the tube can be made from methyl methacrylate. In our group, the equipment used to make the PMMA tube is shown in Figure 2. The general procedure is described in the following text.

A mixture of MMA and initiator of about 1% w/w is placed into a glass tube and half filled. For prepolymerization, the tube is sealed and put into a temperature-controlled oil bath to polymerize, for a certain time. After prepolymerized is accomplished, the tube is put horizontally into the electric furnace shown in Figure 2 and rotated about its center axis. Under this circumstance, the mixture of polymer and monomer in the tube forms a tube under the action of acentric force. When the polymerization is finished, one PMMA tube can be made with controlled molecular weight and geometrical dimensions.

The PMMA tube is sealed on one end and then filled with a solution composed of MMA as the monomer, azobis (isobutyronitrile) (AIBN) as the initiator, dodecyl mercaptan

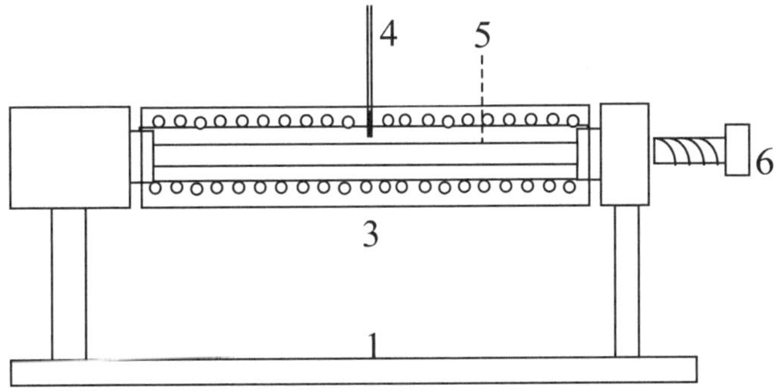

Figure 2. Schematic of the primary elements of the polymerization apparatus for preparing a PMMA tube 1. Base; 2. Electric motor; 3. Electric furnace; 4. Temperature controlling; 5. Glass tube; 6. Adjusting Knob.

as the chain transfer agent, bromobenzene as the unreactive dopant that has a higher refractive index, and a specified amount of rare earth octanoate. The tube is placed into a glycerin bath and held at a constant temperature of 80 °C for 1 h and then at 70 °C until solidification is complete. After solidification, the tube is placed into a vacuum oven at about 100 °C for 2 h and then the preform rod is ready for characterization. Similar procedures were adapted to make graded index preform in other reported works [13, 14].

In contrast to other polymerization processes, interfacial-gel polymerization is carried out in a container composed of polymer that can be dissolved into monomers prior to polymerization. When the tube is filled with a solution mainly composed of the monomer, the inner wall of the tube becomes slightly swollen and a gel phase is formed. It is well known that polymerization is accelerated in the gel phase due to the so-called gel effect. Because the dissolution process comprises a diffusion process of monomer and other molecules, such as dopants and rare earth salts, from the center region to the wall of the tube and of polymer in the opposite direction, a concentration gradient of dopants is formed when the diffusion equilibrium is reached. Generally speaking, the dopant used in interfacial-gel polymerization is of higher refractive index than the polymer. In this way, the diffusion equilibrium results in higher refractive index in the center part of the preform rod that has a higher concentration of dopant and lower refractive index in the periphery of the rod that has lower concentration of dopant. The detailed process is given by the schematic description shown in Figure 3, in which other molecules that occur in smaller amounts in than monomers and dopants are not shown for the reader's convenience. From the point of theory, we can see that this assumption makes theoretical treatment easy and intelligible, but its validity must be proved by experimental work, which we discuss in a later part of this review.

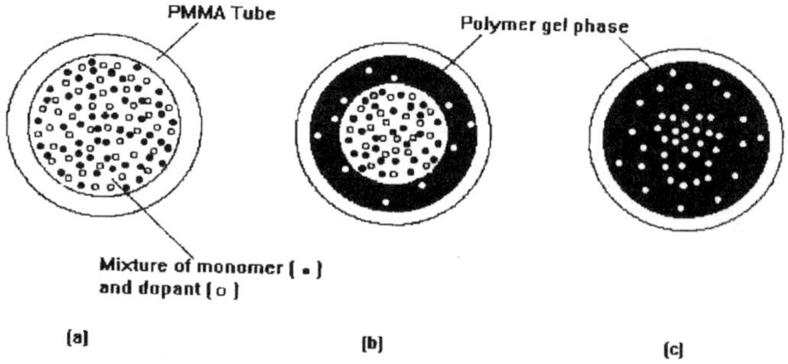

Figure 3. Schematic cross sections of the tube used in interfacial-gel polymerization. (a) Before the diffusion process and polymerization, monomers and dopants are distributed within the tube, (b) Diffusion and polymerization start, and only small amounts of dopant go into the gel layer, (c) Equilibrium between dopant and polymer diffusion is reached. Adopted with permission from Q. Zhang et al., *Macromolecules* 30, 7874. Copyright © 1997, American Chemical Society.

3. DIFFUSION PROCESS GIVEN BY FREE-VOLUME THEORY

3.1. Theoretical Fundamentals for Analysis of the IGP Process

From the preceding qualitative description of interfacial-gel polymerization, it can be seen that the gradient index distribution is formed by the diffusion of dopants in polymer gels. Obviously, a theoretical model that is able to describe the quantitative behavior of the process is highly desirable, not only for understanding the effects of various parameters on the formation of the gradient index distribution, but also for design of laboratory-scaled experiments. Up to now, the only reported theoretical model [15] to describe the process can be traced back to a reexamination of the free-volume theory, which was introduced by Vrentas and Duda [16] in 1977 to explain diffusion in polymer—solvent mixtures. After over 20 years of work based on free-volume theory, a method to predict solvent self-diffusion in polymer–solvent system has been developed. The theoretical model that describes interfacial-gel polymerization is one example of numerous applications concerning migration of small molecules in polymeric materials. In this short review, only the theoretical model concerning interfacial-gel polymerization is presented after a simple introduction to free-volume theory relative to diffusion processes, Detailed theoretical fundamentals can be found in related references [16–20].

The free-volume theory is assumed to be valid when there is a sufficient amount of polymer–polymer contact (for example, when the domains of polymer molecules begin to overlap), on the basis of which the analysis of diffusion behavior in concentrated polymer solution has been established [20]. A general version of the free-volume theory of self-diffusion supposes that the diffusion process depends on the probability that a molecule will obtain sufficient energy to overcome attractive forces and the probability that a fluctuation in local density will produce a hole of sufficient size so that the diffusing molecule can jump. For a polymer—dopant system, the self-diffusion coefficient can be determined using the expressions [18]

$$\ln D_1 = \ln D_0 - \frac{E}{RT} - \left[\frac{(1-\omega)\widehat{V}_1^* + \omega\xi\widehat{V}_2^*}{\widehat{V}_{FH}/\gamma}\right] \tag{3.1}$$

$$\widehat{V}_{FH}/\gamma = (1-\omega)\left(\frac{K_{11}}{\gamma}\right)(K_{21} + T - T_{g1}) + \omega\left(\frac{K_{12}}{\gamma}\right)[K_{22} + \alpha(T - T_{g2})] \tag{3.2}$$

In this review, the dopant, which is used instead of penetrants or solvents as in other papers, is component 1 and the polymer is component 2. In addition, D_0 is an effectively constant preexponential factor, E is active energy, $\widehat{V}_1^*$ is the specific hole free volume of component i required for a jump, ω is the mass fraction of polymer, T is the temperature, T_{g1} is the glass transition temperature of pure component i, ξ is a size parameter that is the ratio of the critical molar volume of the small molecule jumping unit, $\widehat{V}_{FH}/\gamma$ is the free volume of the solution, and (K_{11}/λ), K_{21}, (K_{12}/γ), and K_{22} are free-volume parameters. The parameter α gives a zero polymer free volume at 0 °C according to the work reported previously [18]. The free-volume formulation is based on utilization of the following four assumptions:

1. The partial specific volumes of the dopant and polymer are independent of composition, so that no volume change occurs due to mixing.
2. All thermal expansion coefficients are represented adequately by average values over the temperature interval of interest.
3. The parameter γ remains the same over the complete concentration interval.
4. The energy E remains the same for all dopant concentrations. $E = E_p - E_d$, where E_p and E_d represent energies per mole that a molecule needs to overcome attractive forces that hold it to its neighbors. The quantity E_p describes energy effects for concentrated polymer solutions for which the domains of polymer molecules overlap; E_d describes energy effects near the pure dopant limit.
   Hence, a single value of E is used in Eq. (2.1) over most of the concentration range.

These assumptions also are used in interfacial-gel polymerization when the free-volume theory is applied to calculate the gradient index distribution on the cross section of the preform rods. The second probability is, of course, concerned with the availability of sufficient hole free volume for transport to occur. It generally is assumed that the diffusion process is free-volume driven for temperatures sufficiently close to the glass transition temperature of the system [21]. The concentration dependence of the self-discussion coefficient has been found to be more pronounced in the rubbery polymer above the glass transition temperature (T_g) than it is for the glassy polymer [22]. Note here that, at a given polymerization temperature, the limiting conversion for cessation of polymerization has been identified with the conversion at which the medium turned into glass [23]. This means that the diffusion in the bulk system above and close to the glass temperature is the same as the diffusion in the polymerization system below and close to the limiting conversion. In other words, the higher is the conversion, the smaller is the free-volume drive. Furthermore, it can be expected that diffusing molecules, if the exist, will distribute according to the conversion gradient, if it exists, in a closed system and the distribution of diffusing molecules will be suppressed as the polymerization continues. It is this point that makes predication of the gradient index distribution possible in the interfacial-gel polymerization system. Because there is a gradient of polymer concentration in the direction vertical to the center axis and in the cross section area of the preform rods, and because the dopant adopted in IGP has a larger refractive index than that of the monomer and the polymer, the refractive index distribution can be predicted by the distribution of the dopant's diffusion coefficients along the polymer concentration gradient.

3.2. Evaluation of a Model for Forming a Gradient Index Profile in IGP

Unlike other polymerization processes, IGP is carried out in a container (such as a tube) composed of polymers that can be dissolved into monomers prior to polymerization. When the polymer tube is filled with monomers, the inner wall of the tube becomes slightly swollen and a gel phase is formed. It is well known that polymerization is accelerated in the gel phase in which the polymer contents are between 20 and 80%, in general, due to the gel effect. Therefore, the polymerization is faster in the gel phase than in the polymer content >80% phase, which is closer to the tube inner wall, and in the polymer content <20% phase, which is closer to the center region of the tube, because dissolution is the process of diffusion of monomers and dopants from the center range to the wall of the tube.

Under these circumstances the gel layer gradually thickens toward the center region as polymerization continues, and a gradient polymer concentration distribution forms along the radial direction of the tube. By optimization of polymerization conditions, experience has shown that, at a certain time, the polymer phase reaches the center axis of the tube and an equilibrium state is formed in which the dopant concentration decreases monotonically from the center axis to the periphery [24]. After this time, the radial distribution of the dopant remains unchanged because the available hole free volume is fixed at each point on the cross section. Although unreacted monomers will continue to polymerize in the gel layer, the volume condensation caused by conversion from monomers to polymers can take place only in the longitudinal direction of the reaction tube.

A real polymer concentration distribution at equilibrium along the radial direction of the tube has not been available, because past studies on the kinetics of polymerization often were limited to low conversion, and there is a distinct difference in the kinetics between low and high conversion. However, through numerous studies on intermediate to high conversion (i.e., beyond about 50% conversion of monomer into polymer), it has been experimentally well established that the overall rate of polymerization decreases drastically with conversion [25]. Furthermore, at a given polymerization temperature, the limiting conversion for cessation of polymerization has been identified with conversion at which the medium turned into a glass state [26]. For the polymerization of MMA at 50 °C, the limiting conversion was found to be about 85% [18]. On the other hand, below the limiting conversion, polymerization is not controlled by diffusion of monomers, that is, the rate of monomer diffusion is greater than the rate of polymerization, which results in the observation that the gel layer can thicken quickly to the center of the tube. From this analysis, we can deduce that the polymer concentration gradually increases from the center to the inner wall of the tube. However, a qualitative relationship between polymer concentration and the radial distance of the tube should be assumed to establish a model of IGP. Several functions for this relationship have been published [15], such as

$$\omega = a\exp(r) + b \tag{3.3}$$

in which ω is the weight fraction of polymer, r is the radial distance, and a and b are distribution constants.

To obtain the GI profile, another molecule that is inert to polymerization is mixed with the monomer before polymerization and both are filled into the tube at the same time. As the gradient polymer concentration is shaped, the inert molecule and the monomer gradually distribute with a concentration gradient contrary to the polymer concentration gradient in the gel layer. This concentration distribution is formed by equilibration between the formation of polymers and the diffusion of small molecules, and is fixed when the gel layer reaches the center of the tube. According to the free-volume theory, the diffusion coefficient of the inert molecule at different polymer concentrations can be calculated by Eqs. (3.1) and (3.2). By combining these formulas with Eq. (3.3), the relationship between the diffusion coefficient and radial distance can be obtained.

As stated before, Eqs. (3.1) and (3.2) are used in a binary diffusion system composed of a polymer and a dopant. In the IGP system, two dopants, the inert molecule and the monomer, are considered. To calculate each diffusion coefficient of the two dopants by Eqs. (3.1) and (3.2), first assume that either one of the two dopants is distributed according to its diffusion coefficient in the IGP system. Note that another method was reported previously that interpreted the self-diffusion of both dopants in a ternary system [28]. Detailed discussion about this aspect will be given later. If two diffusion coefficients are obtained for two dopants under a state of equilibrium, the concentration distribution of the inert molecule can be expressed as [29].

$$C_1 = \frac{D_1}{D_1 + D_2}(1 - \omega) \tag{3.4}$$

in which C_1 is the concentration distribution of inert molecules, D_1 and D_2 are the diffusion coefficients of inert molecules and monomers, respectively, and ω is the polymer concentration gradient. From Eqs. (3.1)–(3.4), the concentration distribution of the inert molecule along with the radial direction can be obtained.

For a compatible system, the relationship between refractive index and composition of the system can be correlated theoretically via the Lorentz and Lorenz formulation [30]

$$\frac{n^2 - 1}{n^2 + 2} = \frac{1}{V}[(R_{m_1} - R_{m_2})C_1 + R_{m_2}] \tag{3.5}$$

where n is the refractive index, V is the molar volume, and R_{m_1} and R_{m_2} denote the molar refraction of the dopant molecule and the monomer unit of the polymer, respectively.

If Eq. (3.5) is modified, a relationship between n and C_1 can be obtained as

$$n = \sqrt{\frac{2[(R_{m_1} - R_{m_2})C_1 + R_{m_2}] + V}{V - [(R_{m_1} - R_{m_2})C_1 + R_{m_2}]}} \tag{3.6}$$

It is evident that the refractive index distribution as a function of radial distance can be simulated by Eqs. (3.1)–(3.6).

4. GRADIENT REFRACTIVE INDEX DISTRIBUTION OF UNDOPED POFs

4.1. Description by Self-Diffusion Approaches of Free-Volume Theory in a Binary System

As described in Section 2.2, the manufacture of GI POF starts with making a hollow polymer tube that has a wall that can be swollen by monomer used in the subsequent IGP reaction. The thickness of the tube wall is significant for GI distribution of preforms and is controlled by the quantity of monomer used to make the hollow tube. The thickness has a relationship with prepolymerization time and temperature adopted in the IGP reaction. For a reaction with longer prepolymerization time and higher temperature, a greater extent of swelling results. Sometimes the shape of the tube can be changed by excessive swelling, which has occurred in work performed in my group.

When the tube is made, a mixture composed mainly of monomer and one inert molecule can be poured into the tube for interfacial-gel polymerization. The inert molecule usually is selected for its unreactivity to polymerization, higher refractive index than that of the monomer and the structural unit in the polymer chain, compatibility with the monomers and polymers, and larger mass than that of the monomer. These properties are often the embodiment of the chemical structure and the body size of the inert molecule. Several molecules used as inert molecules have been reported, namely bromobenzene (BB) [13], benzyl *n*-butyl phthalate (BBP), BB-d5 [31], bromonaphthalene (BN), benzyl phenyl acetate (BPAc) [32], dibutyl phthalate (DBP), benzyl benzoate (BEN) [6], diphenyl sulfide (DPS) [14], and triphenyl phosphate (TPP) [33]. Among all these molecules, BB was the first to be used and has been studied fully. Compared with MMA, BB has a higher refractive index (1.56) and density (1.5) [13]. In terms of densities, size calculations for BB and MMA shows that the molecular sizes of the two molecules are approximately the same ($d = 6.9$ Å), but BB is still often used as the inert molecule because of its higher refractive index and good compatibility with MMA and PMMA. The calculated solubility parameters are reported to be 7.814, 8.468, and 9.327 $(\text{cal/cm}^3)^{1/2}$ for MMA, BB, and PMMA, respectively [24]. In our experimental work, we found that BB and MMA can mix at any concentrations [15].

On the other hand, the IGP system of MMA and BB has been treated theoretically, because the parameters required by Eqs. (3.1) and (3.2) can be found in the literature as listed in Table 1 [15].

Table 1. Vrentas-Duda free-volume parameters in a binary system.

Parameters	MMA	BB
$\widehat{V}_1^*$ (cm^3/g)	0.87	0.591
$\widehat{V}_2^*$ (cm^3/g)	0.757	0.757
$(K_{11}/\gamma) \times 10^3$ [cm^3/(g · K)]	0.815	0.64
$K_{21}(K)$	143	184.6
$(K_{12}/\gamma) \times 10^3$ [cm^3/(g · K)]	0.477	0.477
$K_{22}(K)$	52.38	52.38
$T_{g1}(K)$	225	242.4
$T_{g2}(K)$	378	378
α	0.44	0.44
D_0 (cm^2/s)	4.07×10^{-5}	1.92×10^{-5}
E (Cal/mol)	0	0
ξ	0.60	0.60

Using the data listed in Table 1 and Eq. (3.3), the dependence of D_{MMA} and D_{BB} on polymer concentration at 80 °C was calculated. The results show that the changing trends of D_{MMA} and D_{BB} values with PMMA concentration are identical, and that D_{MMA} is slightly larger than D_{BB} as illustrated in Figure 4. Because MMA and BB are nearly same size, this difference mainly comes from the difference of correspondent D_0 values, which are calculated from each viscosity of MMA and BB. Although it also has been verified that D is independent of the molecular weight of the polymer [18], there are still many factors to determine diffusion coefficients in the IGP process that need to be studied intensively.

From the model given in Section 3.2, it is easy to get theoretical results of the gradient refractive index (GRI) distribution on a cross section of preform using the data of D_{MMA} and D_{BB} values. One example is shown in Figure 5, in which experimental results are compared with the curves obtained theoretically under the assumptions of Eq. (3.3) and boundary conditions $\omega = 0.1$ at $r = 0$ and $\omega = 0.9$ at $r = 0.9R$. Absolute values vary slightly between experimental

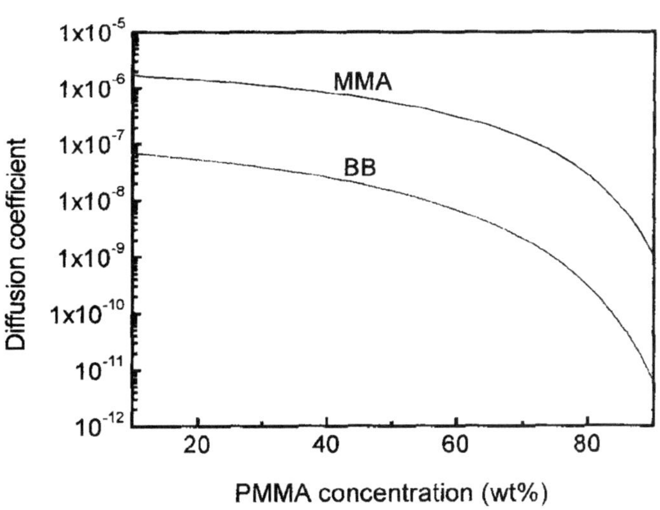

Figure 4. Values of D_{MMA} and D_{BB} as a function of PMMA concentration at 80 °C. Adopted with permission from Q. Zhang et al., *Macromolecules* 30, 7874. Copyright © 1997, American Chemical Society.

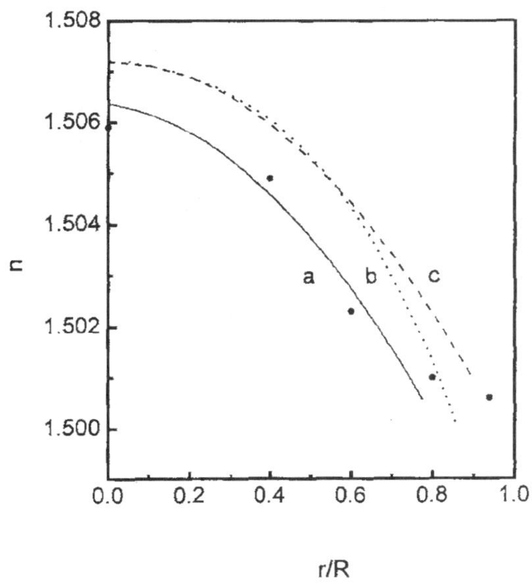

Figure 5. Refractive index distribution of a PMMA preform: a (solid curve), interfacial-gel polymerization of PMMA and dopant BB at 80 °C; b (dotted curve), theoretical simulation; c (dashed curve), result calculated by Eq. (1.1). Adopted with permission from Q. Zhang et al., *Macromolecules* 30, 7874. Copyright © 1997, American Chemical Society.

and theoretical distributions, although the changing trends are identical.

Except the preceding consideration of experimental techniques, when the thickening process ceases, the boundary conditions for the polymer concentration gradient in the tube have been assumed as $\omega = 0.1$ at $r = 0$ and $\omega = 0.9$ at $r = 0.9R$ for Eq. (3.3), in which R is the outer radius of the tube. This polymer concentration gradient can result in a corresponding distribution of inert molecules, because the inert molecules have different diffusion coefficients at different polymer concentration along the radial direction according to Eqs. (3.1) and (3.2). The result of the IGP system for MMA and BB can be seen visually from Figure 3c, in which open circles represent dopants. The inert molecules are distributed in a radial gradient distribution in the cross section. Quantitative results are shown in Figure 6. Figure 6a is the curve of the corresponding distribution of polymer concentration in the same system. It can be seen from Figures 5 and 6 that the result of refractive index distribution is largely dependent on the distribution function of polymer concentration given by Eq. (3.3), which is only an assumption at present, because it is not easy to measure the distribution experimentally. Different assumptions also have been reported, such as [27]

$$\omega = a(r/R)^2 + b \tag{4.1}$$

The difference between results from Eqs. (3.3) and (4.1) under the same experimental conditions and boundary conditions of polymer concentration distribution can be seen in Figure 7(1). Corresponding refractive index distributions are compared with experimental results in Figure 7(2). It can be seen from Figure 7(2) that, within the diameter range of the

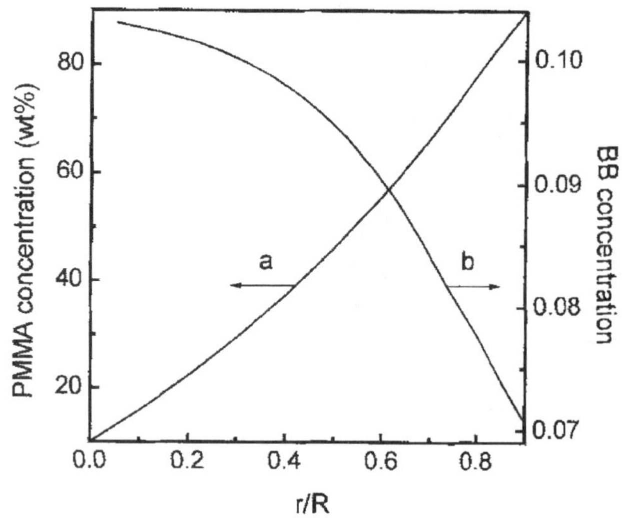

Figure 6. Radial distribution of (a) polymer concentration and (b) BB concentration. Adopted with permission from Q. Zhang et al., *Macromolecules* 30, 7874. Copyright © 1997, American Chemical Society.

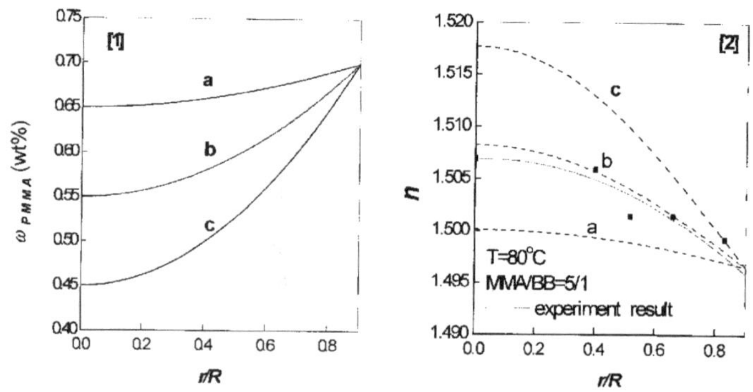

Figure 8. Polymer concentration distribution at different boundary conditions at $r/R = 0.9$, $\omega = 0.7$ and at $r = 0$, $\omega = 0.65$, 0.55, and 0.45 for curves a, b, and c, respectively. [2] The dependence of the refractive index of the PMMA preform of MMA/BB = 5/1 at 80 °C with different boundary conditions corresponding to [1]. The dashed curves denote theoretical simulates; the solid curve denotes the experimental result; the solid boxes denote refractive indices measured by the Becke line method. Adopted with permission from F. Zhang et al., *Polymer* 41, 9155. Copyright © 2000, Elsevier Science.

preform, experimental data are closer to the result obtained from Eq. (4.1) than the result from Eq. (3.3) and the curve is also more similar to the curvature of the experimental result.

In addition to the distribution function of polymer concentration, the boundary conditions of the function also are assumed during theoretical analysis. For different boundary conditions, the theoretical distribution of refractive index changes a lot as shown in Figure 8(2); as much difference can be seen from Figure 8(1) for polymer concentration distributions. Figure 8 shows the result of the IGP system of MMA/BB $= 5/1$ at different boundary conditions. It is clear that the optimized assumption condition is $\omega = 0.55$ at $r = 0$ and $\omega = 0.7$ at $r = 0.9R$ compared with the experiment result. This justification can help to reveal the detailed mechanism of the IGP process. Further justification comes from experimental work for IGP systems with different ratios of MMA and BB, which can be seen in Figure 9. Boundary conditions of the polymer concentration distribution were found not only to have an optimization, but also to be connected with the ratio of monomer to inert molecules [27].

Figure 9(1), easily reveals that the gel layer is thicker for the system with a high monomer concentration (curve c) than for the system with a low monomer concentration (curve a). The results given in Figure 9(2) show that there is also one optimized value of monomer concentration to obtain a suitable refractive index distribution during IGP processes. It can be expected that this optimized value is also dependent on the properties of the polymer tube used in IGP, such as thickness of the tube, molecular weight, and distribution of the polymer that comprises the tube. Much research still needs to be done in studying these factors for this specific technique.

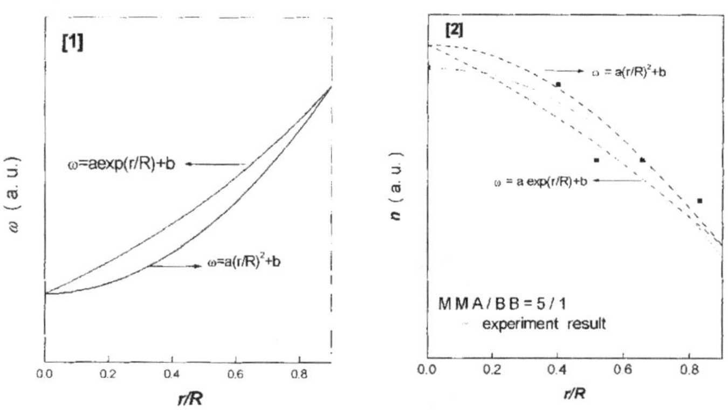

Figure 7. [1] Radial distributions of polymer concentration with different functions of polymer concentration distribution. [2] The dependence of the refractive index distribution of the preform at MMA/BB = 5/1 and 80 °C. The dashed curves denote theoretical simulations; the solid curve denotes experimental results; the solid boxes denote refractive indices measured by the Becke line method.

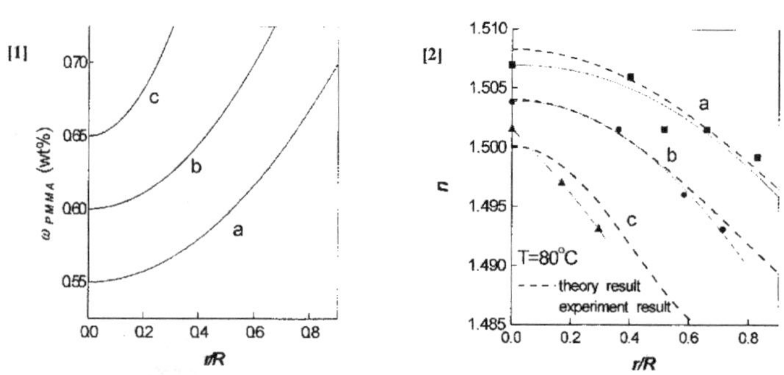

Figure 9. The distribution of the refractive index of a PMMA preform with different MMA/BB of the polymerization system at 80 °C: (a) MMA/BB = 5/1; (b) MMA/BB = 7/1; (c) MMA/BB = 10/1. The dashed curves denote theoretical simulations; the solid curve denote experimental results; the solid boxes, circles, and triangles represent refractive indices measured by the experimental method.

Any compounds that show fluorescence have two characteristic spectra: excitation and emission. Because excitation and emission usually take place at transitions between the ground state and the first excited state of organic dyes, partial overlap between absorption and fluorescence spectra is found in some organic molecules. There are many permitting transitions between energy levels of rare earth ions and, correspondingly, there are many absorption or fluorescence peaks in rare earth ion spectra compared with organic dye molecules. By properly selecting excitation, a segregated emission can be obtained. Furthermore, when rare earth compounds are irradiated at the absorption of an organic coordination group other than the absorption of rare earth ions, the rare earth ions, which are coordination centers, can be excited through energy transfer from the excited organic group to the ground ions and there is no overlap between excitation and emission peaks, because the absorption of organic groups usually is located within the ultraviolet light range.

As described previously, the emitting group often is not identical with the absorption group in rare-earth-doped polymers. The experimentally observed quantum efficiency may, therefore, be quite different from the "true" quantum efficiency which is a characteristic feature of the emitting group, irrespective of the mode of excitation. Some kinds of organic dyes can emit more efficiently than rare earth ions. For example, rhodamine B in ethanol solution has a larger fluorescence cross section ($\sigma = 3.0 \times 10^{-20}$ m^2 [40] than that of Sm^{3+} in acetone-d$_6$ solution ($\sigma = 3.3 \times 10^{-21}$ m^2 at 645 nm) [46]. However, rare earth ions are chosen as an emitting group based on the feature of long fluorescence lifetime [44]. Both the fluorescence lifetime and the emission cross section are known to be strongly influenced by the environment surrounding rare earth ions, and a figure of merit for a rare-earth-doped fiber amplifier is proportional to the emission cross section as well as the observed lifetime [37].

There is no problem of photochemical stability with rare earth ions because ions are the smallest units in chemical reactions. Absorption and fluorescence of rare earth ions come from their f-f electronic transitions, which are more stable than those of organic dyes, because their inner electronic energy levels are less affected by macroscopic actions, except the action of a microscopic coordinating field around the ions. Furthermore, rare-earth-doped POFs exhibit broad emission profiles and favorable implications for signal wavelength tuning ability, because the polymer host is a disordered medium [44].

The initial research work on rare-earth-doped POF was concerned mainly with step index (SI) POF, such as the manufacture of Nd octanoatc-doped SI POF [39], amplified spontaneous emission of the POF [47], and numerical simulation of rare-earth-doped PMMA fiber amplifiers [46]. In all these works, the introduction of organic groups is used mainly to achieve better compatibility between rare earth ions and the polymer matrix. However, it has received more attention if the GI distribution is affected by rare earth compounds in rare-earth-doped GI POF, because the compounds used in IGP usually have a bulk volume.

5.2. Synthesis and Characterization of Organic Rare Earth Salts

The rare-earth-doped preform for GI POFs can be made according to the procedure given in Section 1. The only difference lies in the ingredients of the mixture for polymerization. In addition to initiator, chain transfer agent, and so forth, one of the rare earth compounds, such as organic salts or chelates, is added to the mixture. In reported work, such compounds include rare earth octanoate [48, 49], Eu chelates of trifluoroacetylacetone [Eu(TFAA)] and hexafluoroacetylacetone [Eu(HFAA)] [44], and Nd chelate of deuterated tri(hexafluoroacetylacetone) [Nd(HFA-D)] [50]. To get rare-earth-doped GI POF, the first step is to synthesize rare earth compounds that are compatible with the polymer. A detailed description of the synthesis of rare earth octanoate is shown as

$$Re_2O_3 + 6HNO_3 \rightarrow 2Re(NO_3)_3 + 3H_2O \tag{5.1}$$

$$2Re(NO_3)_3 + 3Na_2CO_3 \rightarrow 6NaNO_3 + Re_2(CO_3)_3 \tag{5.2}$$

$$Re_2(CO_3)_3 + 6C_7H_{15}COOH \rightarrow 2Re(C_7H_{15}COO)_3 + 3H_2O + 3CO_2 \uparrow \tag{5.3}$$

in which Re represents the rare earth ion, such as Eu^{3+} or Nd^{3+}. For example, neodymium octanoate (NOA) was made from neodymium oxide. The oxide was dissolved in concentrated nitric acid by heat to form a solution of neodymium nitrite. Adding the solution to a solution of sodium carbonate that had a concentration of 10%, yielded solid Nd^{3+} carbonate with nearly 100% product. Following desiccation in a vacuum oven at 80 °C for two hours, the solid Nd^{3+} carbonate was titrated by excess octanoic acid at a constant temperature of 80 °C. Then the reacted solution was filtrated and the filtrated solution was put into a large amount of acetone, where a great deal of sedimentation, which is neodymium octanoate according to infrared (IR) and elementary analysis, was obtained.

Rare earth octanoates obtained by the foregoing synthesis route usually contain two kinds of isomers: chelating isomers and bridge isomers. The infrared spectrum of NOA in Figure 11 shows that there are two strong absorption peaks located at 1541.7 and 1454.7 cm^{-1}. They denote absorption of the antisymmetrical (υ_1) and the symmetrical (υ_2) vibration of the carboxyl group, respectively. The frequency difference ($\Delta\upsilon$) between these two peaks is 87 cm^{-1}. Spectroscopic studies on the structure of carboxylate complexes [51–56] have found that a relationship exists between the O-C-O angle and the frequency difference ($\Delta\upsilon = \upsilon_1 - \upsilon_2$). Generally speaking, there are two kinds of structure of Nd^{3+} octanoate: chelating structure and bridging structure. In chelating acetates the O-C-O angle is smaller than in bridging acetate [57], which is confirmed by the structural evidence: chelating acetate angles [58, 59] are 111° and 121°;

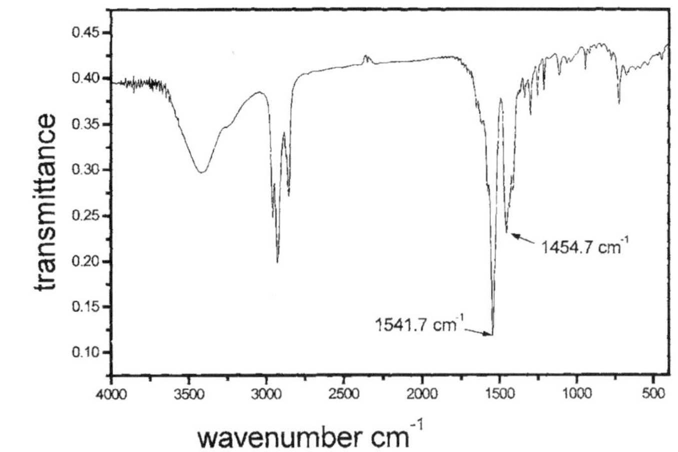

Figure 11. Infrared spectrum of NOA. The numbers in the figure refer to the wave numbers of antisymmetrical and symmetrical vibration absorption of carboxyl group.

bridging angles [60–62] are 125°, 123°, and 124°. Grigor'ev [57] calculated the effect of changing the O-C-O angle without changing the force constant. He found that increasing this angle should increase v_1 and decrease v_2, and hence increase Δv. The classical frequency differences of bridge linkage and chelation linkage between ion and carboxyl group are 140 and 80 cm^{-1}, respectively [63]. Figure 11 shows the IR spectrum of NOA, from which $\Delta v = 87\ cm^{-1}$ can be obtained. These data show that NOA has chelation linkage, as given by the two-dimensional structure in Figure 12, from which it can be seen that one macromolecule is formed for bridge isomers because one carbonyl group connects with two rare earth ions.

Two isomers of rare earth octanoates can be separated by the solubility property of each isomer. The chelation linkage isomer can be dissolved into methanol, ethanol, and acetic acid; the bridge linkage isomer cannot. Detailed analysis of their IR spectra is provided in Table 4. Three chelation isomers also can be dissolved into MMA to a certain extent. Preforms for rare-earth-doped POF can be made from this mixture solution according to the procedure given in Section 2.2. However, the microstructure of rare earth ions doped in solid material is different from that in solution. Some kinds of clusters may be formed during solidification of the mixture solution. On the other hand, the specific clusters that resulted from actions between rare earth ions and the polymer matrix are sensitive to the spectroscopic characteristics of the doped polymer. There has been much work on such clusters, although study is just beginning about their effect on the spectroscopic properties of the doped polymer. Note that doped polymers are classified as ionomers in polymer science, in which field clusters are of specific significance as described in the following section.

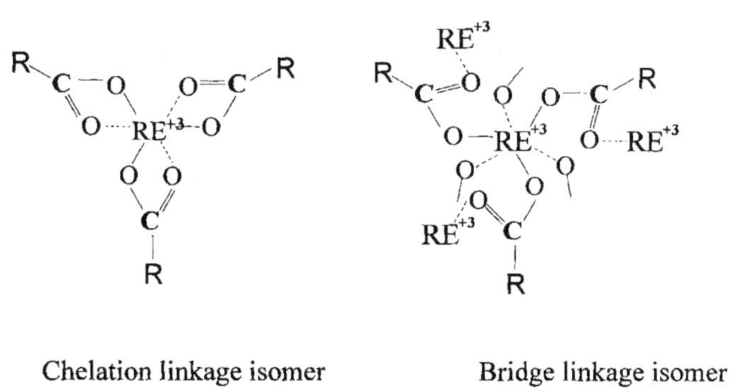

Figure 12. Structural units of two isomers of rare earth octanoates.

Table 4. Difference between ν_{as} and ν_a of Eu^{3+}, Nd^{3+}, and Sm^{3+} octanoates (EOA, NOA, and SOA, respectively)

	Salts					
	EOA		NOA		SOA	
Isomer	Chelate	Bridge	Chelate	Bridge	Chelate	Bridge
ν_{as} (cm^{-1})	1547.3	1566.6	1541.7	1545.4	1545.7	1546.5
ν_a (cm^{-1})	1467.1	1453.3	1454.7	1445.3	1466.4	1446.4
$\Delta\nu$ (cm^{-1})	80.2	113.3	87.0	100.1	79.1	100.1

5.3. Aggregation of Rare Earth Ions in Rare-Earth-Doped POFs

5.3.1. Introduction

In the past years, a considerable amount of research has been devoted to random ionomers due to their unique physical properties [64–68]. Ionomers typically are defined as polymers that have an acidic side group neutralized by cations [69]. These side groups join with the main chain through covalent bonds. The rare-earth-doped polymer is of similar structural character to ionomers because the ions are connected with the polymer chain by coordination bonds, although coordination bonds are much weaker than ionic bonds in ionomers. Furthermore, the cluster model used in ionomers is applicable to the rare-earth-doped polymers because the formation of clusters in both materials results from the incompatibility between inorganic ions and the organic polymer matrix.

A number of models for the morphology of random ionomers have been proposed to explain existing experimental observations. The earliest and most direct evidence for the existence of ion aggregation in ionomers was the appearance of a peak in the small angle X-ray scattering (SAXS) pattern of an ethylene–methacrylic acid copolymer neutralized with Na^+ [70]. Bononto and Bonner [71] developed the first qualitative model for the morphologies of random ionomers in 1968. In the same year, Longworth and Vaughan [70] proposed a model based on the analysis of SAXS data, where a Bragg peak spacing of ca. 50 Å was observed in the scattering profiles. This "ionic" peak was interpreted as being due to scattering from ordered hydrocarbon chains between ionic aggregates rather than the aggregates themselves.

In 1970, Eisenberg [72] presented a theoretical model in which the electrostatic interactions of ionic pairs, steric effects, and the dielectric constant of ionomers were proposed to explain the formation of multiplets and clusters. This model was able to account for the mechanical data by citing the case of an ethylene–sodium methacrylate copolymer that contained 4.5 mol% of the salt.

Marx et al. [73] proposed a model in which the ionic SAXS peak was ascribed to the electron density difference between the metal cations and the hydrocarbon, where the scattering moieties were treated as points on a paracrystalline lattice. In this model, only multiplets were postulated to exist, with no occurrence of clustering or phase separation when the ion concentration is within 5.5 mol%. When the ion concentration was above 7.9 mol%, a low-angle X-ray peak was observed for cesium salts.

MacKnight et al. [74] proposed a model derived from the radial distribution function of the scattered X-rays for partially neutralized ionomer of poly(ethylene-*co*-cesium methacrylate). This model, known as the core–shell model, postulates the existence of clusters with a radius of 8–10 Å, each containing ca. 50 metal ions. The cluster is shielded from the surrounding matrix ions by a shell of hydrocarbon chains that is on the order of 20 Å thick. The considerable number of ions per aggregate and the relatively large dimensions of the clusters postulated in this model might provide an explanation for the observed high T_g. A peak was observed in the copolymers of cesium salts of ethylene–methacrylic acid and ethylene–acrylic acid that had cesium ion concentrations ranging from 2.1 to 6.3 mol%.

In 1980, Roche et al. [75] suggested a model that is conceptually similar to the core–shell model, but differs in the proposed geometry of the clusters. In this model, a lamellar, instead of spherical, central ionic core is proposed that has a region of depleted ions on either side. In this model, the spacing between the core and the shell accounts for the SAXS peak, and the size of the clusters may be sufficient to explain the observed two-phase behavior.

In 1983, Yarusso and Cooper [76] proposed a modified hard-sphere model in which the multiplets are proposed to have liquidlike order and to have a distance of closest approach that is determined by the layer of hydrocarbon chains attached to and surrounding each multiplet. This model is in excellent agreement with the experimentally determined "ionic" peak in the SAXS profiles of sulfonated polystyrene (SPS) that has a composition of about 6% methacrylic ions [77].

A model proposed by Ding et al. [78], based on anomalous SAXS from SPS ionomers, accurately described the entire SAXS profile. This model also attributes the SAXS ionomer peak to interparticle scattering of hard spheres with liquidlike order, but includes the effects of a nonrandom distribution of lone metal ions, thereby accounting for the small-angle upturn. In their report, 5.6 mol% of the styrene units were sulfonated and the scattering of this polystyrene ionomer showed a clear peak.

In 1990, Eisenberg et al. [79] proposed a new model of clustering in random ionomers that originated from the multiplet–cluster model proposed in 1970. In this model, the formation of multiplets is thought to be affected by the strength of the electrostatic interactions between the metal ions, the elastic forces of the chains, the ion content of the ionomer and the dielectric constant of the polymer. Each ion pair in a multiplet effectively anchors the polymer chain at the point to which it is attached. Hence, the mobility of the polymer chain in the immediate vicinity of a multiplet is expected to be greatly reduced relative to that of a chain in the bulk polymer. A region of restricted chain mobility or "skin" therefore surrounds each multiplet, and there is no definite boundary between the region of restricted mobility and the rest of the polymer. The thickness of the region of restricted mobility surrounding each multiplet is determined largely by the flexibility of the polymer backbone.

As the ion content is increased to a critical point, some overlap is encountered among the regions of restricted mobility surrounding each multiplet (the multiplets themselves remain intact). As the overlap becomes more frequent, large contiguous regions of restricted mobility that have their own T_g, must form. Such regions constitute clusters and exhibit behavior characteristic of a phase-separated region. An important feature of this model is that it does not require the clusters to be of any particular geometry and, in fact, highly irregular shapes are probable. The presence of clusters is determined by the existence of sufficiently large regions of material that has restricted mobility.

This model accounts for wide ranges of experimentally observed phenomena. The clusters behave as if they were phase–separated from the regions of more mobile segmental motion in that they exhibit their own T_g, which is significantly higher than the T_g of the unclustered components. Also in these regions a very prevalent intermultiplet spacing would be expected. This spacing accounts for the characteristic ionic peak in the SAXS profiles of these materials above a certain ion content.

In all these studies all of the models are based on materials that have ion concentrations ranging from 1.0 [80] to 18 mol% [72]. For the rare-earth-doped material used for doped POF, generally speaking, only very low ion concentration is needed, usually within 100–1000 ppm. Under this circumstance, new methods for characterization of such material need to be developed. In this short review, methods newly developed by my group will be discussed in the following sections.

5.3.2. Internal Friction Method

The mobility of the main and side chains of macromolecules is one of the factors that determine the mechanical and other physical properties of solid polymers. The temperature dependence of molecular mobility is characterized by various transitions in which a certain mode of chain motion sets in. The most important transition in amorphous polymers is the glass transition temperature, T_g, at which the micro-Brownian motion of segments of the main chains becomes active. It is well known that ion-containing polymers perhaps provide an ideal system for the study of intermolecular force, because the interaction between ions can be quantified most precisely. For organic polymer salts, an initially rapid rise of the glass transition temperature with anion content, is observed, followed by a sigmoidal shaped accelerated rise of the glass transition, and then another relatively lower rate of increase at still higher ion concentrations [81]. On the other hand, ion-containing polymers can be plasticized effectively with respect to backbone and ionic domain plasticizers. The former is the more conventional type in which the plasticizing arises from the free-volume effect (backbone plasticizers); the latter can act ideally by diminishing the interchain association of the ionic groups on the polymer chain (ionic domain plasticizers) [82].

The internal friction method is a sensitive method that has been used widely in recent decades to investigate the various low-frequency relaxation processes of solid materials [83, 84]. Under certain conditions, the internal friction method can be used to determine the glass transition, which is often called the α transition, of amorphous solid materials. The α transition can provide much information concerning the microstructure of polymers [85].

The investigation on the glass transition of gadolinium-octanoate-containing poly(methyl methacrylate) (GOA-PMMA) was performed by the internal friction method [86]. The low-frequency internal friction probe (internal friction = IF, which has the same physical meaning as tan δ adopted in other mechanical spectroscopy) gave several different frequency measurements with forced oscillation mode within certain temperature measurement ranges for GOA-PMMA samples with various Gd^{3+} concentrations. The result obtained shows that in the temperature range studied, there is only one IF peak for each sample and, with increasing Gd^{3+} content, the height of the peak decreases. When Gd^{3+} was doped into PMMA, the chain length or domains for the relaxation at given temperature were further restricted by coordination actions of Gd^{3+} and became shorter or smaller, which resulted in a decrease in IF. This phenomenon, which results from the stiffening of the matrix imparted by ionic domains, also has been found in other systems [87]. On the other hand, the novel changing tendency of T_g also has been found for GOA-PMMA samples: T_g increases with increasing ion concentration when ion concentration is lower than 0.3 mol% beyond this ion concentration, T_g decreases to a level of about 105 °C, which is nearly 10 °C lower than the T_g of the control sample [86].

Generally speaking, plasticization of ion-containing polymers includes two aspects: one is the more conventional type where the plasticizing effect arises from free-volume effects and is usually called backbone plasticization; the other acts ideally by diminishing the interchain association of the ionic groups on the polymer chain and is called ionic domain plasticization. It has been found that ionic domain plasticization cannot cause a reduction of T_g and tan δ of polymers [88]. Although coordination also can take place between Gd^{3+} and the free carbonyl or ester groups on the flexible polymer chains, because Gd^{3+} can have multiple (seven to nine) coordinated types of sites [89], the backbone plasticization of GOA can be expected to a greater extent because GOA is a long chain aliphatic ester, which is not compatible with PMMA chains. This structure characteristic results in small phases composed of GOA molecules distributed in the PMMA matrix. According to the observation of IF measurement [86], the phase can be considered as multiplets rather than clusters, because the concentration of GOA is low. An increase in the free volumes of polymers imparted by incompatibility between two phases causes a reduction in the T_g of GOA-PMMA.

Taking all the tendencies for change into consideration, it can be deduced that coordination and plasticization both exist in GOA-PMMA samples. The maximum T_g is reached by equilibrium between two opposite effects. Based on this phenomenon, a new multiplet structure can be expected: ions are on the outer surface of the multiplet and long aliphatic chains of GOA compose the core of the multiplet, because they are not compatible with the polar matrix of PMMA. A diagram of the multiplet model is shown in Figure 13.

The model given in Figure 13 is relevant only to the samples that have low doping concentration. When the doping concentration increases, clusters that contain more than two rare earth ions will be formed. A detailed analysis of this phenomenon can be found in the reviews cited in Section 5.3.1. Because energy transfer between ions within one cluster easily takes place, rare-earth-doped polymers with high ion doping concentration are less frequently adopted to manufacture rare-earth-doped POF.

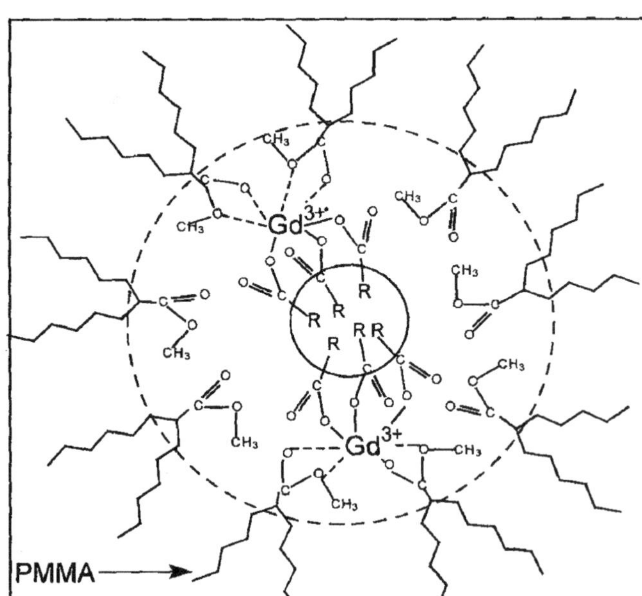

Figure 13. Schematic representation of the multiplet model of GOA-PMMA. Adopted with permission from Q. Zhang et al., *Polymer* 39, 3787, Copyright © 1998, Elsevier Science.

5.3.3. Atomic Rate Equations for Rare-Earth-Doped POFs

The optical properties of most trivalent lanthanide compounds in the near-ultraviolet, visible, and near-infrared spectral regions can be accounted for in terms of intraconfigurational 4f-4f radiative transitions [70, 90]. Along with these radiative transitions, energy transference occurs. Cross relaxation takes place between close ions such as occurs in multiplets. All of these processes can be described by atomic rate equations that correspond to ion density and populations of atoms (ions) in each energy states, respectively [91]. This principle can be used to characterize the multiplet structure of rare-earth-doped polymer through a model, which is given for a Nd^{3+}-doped PMMA (NOA-PMMA) as an example.

For NOA-PMMA, these energy transferences occur between an excited ion and a ground ion. The ion is pumped by absorbed light to a excited energy level, $^4F_{3/2}$, and then, the excited state $^4F_{3/2}$ (donor) gives half its energy to the acceptor in the ground state $^4I_{9/2}$, so that both ions end up in the intermediate state $^4I_{15/2}$ from which they both relax nonradiatively to the ground state. All of these processes can be seen in the diagram in Figure 14.

For the sample of Nd^{3+}-doped POF, the model uses the length dimension of the optical fiber as the direction of light transmission (z axis). Due to the low rare earth ion concentration in Nd^{3+}-doped POF, we assumed that (1) a multiplet contains two Nd^{3+} ions, which we express as the Nd^{3+} pair, and (2) there is no interaction between ions of different multiplets. In this case, the rate equations for Nd^{3+} ions in the fiber to transfer between excited and unexcited (ground) energy states can be written as [92]

$$\frac{dN_1^p}{dt} = \frac{N_1^p}{\tau} - \frac{N_1^p}{\tau_{rad}} + 2\frac{N_2^p}{\tau_{rad}} + 2\omega_p N_0^p - \omega_p N_1^p \tag{5.4}$$

$$\frac{dN_2^p}{dt} = -2\frac{N_2^p}{\tau_{rad}} + \omega_p N_1^p \tag{5.5}$$

$$\frac{dN_c^s}{dt} = -\frac{N_c^s}{\tau_{rad}} + \omega_p N_g^s \tag{5.6}$$

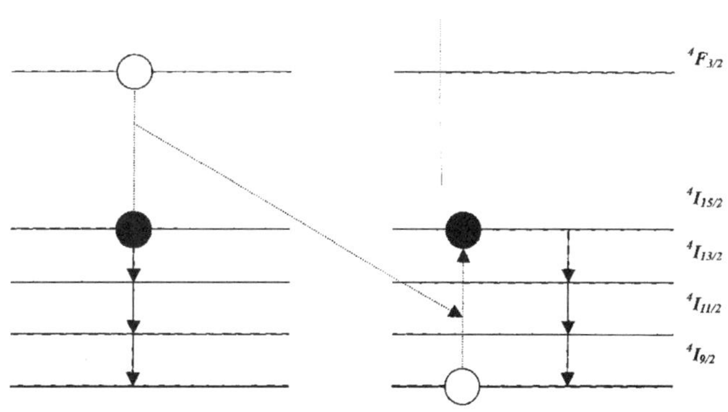

Figure 14. Cross relaxation process for Nd^{3+} ions within one multiplet. The open circle indicates the initial position of the electrons; the closed circle indicates the final position; dashed lines denote energy transfers; solid lines denote nonradiative transitions.

where N_0^p, N_1^p, and N_2^p are concentrations of the unexcited Nd^{3+} pair, single excited Nd^{3+} pair, and doubly excited Nd^{3+} pair, respectively, N_g^s and N_e^s are concentrations of unpaired Nd^{3+} in the ground state and excited state, respectively, τ_c is the time constant of cross relaxation, and τ_{rad} is the lifetime of the metal state. In Eq. (5.6), ω_p is the pump rate and is of the form

$$\omega_p = \frac{p\sigma_p}{h\nu_p A_{eff}} \tag{5.7}$$

in which p is the pump power, σ_p is the absorption cross section, h is Planck's constant, ν_p is pump frequency, and A_{eff} is the effective mode square of pump light. Whereas each pair has two ions, the total excited ion concentration, the unexcited ion concentration, and the total ion concentration are, respectively,

$$N_c = 2N_2^p + N_1^p + N_c^s \tag{5.8}$$

$$N_g = 2N_0^p + N_1^p + N_g^s \tag{5.9}$$

$$N_{tot} = N_c + N_g \tag{5.10}$$

The ratio of "multipleted" ions to the total ions is defined as ρ. Then the steady solution of Eq. (5.4)–(5.6) can be worked out as

$$N_g = \frac{N_{tot}\omega_{rad}}{\omega_p + \omega_{rad}}(1-\rho) + \rho N_{tot}\frac{\omega_c + \omega_p + W_r}{(\omega_e + 2\omega_p + \omega_p^2/\omega_{rad} + \omega_{rad})} \tag{5.11}$$

In Eq. (5.11), the last item was imputed to energy transfer between multipleted ions, ω_{rad} is the transition rate from the metastable state, and ω_c is the cross relaxation rate, that is,

$$\omega_{rad} = \frac{1}{\tau_{rad}} \qquad \omega_c = \frac{1}{\tau_c} \tag{5.12}$$

When a stimulating energy power (p, pump power) was launched into the Nd^{3+}-doped POF and after steady state was established, the pump power changes along the z direction in the form

$$\frac{dp}{dz} = -\sigma_p N_g p \tag{5.13}$$

In Eq. (5.13), amplified stimulated emission and the background absorption of POF were ignored, because the infrared fluorescence caused by stimulated emission is strongly absorbed by the host of Nd^{3+}-doped POF and because of the considerably short length of POF used as samples in the experiments.

Equation (5.13) was integrated numerically along the fiber length (z axis) to obtain the fiber transmission $T_f = p(l)/p(0)$, where $p(0)$ is the launched pump power and $p(l)$ is the output pump power at fiber length l. It is obvious that the transmission was closely related to the basic multiplet parameter, ρ, because the other parameters are fixed under experimental conditions. So by using the fiber transmission data, ρ can be enumerated via Eq. (5.13).

For a POF sample that has several hundreds parts per million Nd^{3+} ion concentration, it is also reasonable that the POF sample made from bulk samples has the same morphology as the bulk sample. When a light is introduced into the fiber and a steady state is established, various rate processes described by Eq. (5.4)–(5.6) reach equilibrium. Under this circumstance, N_g, the unexcited ion concentration of Nd^{3+} in the ground state, can be calculated according to Eq. (5.11), and can be affected by N_{tot}, ρ, and light power P. N_g in Eq. (5.11) can be divided into two parts. The first part is contributed to the isolated Nd^{3+} ions (N_{g1}) and the second to multipleted Nd^{3+} ions (N_{g2}). Figure 15 shows how each item changes as P and ρ increase. From the Figure 15, it can be deduced that all of N_g, N_{g1}, and N_{g2} decrease dramatically with increasing pump power. In Figure 15(a), values of N_g for each ρ value remain unchanging at about 0.05 of N_g/N_{tot} when pump power P approaches 0.8 W. This resulted mainly from the item of N_{g2}. If there were no multiplets in the sample, N_g, which only comprises N_{g1}, would disappear when P is large enough as shown in Figure 15(b). N_{g1} contributes the most to N_g when the value of ρ is small. When the value of ρ increases, N_{g1} decreases, but both N_g and N_{g2} increase. This can be explained by the fact that excited Nd^{3+} ions in multiplets can reach ground state more easily and more rapidly because the cross relaxation between Nd^{3+} ions takes place in a short time: $\tau_c = 10^{-7}$ s for a couple of Nd^{3+} ions within one multiplet. When multiplet concentration ρ increases, Nd^{3+} ions in the ground state, N_g, also increase.

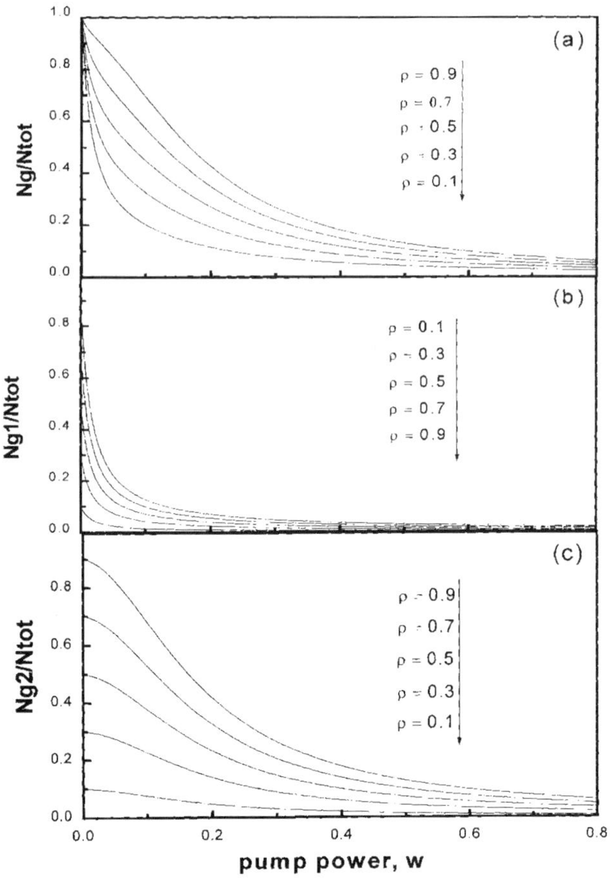

Figure 15. Relationships between items of Eq. (5.11) and pump power. (a) The relationship between combined unexcited concentration and pump power, (b) The relationship between unexcited concentration of isolated Nd^{3+} ions and pump power; (c) The relationship between unexcited concentration of multipleted Nd^{3+} ions and pump power.

Assuming a POF sample with 200 ppm Nd^{3+} ion concentration and given different ρ values, theoretical curves of the relationship between transmittance and pump power that result from Eq. (5.13) are as shown in Figure 16. The curves are similarly shaped but have different transmittance at the same pump power. A small difference can be found in curve shape between curves with $\rho > 0.5$ and $\rho < 0.5$. For curves with $\rho > 0.5$, a higher rate of transmittance over pump power can be observed when p is greater than 0.3 W. This observation can explained in terms of the existence of more multiplets and of cross relaxation in the system. Another explanation comes from the limitation of this theoretical model. During integrating of Eq. (5.13), one important assumption is that only multiplets composed of two Nd^{3+} ions are in the samples. This is true when the concentration of Nd^{3+} ions is low, but when the concentration increases, that is, when ρ increases, multiplets composed of more Nd^{3+} ions or clusters may form in the sample. Under this circumstance, a more complex model that contains more rate processes or another model that deals with high ion content ionomers will be considered. For the $\rho <$ 0.5 system, it is interesting to note that a curve can be made for each ρ value, which means that each Nd^{3+}-doped POF sample would has its own characteristic transmittance curve, even though it has very low ion concentration. Therefore, from this deduced model, POF samples with different Nd^{3+} ion concentrations can be expected to have different curves as shown in Figure 16.

Fiber transmissions of SI-type POFs doped with 70, 100, and 200 ppm Nd^{3+} ion concentrations were measured under

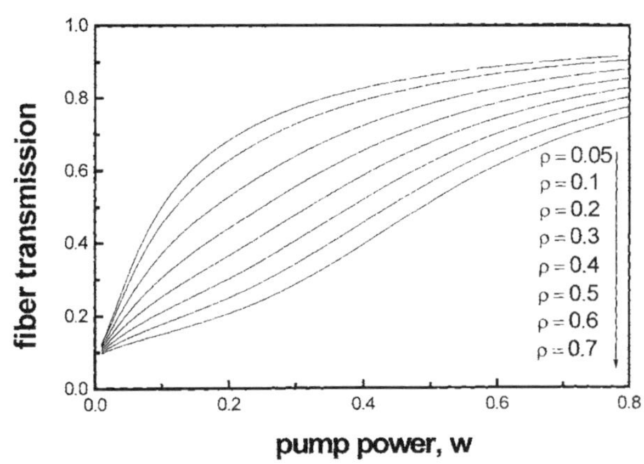

Figure 16. Theoretical relationship between fiber transmission and pump power with different values of ρ in NOA-PMMA bulk that has 200 ppm Nd^{3+} ion concentration. In the calculation, the following parameters were used: $\sigma_p = 8 \times 10^{24}$ m^2, $h\nu_p = 247 \times 10^{-19}$ J, $A_{eff} = 1.2 \times 10^{-11}$ m^2, $\tau_{rad} = 8 \times 10^{-4}$ s, and $\tau_c = 10^{-7}$ s.

different pump power and the result are shown in Figure 17, in which experimental data points are denoted by □, △, and ○ for POF with Nd^{3+} ion concentrations of 200, 100, and 70 ppm, respectively. It can be Observed in Figure 17 that for each Nd^{3+} concentration, the fiber transmissions increase as the pump power increases. At high pump power of nearly 500 mW, the fiber transmissions reach a constant value and a residual absorption could not saturate. For different Nd^{3+} ion concentrations, the value of the residual absorption is different. This can be attributed to energy transference between two Nd^{3+} ions within a multiplet.

By adopting the preceding theoretical model, three ρ values for Nd^{3+}-doped POF samples with concentrations of 70, 100, and 200 ppm Nd^{3+} ions were estimated. The solid lines in Figure 17 are theoretical curves obtained by selecting a fitting parameter ρ. The value of ρ was 0.05 for the fiber with a concentration of 70 ppm, 0.1 for a concentration of 100 ppm, and 0.3 for a concentration of 200 ppm. As mentioned, Eisenberg's model proposed in 1990 that a multiplet may contain only two metal ions as well as a quartet when the ion concentration is low enough [79]. However, the degree to which the ion concentration decreases while the multiplet that contains two Nd^{3+} ions is formed is dependent on many factors. The foregoing result provides a quantitative description of the probability of the existence of a multiplet that contains two Nd^{3+} ions at certain low ion concentration and the critical Nd^{3+} ion concentration for forming the multiplet is within 100 ppm.

The data presented for ρ versus Nd^{3+} ion concentration engender the deduction that the larger is the ion concentration, the greater is the value of ρ. From the results we also can see that the fraction of multipleted ions, ρ, is lower than 0.3 when the ion concentration is within 200 ppm. As already defined, ρ is the ratio of multipleted ions to the total ions and a multiplet is an aggregate that contains two Nd^{3+} ions. Therefore it can be inferred that a large part of the

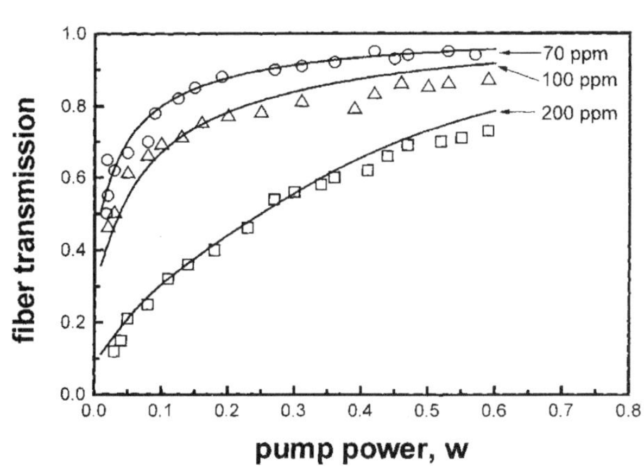

Figure 17. Transmission data for the three different fiber and best fits of the model with Nd-Nd pairs (solid lines). The circles, triangles, and boxes refer to experimental data points of the three fibers at 70, 100, and 200 ppm, respectively. The theoretical simulation used the following parameters: $\sigma_p = 8 \times 10^{24}$ m^2, $h\nu_p = 2.47 \times 10^{-19}$ J, $A_{eff} = 1.2 \times 10^{-11}$ m^2, $\tau_{rad} = 8 \times 10^{-4}$ s, and $\tau_c = 10^{-7}$ s.

Nd^{3+} ions are distributed evenly in the polymer bulk when the ion concentration is below 200 ppm and only a small portion of Nd^{3+} ions join together to form a multiplet. Each multiplet links two or more than two PMMA chains. These multiplets behave like a conventional cross-linking point. So NOA-PMMA with such ion concentration can be viewed as a conventional ion cross-linked polymer and these multiplets may influence the properties of the material in a fashion similar to their effect on ion cross-linked polymers. In addition, because the multiplets formed in rare-earth-doped POF will affect the photonic properties, the result given herein are only a guideline for designing active POFs.

6. REFRACTIVE INDEX DISTRIBUTION OF RARE-EARTH-DOPED POFS

As described in Section 1, GI POFs are generally obtained through two separate processes: first, GI preforms are made by the IGP technique and, second, the preforms are drawn into GI POFs. Fiber made by the second process maintains the properties of the preform, so the preform can be characterized before it is drawn into fibers and, in this way, the fiber properties can be controlled better. This procedure can be adopted easily to make rare-earth-doped GI POFs.

During the preparation of performs of rare-earth-doped GI POF, the reaction mixture is composed of monomer, initiator, chain transfer agent, and an unreactive high refractive index compound, as well as a rare earth compound. Compared with the composition in Section 2.2, the rare earth compound is introduced in the interfacial-gel polymerization step. Usually, the rare earth compound has a large volume, which may cause a change in the refractive index distribution in the cross section of the preform in terms of the diffusion model of free-volume theory. This problem has received attention in reported works on rare-earth-doped GI preforms and GI POFs. For rare earth octanoates, such as neodymium octanoate [48] and samarium octanoate [49], interfacial-gel polymerization was carried out in a PMMA tube using BB as the unreactive high refractive index compound. The distribution and difference of refractive index between the center of the preform and a cladding region were found to be identical to other interfacial-gel polymerizations without rare earth compounds. This result was found in a rare-earth-doped GI preform that had a low rare earth concentration of 60–700 ppm [48, 49]. For low concentrations, the main contribution to refractive index comes from the unreactive high refractive index BB compound, although rare earth compounds can diffuse along with the BB and have large refractive index.

For high rare earth concentrations, such as 1 wt% of deuterated tris (hexafluoroacetylacetone) theodymiom [Nd(HFA-D)$_3$), the rare earth compound has the same GI distribution along the radius of the preform [50]. At the center of the preform, the concentration reaches its highest value and then gradually decreases along with the radius and reaches its lowest value at the normalized radius of 0.6. At the same time, rare earth chelates have no effect on the distribution of refractive index of the POF at such a concentration as 1 wt%. The preform was heat-drawn into a fiber at 200 °C. The fiber diameter and the corediameter were 500 and 300 mm, respectively. The index difference between the center of the core and the clad region was 1.1% [50]. Note that rare-earth-doped preforms can be drawn easily into fibers, although there is relatively strong action between rare earth ions chains of polymer matrix. This action comes from coordination between rare earth ions and polar groups in polymer chains, such as ester groups in PMMA. When the temperature is enough high, this pre-ionic cross linking dissociates and polymer chains can move freely. However, there are no reports on whether this dissociation affects the distribution of refractive index of a GI POF obtained by this procedure.

Furthermore, there is still a need to develop amplifiers or lasers of rare-earth-doped POF and/or GI POF. Nearly all the host materials used so far have been inorganic glasses and crystals. For rare-earth-doped POF, eliminary work has been reported on superfluorescence [44], amplified spontaneous emission [47], and the effects of deuterated or fluorinated matrices on absorptions and fluorescence [50]. As for amplification of rare-earth-doped POF, it is thought that the observed metastable-state lifetime of a rare earth ion transition should be maximized [46], which is most influenced by the surrounding environment.

The electronic configuration of a trivalent rare earth is $[Xe]4f^{N-1}5s^2 5p^6 6s^0$. In this configuration, [Xe] represents the closed shell of the electronic configuration of xenon [93], and one and two electrons are removed from the 4f and 6s shells, respectively. The known 4f-4f transitions of a rare earth ion can be characterized by electric dipole and electric quadrupole as well as magnetic dipole contributions. If the atomic states were pure 4f states, parity would forbid any electric dipole transitions. However, if these states mix with higher lying opposite-parity states, such transitions are possible and, in fact, are dominant. This mixing usually takes place under a coordination field forwarded by the host matrices and can be examined in terms of the optical intensity parameters Ω ($\lambda = 1, 2, 3$) [94, 95] and other parameters [96]. On the other hand, the N-1 inner electrons of the 4f shell are shielded from the external field by the outermost 5s, 5p shell. Under this circumstance, to provide an adequate basis for carrying out quantitatively reliable 4f-4f intensity calculations, lots of work has concentrated on electric dipole intensity parameters for lanthanide 4f-4f transitions because of both the complexity in spectroscopic characteristics of rare-earth-doped materials and the magnetic dipole oscillator strengths that are orders of magnitude smaller than that of the electric dipole [97, 98]. Detailed theoretical treatment of this field is beyond the scope of this short review. Interested readers may refer to [94–98] and references cited therein.

A known theoretical result is that the electric dipole oscillator strengths of one rare earth ion are related to a series of field parameters that characterize the environment of the ion, such as Ω ($\lambda = 1, 2, 3$), which represents the optical intensity parameters. For rare-earth-doped POFs, where the host material is amorphous polymer (PMMA, in general), polymer chains would have actions on rare earth ions, especially because there are polar groups in polymer chains, such as the coordination between rare earth ions and ester groups in PMMA. Three samples were used to elucidate the coordination [99]: A, a NOA solution in acetic acid; B, a solution of NOA and PMMA in toluene; C, a PMMA disk doped with NOA.

Table 5. Intensity parameters for Nd^{3+} in the systems A, B, and C ($\times 10^{-20}$ cm^2).

System	Ω_2	Ω_4	Ω_6
A	0.26	0.52	0.76
B	0.48	0.73	1.28
C	0.62	0.78	1.48

From absorption spectra of A, B, and C samples, the optical intensity parameters of Nd^{3+} can be obtained by Judd–Ofelt theory [99], and the results can be seen in Table 5. It is well known that the $^4I_{9/2} \rightarrow {}^4G_{5/2}$ transition in Nd^{3+} is hypersensitive and, furthermore, the parameter will be determined almost entirely by the strength of this band. The largest Ω_2 parameter shown in Table 5 for system C is due to its relatively stronger $^4I_{9/2} \rightarrow {}^4G_{5/2}$ absorption band, which shows that stronger coordination exists in system C than in the other systems. On the other hand, compared with Ω_4 and Ω_6, the small values for Ω_2 indicate the presence of ionic bonding between the rare earth ions and surrounding ligands [100]. For rare earth complexes, covalent bonding was present with large Ω_2 values [46].

Study on Er-doped glass fiber amplifiers (EDFA) has shown that the actual Er doping distribution can be used for accurate modeling of EDFA [101]. Unlike Er-doped glass fiber, the absolute rare earth concentration and the distribution can be obtained only in the system of rare-earth-doped GI preforms [50]. This shows that research on rare-earth-doped GI POFs is far behind that of glass fibers. A similar relationship between structure and property can be expected for two kinds of rare-earth-doped fibers, but more work needs to be done in the field of rare-earth-doped GI POFs.

ACKNOWLEDGMENTS

The investigations on rare-earth-doped plastic optical fiber have been carried out since 1994 in the Department of Polymer Science and Engineering, University of Science and Technology of China and supported by the National Natural Science Foundation of China (grants 59473017, 59773013, and 50025309), which the author gratefully acknowledges.

REFERENCES

1. T. Kaino, in "Polymers for Lightwave and integrated Optics" (L. Hornak, Ed.), pp. 1–38. Dekker, New York, 1992.
2. D. Marcuse, "Principles of Optical Fiber Measurements" Academic Press, New York, 1981.
3. S. Kawakami and T. Nishizawa, *IEEE Trans. Microwave Theory Tech.* MTT-16, 814 (1968).
4. F. P. Kapron, *J. Opt. Soc. Am.* 60, 1433 (1970).
5. R. Olshansky and D. B. Keck, *Appl. Opt.* 15(2), 483 (1976).
6. T. Ishigure, E. Nihei, and Y. Koike, *Appl. Opt.* 35(12), 2048 (1996).
7. B. C. Ho, J. H. Chen, W. C. Chen, Y. H. Chang, S. Y. Yang, J. J. Chen, and T. W. Tseng, *Polym. J.* 27(3), 310 (1995).
8. J. H. Liu, H. T. Liu, and Y. B. Cheng, *Polymer* 39, 5549 (1998).
9. Y. Ohtsuka, *Appl. Phys. Lett.* 23, 247 (1973).
10. Y. Ohtsuka and I. Nakamoto, *Appl. Phys. Lett.* 29, 559 (1976).
11. Y. Koike, Y. Takezawa, and Y. Ohtsuka, *Appl. Opt.* 27, 486 (1988).
12. Y. Koike, E. Nihei, N. Tanio, and Y. Ohtsuka, *Appl. Opt.* 29, 2686 (1990).
13. T. Ishigure, E. Nihei, and Y. Koike, *Appl. Opt.* 33, 4261 (1994).
14. T. Ishigure, H. Horibe, E. Nihei, and Y. Koike, *Electron. Lett.* 30, 1169 (1994).
15. Q. Zhang, P. Wang, and Y. Zhai, *Macromolecules* 30, 7874 (1997).
16. J. S. Vrentas and J. L. Duda, *J. Polym. Sci., Polym. Phys.* Ed. 15, 403 (1977).
17. R. D. Ferguson and E. Von Meerwall, *J. Polym. Sci., Polym. Phys.* Ed. 18, 1285 (1980).
18. A. Faldi, M. Tirrell, T. P. Lodge, and E. Von Meerwall, *Macromolecules* 27, 4184 (1994).
19. J. S. Vrentas, C. M. Vrentas, and N. Faridi, *Macromolecules* 29, 3272 (1996).
20. J. S. Vrentas and C. M. Vrentas, *Eur. Polym. J.* 34, 797 (1998).
21. J. S. Vrentas and C. M. Vrentas, *Macromolecules* 26, 1277 (1993).
22. J. S. Vrentas and C. M. Vrentas, *Macromolecules* 27, 5570 (1994).
23. K. Horie, I. Mita, and H. Kambe, *J. Polym. Sci., Part A-1* 6, 2663 (1968).
24. Y. Koike, T. Ishigure, and E. Nihei, *J. Lightwave Technol.* 13, 1475 (1995).
25. C. H. Bamford and C. F. H. Tipper (Eds.), "Comprehensive Chemical Kinetics", Vol. 14A. Elsevier, Amsterdam, 1976.
26. K. Horie, I. Mita, and H. Kambe, *J. Polym. Sci., Part A-1* 6, 2663 (1968).
27. F. Zhang, X. Wang, and Q. Zhang, *Polymer* 41, 9155 (2000).
28. R. D., Ferguson and E. Von Meerwall, *J. Polym. Sci., Polym. Phys.* Ed. 18, 1285 (1980).
29. V. Field, H. F. Hameka, and U. Blukis (Eds.), "Physical Chemistry," Macmillan, Inc., New York, 1977.
30. D. W. Van Krevelen, "Properties of Polymers," 3rd ed., p. 291. Elsevier, New York, 1990.
31. E. Nihei, T. Ishigure, and Y. Koike, *Appl. Opt.* 35, 7085 (1996).
32. J.-H. Liu and H.-T. Liu, *Polymer* 38, 1251 (1997).
33. K. Kuriki, T. Kobayashi, N. Imai, T. Tamura, S. Nishihara, A. Tagaya, Y. Koike, and Y. Okamoto, *IEEE Photon. Technol. Lett.* 12, 989 (2000).
34. J. S. Vrentas, J. L. Duda Duda, and H. C. Ling, *J. Polym. Sci. Polym. Phys.* Ed. 22, 459 (1984).
35. M. J. F. Digonnet, *IEEE J. Lightwave Technol.* 4, 1631 (1986).
36. R. J. Mears, L. Reekie, S. B. Poole, and D. N. Payne, *Electron. Lett.* 22, 159 (1986).
37. M. Digonnet (Ed.), "Rare Earth Doped Fiber Lasers and Amplifiers," Dekker, New York, 1993.
38. E. Desurvire, "Erbium Doped Fiber Amplifiers," Wiley, New York, 1994.
39. Q. J. Zhang, H. Ming, and Y. Zhai, *J. Appl. Polym. Sci.* 62, 887 (1996).
40. A. Tagaya, Y. Koike, T. Kinoshita, E. Nihei, T. Yamamoto, and K. Sasaki, *Appl. Phys. Lett.* 63, 883 (1993).
41. N. E. Wolff and R. J. Pressley, *Appl. Phys. Lett*, 2, 152 (1963).
42. Y. Okamoto, Y. Ueba, N. F. Dzhanibekov, and E. Banks, *Macromolecules* 14, 17 (1981).
43. W. Y. Xu, Y. S. Wang, D. G. Zheng, and S. L. Xia, *J. Macromol. Sci.-Chem.* A25(10&11), 1397 (1988).
44. T. Kobayashi, S. Nakatsuka, T. Iwafuji, K. Kuriki, N. Imai, T. Nakamoto, C. D. Claude, K. Sasaki, Y. Koike, and Y. Okamoto, *Appl. Phys. Lett.* 71(17), 2421 (1997).
45. R. B. Seymour and C. E. Carraher, in Jr., "Polymer Chemistry, An Introduction" (J. J. Lagowski, Ed.), Dekker, New York, 1981.
46. C. Koeppen, S. Yamada, G. Jiang, and A. F. Gario, *J. Opt. Soc. Am. B* 14(1), 155 (1997).
47. Q. J. Zhang, P. Wang, X. F. Sun, Y. Zhai, P. Dai, B. Yang, H. Ming, and J. P. Xie, *Appl. Phys. Lett.* 72(4), 407 (1998).
48. Q. J. Zhang, P. Wang, and Y. Zhai, *J. Appl. Polym. Sci.* 67, 1431 (1998).
49. Q. J. Zhang, X. F. Sun, and P. Wang, *Polym. Int.* 45, 185 (1998).
50. K. Kuriki, T. Kobayashi, N. Imai, T. Tamura, S. Nishihara, A. Tagaya, Y. Koike, and Y. Okamoto, *IEEE Photon. Technol. Lett.* 12(8), 989 (2000).
51. F. A. Cotton, in "Modern Coordination Chemistry" (J. Lewis and R. G. Wilkins, Eds.), Interscience, New York, 1960.
52. K. Nakamoto, "Infrared Spectra of Inorganic and Coordination Compounds," p. 198. Wiley, New York, 1963.
53. Spectroscopic Properties of Inorganic and Organometallic Compounds, Specialist Periodical Reports, Chem. Soc., London 1, 199(1968); 2, 333 (1969).
54. G. B. Deacon and R. J. Phillips, *Coord. Chem. Rev.* 33, 227 (1980).
55. J. A. Faniran and K. S. Patel, *Spectrochim. Acta Part A* 32, 1351 (1976).
56. N. W. Alcock, V. M. Tracy, and T. C. Waddington, *J. Chem. Soc. Dalton Trans.* 2243 (1976).
57. A. I. Grigor'ev, *Russ. J. Inorg. Chem.* 8, 409 (1963).
58. J. H. Talbot, *Acta Cryst.* 6, 720 (1953).
59. W. H. Zachariasen and H. A. Plettinger, *Acta Cryst.* 12, 526 (1995).
60. H. Koyama and Y. Saito, *Bull. Chem. Soc. Jpn.* 127, 113 (1954).
61. W. H. Bragg and G. T. Morgan, *Proc. Roy. Soc. London, Ser. A* 104, 437 (1923).
62. A. D. Negro, L. Ungaretti, and A. Perotto, *J. Chem. Soc. Dalton Trans.* 1639 (1972).
63. K. Nakamoto, "Infrared and Raman Compounds," p. 222. Wiley, New York, 1977.
64. A. Eisenberg and M. King, "Ion-Containing Polymers, Physical Properties and Structure." Academic Press, New York, 1977.
65. K. A. Mauritz, *J. Macromol. Sci., Rev. Macromol. Chem. Phys. C* 28, 65 (1988).
66. M. R. Tant and G. L. Wilkes, *J. Macromol. Sci., Rev. Macromol. Chem. Phys. C* 28, 1 (1988).
67. W. J. MacKnight and T. R. Earnest, Jr., *J. Polym. Sci., Macromol. Rev.* 16, 41 (1981).
68. J. J. Fitzgerald and R. A. Weiss, *J. Macromol. Sci., Rev. Macromol. Chem. Phys. C* 28, 99 (1988).
69. H. P. Brown, *Rubber Chem. Technol.* 30, 1347 (1957).
70. R. Longworth and D. J. Vaughan, *Nature* (*London*) 218, 85 (1968).
71. S. Bononto and E. F. Bonner, *Polym. Prepr.* 9, 525 (1968).
72. A. Eisenberg, *Macromolecules* 3, 147 (1970).
73. C. L. Marx, D. F. Caulfield, and S. L. Cooper, *Macromolecules* 5, 344 (1973).
74. W. J. MacKnight, W. P. Taggart, and R. S. Stein, *J. Polym. Sci., Polym. Symp.* 45, 113 (1974).
75. E. H. Roche, R. S. Stein, T. P. Russell, and W. J. MacKnight, *J. Polym. Sci., Polym. Phys.* Ed. 18, 1497 (1980).
76. D. J. Yarusso and S. L. Cooper, *Macromolecules* 16, 1871 (1983).
77. A. Eisenberg and M. Navratil, *Macromolecules* 7, 90 (1974).

78. Y. S. Ding, S. R. Hubbard, K. O. Hodgson, R. A. Register, and S. L. Copper, *Macromolecules* 21, 1698 (1988).
79. A. Eisenberg, B. Hird, and R. B. Nirem, *Macromolecules* 23, 4098 (1990).
80. R. A. Weiss and J. A. Lefelar, *Polymer* 27, 3 (1986).
81. A. Eisenberg, "Physical Properties of Polymers," p. 77. American Chemical Society, Washington, DC, 1982.
82. R. D. Lundberg, H. S. Makowski, and L. Westerman, "Ions in Polymers," p. 67. American Chemical Society, Washington, DC, 1980.
83. A. S. Nowick and B. S. Berry, "An Elastic Relaxation in Crystalline Solids," Academic Press, New York, 1972.
84. W. Liu, Y. T. Wen, and Y. T. Qian, *Chinese Scie. Bull.* 39, 908 (1994).
85. N. G. McCrum, "Elastic and Dielectric Effects in Polymeric Solids," Wiley, London, 1967.
86. Q. Zhang, W. Liu, and P. Dai, *Polymer* 39, 3787 (1998).
87. P. P. Smit, *Rheol. Acta* 5, 277 (1966).
88. T. Kurian, D. Khastigir, D. K. Tripathy, S. K. De, and D. G. Peiffer, *Polymer* 37, 4865 (1996).
89. J. Felsche, *Struct. Bonding* 13, 99 (1973).
90. B. G. Wybourne, "Spectroscopic Properties of Rare Earths," Interscience, New York, 1965.
91. E. Desurvire, "Erbium Doped Fiber Amplifiers," p. 6. Wiley, New York, 1994.
92. E. Delevaque, T. Georges, M. Monerie, P. Lamouler, and J. F. Bayon, *IEEE Photon. Technol. Lett.* 5, 73 (1993).
93. H. Haken and H. C. Wolf, "Atomic and Quantum Physics," 2nd ed., Chap. 5, 15, and 19. Springer-Verlag, New York, 1987.
94. B. R. Judd, *Phys. Rev.* 127, 750 (1962).
95. G. S. Ofelt, *J. Chem. Phys.* 37, 511 (1962).
96. M. F. Reid, J. J. Dallara, and F. S. Richardson, *J. Chem. Phys.* 79, 5743 (1983).
97. M. F. Reid and F. S. Richardson, *J. Chem. Phys.* 79, 5735 (1983).
98. P. C. Mehta and S. P. Tandon, *J. Chem. Phys.* 53, 414 (1970).
99. Q. Zhang and P. Wang, *J. Mol. Struct.* 440, 35 (1998).
100. R. D. Peacock, in "Structure and Bonding," Vol. 22, Springer-Verlag, New York, 1975.
101. E. Desurvire, "Erbium-Doped Fiber Amplifiers, Principles and Applications," Chap. 1. Wiley, New York, 1994.

CHAPTER 4

Polymer Optical Fiber Gratings

G. D. Peng and P. L. Chu

Photonics and Optical Communications Group, School of Electrical Engineering and Telecommunications, University of New South Wales, Sydney, Australia

CONTENTS

1. INTRODUCTION

Fiber Bragg grating is characterized by a periodic index change induced by ultraviolet (UV) irradiation in the core along the fiber. Light propagating along the fiber grating is periodically and partially reflected in accordance with the pattern of the index change. When these reflections are in phase, they can constructively produce an enhanced near 100% reflection. Whereas this in-phase condition is sensitively wavelength dependent (dispersive), the reflection as well as the transmission of a fiber Bragg grating are highly wavelength selective. This particular property of fiber Bragg grating results in many interesting effects that lead to generic applications in optical communication and fiber sensing systems.

ISBN: 1-58883-012-8
Copyright © 2004 by American Scientific Publishers
All rights of reproduction in any form reserved.

Fiber Bragg gratings have become essential photonic components since their inception in 1978 when Hill and his colleagues [1] first observed a photoinduced grating in a germanium-doped silica optical fiber. They have been employed for dispersion compensation in optical fiber transmission lines [2], for optical filtering and optical add–drop multiplexing in optical dense wavelength division multiplexing (DWDM) systems [3, 4], for frequency stabilization of optical fiber lasers or semiconductor diode lasers [5], and for gain flattening of erbium-doped fiber amplifiers [6], They are also increasingly popular in various sensors for measuring temperature [7], pressure and strain [8], hydrosound [9], and industrial process [10].

Fiber Bragg gratings have many attractive attributes versus their conventional bulk counterparts such as compactness, stability, portability, high power efficiency (low insertion loss and low connection loss) and, hence, overall better system performance. In particular, they are essentially made of the same fiber that is used for optical signal transmission. In fact, they can be written directly into the transmission fiber or made separately and then fusion-spliced into the fiber link without appreciable loss.

Research and development of fiber gratings have been carried out in many research groups and companies around the world in recent years. So far, work in this regard has predominantly concentrated on silica optical fiber gratings [11]. Most studies of photosensitivity have concentrated on germanium-doped or rare-earth-doped silica optical fibers [12, 13]. Significant advances have been made in fabricating photosensitive silica optical fibers. In addition, the fabrication techniques for silica fiber grating have been rapidly developed for commercial production, and a range of silica optical fiber Bragg gratings are now commercially available.

There is no doubt that Bragg gratings based on Silica fibers have many attractive and unique features. Nevertheless, they also have limitations. The most serious disadvantages seems to be limited wavelength tunability and stress responsivity. Tunability could be a fundamental disadvantage that excludes silica fiber gratings from being used in some important applications, because wavelength tunability is a crucial feature required for future broadcasting and reconfigurable DWDM systems [14]. On the other hand, high stress responsivity is essential in various fiber sensor applications where strain related effects are utilized [15]. In fact, although considerable efforts are being made to improve the tenability of glass fiber gratings, their success has been limited [16–22]. For example, a report on a magnetically tunable fiber Bragg grating (FBG) for fast reconfigurable optical add–drop multiplexers (OADM) demonstrated a 4 nm shift of the Bragg wavelength [18]; a 0.4 nm tunable range was achieved for an eight-channel tunable drop device by thermal tuning of the silica fiber grating [19]; and, using a PZT (Piezoelectrics) for fiber grating tuning, the tuning range obtained was 1.68 nm [20]. A larger tuning range was reporter in a chemically etched cladding fibre grating, where wavelength tuning up to a failure point of 21.2 nm was demonstrated along with repeated cycling over a 9 nm tuning range [21].

The main reason for poor tunability is clearly linked to the rigidity of silica glass. Likewise, the poor stress responsivity is related to a relatively high Young's modulus ($\sim$100 GPa). On the other hand, polymer optical fiber (POF) is highly flexible and elastic compared with silica fiber and, as far as flexibility and elasticity are concerned, POFs (typically $\sim$3 GPa) have quite complementary properties to their silica counterparts. Thus, working on POFs may provide opportunities to develop optical components that have high tunability and high stress responsivity components. This is particularly so if POF can be made significantly photosensitive and POF Bragg gratings can be fabricated. Based on these considerations, work on the photosensitivity of POF and its application for fiber Bragg gratings has begun [23–26].

POF was first developed in the early 1970s, about the same time as silica fiber. In the middle of the 1970s, Schleinitz [27] at DuPont demonstrated that poly(methyl methacrylate)- (PMMA) based POFs could have attenuation as low as 300 dB/km. Mitsubishi then developed a continuous casting process to manufacture POFs based on high purity acrylics [28]. This has since led to the development of various POFs at Mitsubishi Rayon in Japan. However, POF has long been overshadowed by the amazing and rapid development of low loss silica fibers. Only recently have research and development of polymer optical fibers been greatly boosted by an emerging and important market of high-speed (broad-band) local area networks. In these networks, polymer optical fiber is deemed to be able to compete with conventional silica optical fiber. POFs are attractive for their excellent flexibility, low cost, and ease of handling and connection [29, 30]. Moreover, because of the excellent compatibility of base polymer optical fiber materials with many functional organics, special POFs are being targeted and developed for various active and passive photonic devices, such as all-optical switches [31–33], amplifiers [34–37], electro-optic modulators [38], fiber Bragg gratings [23–26], and fiber lasers [39–41]. However, this chapter is concerned mainly with the photosensitivity and grating fabrication of polymer optical fibers and their potential applications.

This chapter is organized as follows: In Section 2, the fundamental principles and basic relationships of fiber Bragg grating are described. In Section 3, the research results on photosensitivities of PMMA-based polymer fiber preforms and polymer fibers are reviewed. In Section 4, the fabrication techniques, mainly the phase mask technique and its associated zeroth-order diffraction issue, are described. In Section 5, the tuning mechanisms and tunability of the center wavelength of POF Bragg gratings are discussed and research results are presented. In Section 6, work on the photosensitivity of CYTOP and CYTOP POF is discussed. In Section 7, the application prospects of POF grating in telecommunication systems and fiber sensor systems are briefly discussed. In Section 8, we briefly address the remaining challenging issues and topics in research and development of POF gratings.

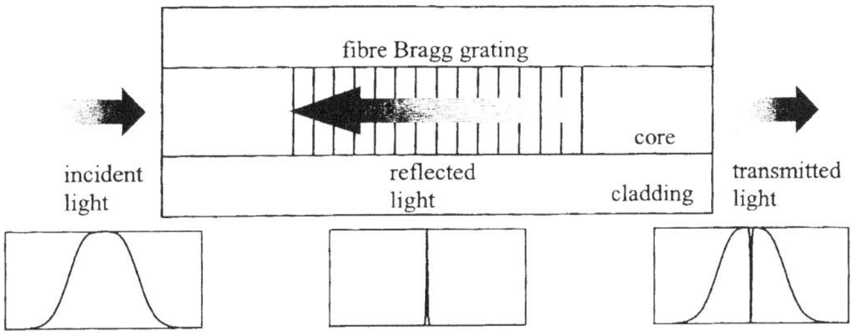

Figure 1. Diagram of a fiber Bragg grating.

2. BASIC PRINCIPLE OF OPTICAL FIBER GRATINGS

A diagram of a fiber Bragg grating is shown in Figure 1. Here an incident light launched into the fiber section written with a Bragg grating in the core is highly reflected within a narrow band around a Bragg wavelength. The Bragg wavelength is determined by the constructive interference, that is, the in-phase condition, of partial reflections from each part of the grating.

Within a simple model of periodic index change, the in-phase condition can be linked directly to the condition that gives the Bragg reflection of a grating. The reflection is the famous Bragg condition that is widely used, and it can be obtained from the usual energy and momentum conservation laws that involve light and matter interaction.

The momentum conservation related to the grating reflection can be expressed explicitly as

$$\mathbf{K} + \mathbf{k}_i = \mathbf{k}_r \tag{1}$$

where $\mathbf{K}$, $\mathbf{k}_i$ and $\mathbf{k}_r$ are the wave vectors of the grating, the incident light, and the reflected light, respectively. This relationship is depicted in Figure 2. The vector $\mathbf{K}$ has a magnitude given by

$$K = \frac{2\pi}{\Lambda} \tag{2}$$

where Λ is the grating period. For optical fiber gratings, the magnitudes of $\mathbf{k}_i$ and $\mathbf{k}_r$ are usually referred to as propagation constants, and are, respectively, given by

$$\beta_i = \frac{2\pi}{\lambda_1} n_{eff} \quad \text{and} \quad \beta_r = \frac{2\pi}{\lambda_r} n_{eff} \tag{3}$$

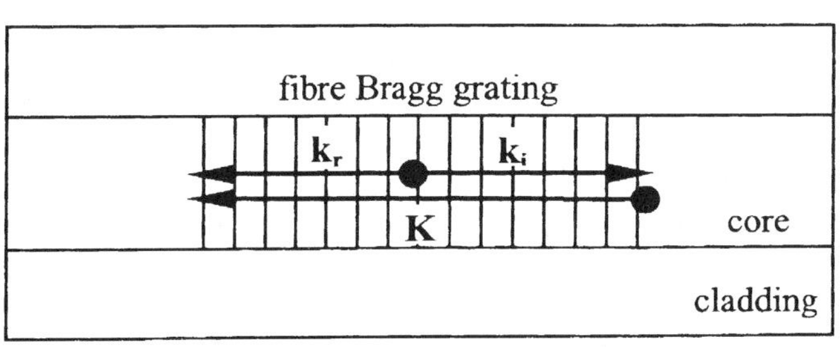

Figure 2. The wave vector relationship on a fiber grating.

where n_{eff} is the effective refractive index of the optical fiber mode (usually the fundamental mode). The incident and reflected waves are both along the fiber axis and in opposite directions to each other. To satisfy the momentum conservation condition, the direction of $\mathbf{K}$ is also along the fiber axis and in the direction of the reflected wave vector. Under the energy conservation law, the incident wave and reflected wave should have the same energy, that is,

$$\eta\omega_i = \eta\omega_r \tag{4}$$

for the same wavelength and we assume there is no energy transfer from one wavelength to another and no radiation loss; $\eta = 2\pi h$, where h is Planck's constant. Hence we can write for frequency

$$f_i = f_r = f_B \tag{5}$$

and for wavelength

$$\lambda_i = \lambda_r = \lambda_B \tag{6}$$

with

$$f_B \lambda_B = c \tag{7}$$

where c is the light velocity and λ_B denotes the Bragg wavelength. Thus from both the energy and momentum conservation conditions, we arrive at the Bragg condition

$$N\lambda_B = 2n_{eff}\Lambda \tag{8}$$

where N is an integer that represents the order of the interaction. In most practical cases of our interest, $N = 1$. The Bragg condition provides the simple and yet very important relationship between the Bragg wavelength and the grating parameter—the period Λ of the index variation.

2.1. Basic Relationships of Fiber Grating

The reflection and transmission in a Bragg grating of an optical fiber are described by the coupled mode theory (see, e.g. [42]). Following the simple analysis of [43], we consider a single fiber with a Bragg grating of length L. We eliminate the common propagation constant (say β_0) and consider only the difference δ between β_0 and the actual propagation constant. We let A and B represent the complex amplitudes of the forward and backward propagating modes and then set

$$A(z) = R(z)\exp(j\delta z) \tag{9}$$

and

$$B(z) = S(z)\exp(-j\delta z) \tag{10}$$

This leads to

$$R'[z] + j\delta R[z] = -j\kappa S[z] \tag{11}$$

and

$$S'[z] - j\delta S[z] = j\kappa R[z] \tag{12}$$

Reorganizing the coupled equations, we have

$$R''[z] + (\delta^2 - \kappa^2)R[z] = 0 \tag{13}$$

and

$$S''[z] + (\delta^2 - \kappa^2)S[z] = 0 \tag{14}$$

Using these equations, it is easy to prove that

$$|R(z)|^2 - |S(z)|^2 = \text{ constant} \tag{15}$$

The modulation Δn in the refractive index determines the Bragg (forward-to-backward) coupling coefficient. For sinusoidal index change along the fiber axis, the coupling coefficient is given by

$$\kappa = \frac{\pi \Delta n}{\lambda} \tag{16}$$

where λ is the signal light wavelength. The frequency offset is $\delta = k - k_B$, where $k = 2\pi/\lambda$ is the signal wave number in free space, $k_B = 2\pi/\lambda_B$ is the center Bragg wave number, and λ_B is the Bragg wavelength.

The range of frequencies where the light is reflected (i.e., the bandgap) has a width proportional to κ. Within the bandgap, that is, for frequency offsets δ that satisfy $\delta^2 < \kappa^2$, almost all the light is reflected, but outside this range (i.e., $\delta^2 > \kappa^2$), the fraction reflected is small.

Physically, this means that each element of power decrease in the forward propagating mode is added to the reflected mode, thus ensuring that the difference remains constant. We have the boundary conditions $R(0) = 1$ and $S(L) = 0$. It can be shown that $\arg[S(z)]$ is also a constant. The result is that, inside the bandgap, both $|R|$ and $|S|$ decrease monotonically as z increases and the curves are similar, because almost all the light is reflected.

However, outside the bandgap, $|R|^2$ and $|S|^2$ are oscillating functions of z, because standing waves are set up in the grating. In this range, $|R|$ remains high for all values of z, while $|S|$ is always fairly small. This means that the constant value of $(|R(z)|^2 - |S(z)|^2)$ is now larger than it was when the parameters were within the bandgap.

2.1.1. Within the Bandgap

Within the bandgap ($\delta^2 < \kappa^2$), we let $^2 = \kappa^2 - \delta^2 > 0$. The exact solution of $R(z)$ to the preceding system can be found easily. We write $R(z)$ in the form

$$R = R_r + j\mathrm{R}_i \tag{17}$$

The exact solution can be expressed as

$$R_r = H\{\kappa^2 \cosh[q(z - 2L)] + (q^2 - \delta^2)\cosh[qz]\} \tag{18}$$

$$R_i = -2q\delta H \sinh[qz] \tag{19}$$

where the constant

$$H = \frac{1}{2(q^2 \cosh^2(qL) + \delta^2 \sinh^2(qL))} \tag{20}$$

The exact solution of $S(z)$ to the foregoing system within the bandgap ($\delta^2 < \kappa^2$) also can be found in the form

$$S = S_r + jS_i \tag{21}$$

with

$$S_r = -2\kappa q H \cosh[qL] \sinh[q(z - L)] \tag{22}$$

$$S_i = -2\kappa\delta H \sinh[qL] \sinh[q(z - L)] \tag{23}$$

The reflectivity from a fiber Bragg grating can be derived as

$$\text{reflectivity} = \left|\frac{S(0)}{R(0)}\right|^2 = \frac{\kappa^2 \sinh^2(qL)}{\delta^2 \sinh^2(qL) + q^2 \cosh^2(qL)} \tag{24}$$

As a simple example, we consider the case when the frequency offset is $\delta = 0$, that is, the incident wavelength is at the center of the bandgap. A simple expression for reflectivity is

$$\text{reflectivity} = \tanh^2[\kappa L] \tag{25}$$

It is obvious that the reflectivity increases with the increase of both the refractive index change Δn (since $\kappa = \pi \Delta n/\lambda$) and grating length L.

2.1.2. Outside the Band-gap

The exact solution of $R(z)$ to the preceding system outside of the bandgap ($p^2 = \kappa^2 - \delta^2 > 0$), can be obtained in a manner similar to Eq. (17):

$$R = R_r + jR_i$$

where

$$R_r = \cos[pz] - Gp \sin[pz] \cos[pL] \tag{26}$$

$$R_i = G\delta \sin[pz] \sin[pL] - (\delta/p) \sin[pz] \tag{27}$$

where $p^2 = \delta^2 - \kappa^2 > 0$ and the constant

$$G = \kappa^2 \sin(pL) pp^2 \cos^2(pL) + \delta^2 \sin^2(pL) \tag{28}$$

We may define the effective change in the local propagation constant caused by the grating as

$$\beta(z) = \frac{d}{dz} \mathrm{arc}[R(z)] = \frac{d}{dz} \arctan\left[\frac{R_i}{R_r}\right] \tag{29}$$

The propagation constant is related to the dispersion of the grating. This is very important for fiber grating in a number of applications such as dispersion compensation in optical fiber transmission lines. Here it is an oscillating function that has one or more cycles over the grating length, depending on the parameters. Its minimum value is $-\delta$, so the effective change in propagation constant of the initial complex mode $[A(z)]$ [viz. $\beta(z) + \delta$] has a minimum of 0. Its maximum is on the order of κ. Thus the propagation constant is periodic and the magnitude of its modulation is on the order of the grating coupling coefficient.

3. PHOTOSENSITIVITY OF PMMA POF

Irradiation of the core of an optical fiber with ultraviolet light introduces a permanent change in the refractive index. This photorefractive effect, referred to as photosensitivity, has great practical significance because it makes possible an ever-increasing number of passive and active fiber devices [44]. It is well known that various single-mode silica optical fibers, especially germanium-doped or rare-earth-doped silica fibers, exhibit a certain photosensitivity. Photorefractive index changes as much as 10^{-2} have been obtained in germanium-doped optical fibers. It is also well known that the photosensitivity of these fibers can be significantly enhanced by various postdrawn photosensitization techniques such as flame brushing or hydrogen loading.

The photosensitivity of PMMA has been studied for about 30 year if not longer. The earliest report we found on the photosensitivity of PMMA is the work of Tomlinson et al. [45]. In their work, they found that properly prepared PMMA (through oxidation of monomer) exhibited a substantial increase in refractive index after irradiation with UV light at 325 (He-Cd^+ laser) or 365 nm (Hg arc). A subsequent report related to the photosensitivity of PMMA is a paper on a dye-doped polymer laser [46]. The solid-state dye laser has a pair of Bragg gratings inside a bulk sample of PMMA doped with rhodamine 6G. These gratings are created by the irradiation of via UV laser beams. The gratings had a dimension of $2 \times 2 \times 2$ mm^3 and a maximum reflectivity of 20%. The material preparation and fabrication processing of current POFs are considerably different from these reports. Nevertheless, these early works are very helpful references.

In the recent research carried out at the University of New South Wales, the photosensitivities of various PMMA-based POFs were investigated. These include PMMA POFs made from undoped, dye-doped, or oxidated preforms under various irradiation wavelength, intensity, and time.

In the initial photosensitivity experiment, a simple interferometry setup is arranged as shown in Figure 3. In the first experiment, a dye-doped, single-mode, twin-core polymer optical fiber was tested. Both cores of the fiber were doped with a typical laser dye, fluorescein (170 ppm). The near-field pattern (end face) and far-field pattern (interference fringe) of this polymer fiber are shown in the insets of Figure 3. The near-field pattern directly shows the two cores of the fiber. The far-field fringe pattern is formed when the fundamental mode in each core is launched equally with a He-Ne laser. Because the core-to-core spacing is small, the fringes are very well separated. They can be seen clearly on a screen a few centimeters away from the end face of the fiber. The Ar laser at wavelengths of 488 or 514 nm is launched into one of the cores of the fiber to induce the photosensitivity effect. To observe this effect, a He-Ne laser of with 1 mW output power at 633 nm is launched equally into each core via a defocussing lens. Because the fiber cutoff wavelength is at 780 nm, care must be taken to launch this 633 nm He-Ne light into the cores so that the only fundamental mode is excited. The fiber length is selected to be much shorter than the coupling length. Hence the power coupling between the two cores is negligible and a stable and uniform output far-field interference fringe pattern can be obtained. The fringe (phase) shift as a function of the exposure time (at a certain Ar laser power level) is directly related to the photosensitivity.

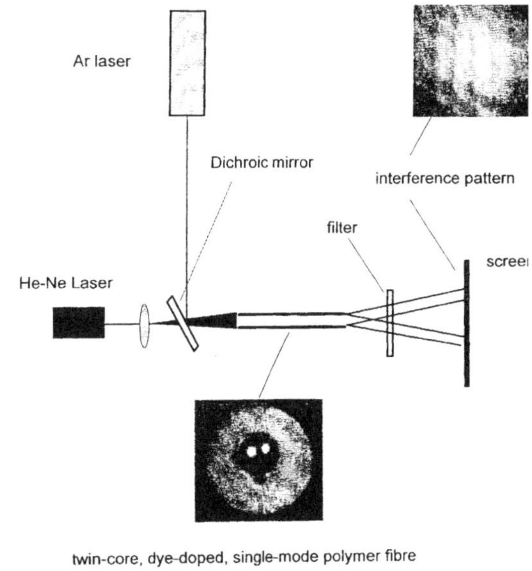

Figure 3. Schematic diagram of the photosensitivity experimental setup, where the insets are the near-field (end face image) and far-field patterns (interference fringe) of a twin-core polymer fiber [24].

In this first experiment, two types of photoinduced phase shifts were observed. They are of quite different nature. The first type of phase shift is nonpermanent and shows fast response. In this case, the phase shift is introduced when the Ar output is launched. It is lost when the Ar output is blocked. This fast response to the Ar irradiation is on a time scale of less than 0.1 ms. This response may come from the thermal nonlinearity linked to the optical absorption of the dye molecules in the core—a factor that is not relevant to the present investigation. The second type of phase shift is permanent and has a fairly slow response (it takes time to build up). Thus the phase shift is not instantaneous and may, in fact, take tens of minutes to build up. However, this phase shift remains when the Ar laser is switched off. This is exactly the photosensitivity that we were looking for.

Figure 4 shows the photosensitivities as a function of exposure time at an Ar irradiation wavelength of 514 nm for the two cores. The experiment on core 1 was carried out in two steps: (1) a 0.36 mW Ar laser power was launched into the fiber for 9 min and then (2) the Ar laser power was increased to 1.08 mW and the fiber was illuminated for 53 min. The rate of phase shift was gradually reduced as the exposure time increased. In the second step, an explicit saturation behavior can be seen. This behavior can be understood from the material system. The photosensitivity must be associated with photochemical processes (e.g., photobleaching) that involve the active molecules of the laser dye. This theory was verified by our observation. It was found that the

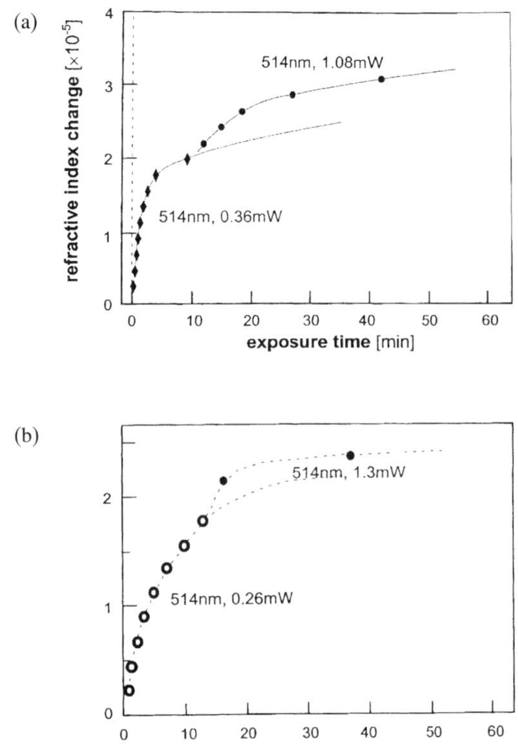

Figure 4. Photosensitivity of a fluorescein-doped twin-core POF Under 514 nm exposure: (a) result from core 1; (b) results from core 2 [24].

absorption loss of the fiber at 514 nm was reduced by 12 dB after the first 9 min illumination in step 1 of the experiment. At the end of step 2, the absorption loss was further reduced by 6 dB. This saturation indicates that the active dye molecules were depleted. In this experiment, a total phase shift of 15π was reached. For a fiber length of 14.5 cm, this shift corresponds to a photorefractive index change of 3.3×10^{-5}.

Similarly, the experiment on core 2 was carried out in two steps: (1) 0.26 mW was launched into the fiber for 13 min, and then (2) the power was increased to 1.3 mW and the fiber was illuminated for 24 min. Similar saturation behavior was observed. In the first 13 min irradiation, the absorption loss was reduced by 12 dB. The absorption loss was further reduced by 3 dB under 1.3 mW and 24 min irradiation. For the same length of fiber, a phase shift of 12π was observed, which corresponds to a photorefractive index change of 2.6×10^{-5}.

A second experiment was carried out using the 488 nm output of the Ar laser. This experiment was carried out in one step and under a constant launch power of 1.84 mW. The maximum phase shift observed is 17π, which gives a photorefractive index change of 3.4×10^{-5} for a fiber length of 16 cm The experimental result is shown in Figure 5.

It is more important to examine the photosensitivity of POF under UV irradiation. Hence another experiment was carried out using the UV output from a tunable optical parametric oscillator (OPO). The experiment was set in a different configuration as shown in Figure 6, where the fiber

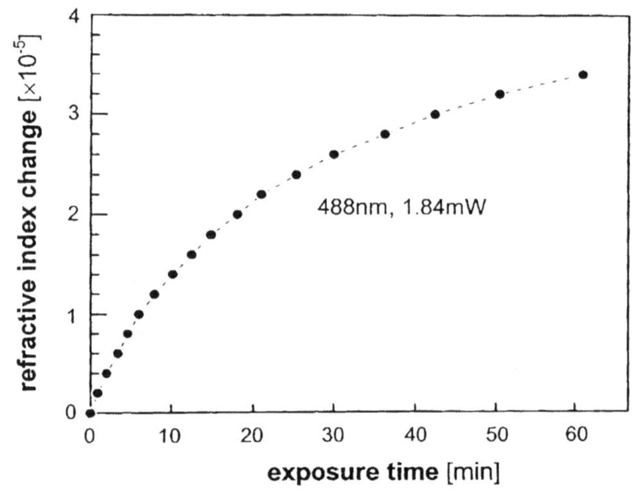

Figure 5. Photosesitivity of flourescein-doped twin POF under 488 nm exposure [24].

under test is illuminated transversely. The OPO generates 5 ns pulses at a rate of 10 Hz. Different UV wavelengths were used, namely 248, 280, and 325 nm. In this experiment, the net effect of photosensitivity depends on the orientation of the twin-core fiber with respect to the illuminating light. Figure 7 shows the phase shifts as a function of the product of pulse energy-exposure time for two different orientations. Noted that the phase shift measured is the relative phase shift, because now both cores are illuminated. With the UV at 248 nm, a (relative) phase shift as large as 10π is obtained when the two cores are in the same plane as the illuminating light. When the two cores are perpendicular to the incident light, the phase shift is much smaller. The reason for this is quite simple: now the cores are exposed at nearly the same level and the photosensitivities produced in each core must be similar. The nonzero phase shift in Figure 7 for the latter case could come from the asymmetry of the cores and/or the uncertainties in orientation of the fiber. From the experiment, the difference of the photoinduced index change between the

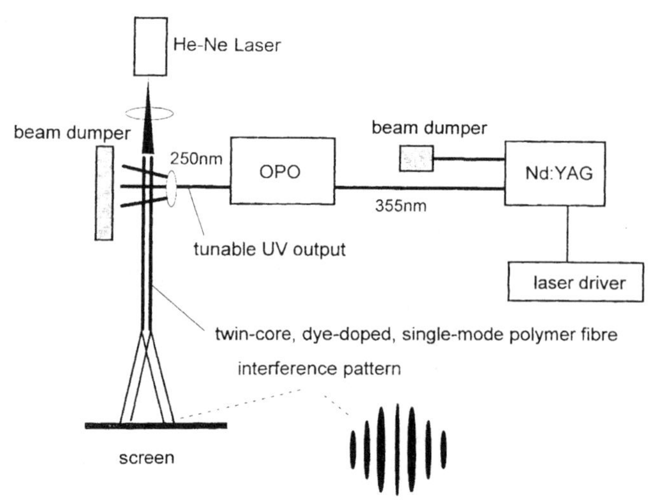

Figure 6. Diagram of the photosensitivity experimental setup for UV wavelength [24].

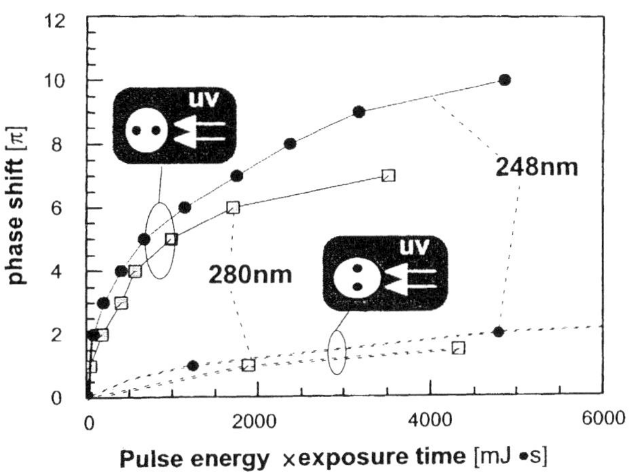

Figure 7. Photosensitivity of a twin-core POF under side-illumination at 248 and 280 nm [24].

two cores can be evaluated. For an effective 3 mm spot size of the UV beam and a phase shift of 10π, the photoinduced index difference is 1.1×10^{-3}. Although accurate photoinduced index change could not be directly worked out from this value, it could be concluded that the net photoinduced index change must be larger, 1.1×10^{-3} under 248 nm irradiation and 8×10^{-4} under 280 nm irradiation, respectively. It is obvious that photosensitivity is wavelength dependent and has a larger value at shorter wavelength.

From these experiments, it is revealed that significant photosensitivity can be introduced to POF with appropriate doping materials, for example, laser dyes at relevant wavelengths. Later it was also demonstrated that the base material (i.e., undoped POF) could have significant photosensitivity at UV wavelengths [24]. These findings provide important evidence that POF grating can be fabricated by making use of these photosensitivities.

4. FABRICATION OF POF BRAGG GRATINGS

Now the immediate task is to create a grating in POF and to evaluate its performance. In general, there are four mechanisms that can lead to the creation of gratings in polymer by means of light:

1. Ablation. This involves melting and burning the polymer surface.
2. Chain scission. This involves breaking the polymer chains, thus decreasing the chain density, which in turns reduces the refractive index.
3. Cross linking. This involves inducing free radicals and then giving rise to the combination of polymer chains. This has the effect of increasing the density of the chains, thus increasing the refractive index of the polymer.
4. Photopolymerization. The incident light generates free radicals, thus causing the polymerization of unreacted monomer within the polymer. As a result, the polymer density increases and the refractive index increases accordingly.

The first two mechanisms, ablation and chain scission, are responsible for the creation of surface gratings in the polymer. The last two mechanisms, cross linking and photopolymerization, are used to create bulk gratings and fiber gratings in the polymer.

Along with the development of silica fiber Bragg gratings, various grating writing techniques have been introduced. Bragg grating was originally fabricated in optical fiber using internal writing [1]. The improved schemes include an interferometer type proposed by Meltz et al. [47] and a simpler technique based on phase mask made from holographic relief grating [48].

4.1. Phase Mask Technique

Many techniques have been developed to write gratings in optical fibers such as the use of a prism. However, the phase mask technique is popular with fiber grating manufactures due to its great simplicity and superior stability [49]. The experimental arrangement is shown in Figure 8. A π phase mask is placed close to the fiber sample and the assembly is illuminated by a UV laser beam of suitable wavelength that can create a photosensitivity effect in the polymer. The positive and negative first-order diffraction beams created by the phase mask form dark and bright fringes within the fiber core. The regions of bright fringes cause a change of refractive index through the photosensitivity effect of the polymer, whereas the regions of dark fringes leave the refractive index of the fiber core unchanged. Thus a grating is created.

An ideal phase mask for this purpose is designed to produce maximum first-order (-1 and $+1$) diffractions without zeroth-order diffraction at a specific wavelength. This can be achieved by making the phase mask with a surface relief depth equal to one-half of the wavelength of the UV writing laser. This means that the depth d satisfies

$$d = \lambda/2(n-1) \tag{30}$$

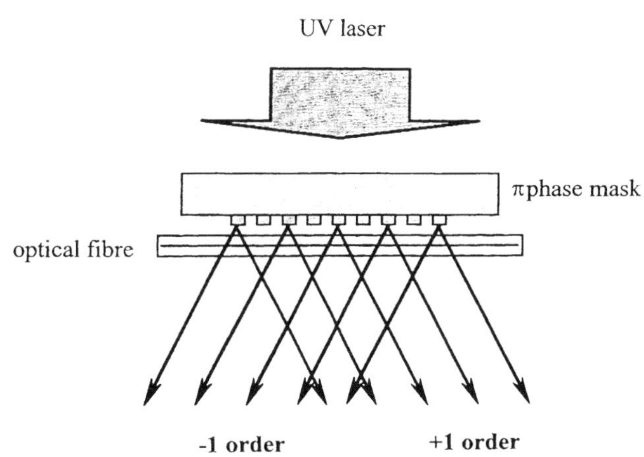

Figure 8. Diagram of grating writing with the phase mask technique.

where n is the refractive index of the phase mask material (typically fused silica, $n = 1.46$). The phase modulation amplitude is thus exactly π and thus the zeroth-order diffraction will be zero due to totally destructive diffraction.

A basic limitation of the phase mask technique is that each mask is fabricated to operate only at a specific wavelength. Any change in the UV wavelength away from the specified value results in an increase in the zeroth-order diffraction at the expense of the first-order diffractions. Moreover, due to practical limitations in mask manufacture, for example, undesirable uniformity and phase deviation from π, there is always a small fraction of the incident writing beam power in the zeroth order. State-of-the-art techniques usually can produce masks with 1% or more zeroth-order diffraction.

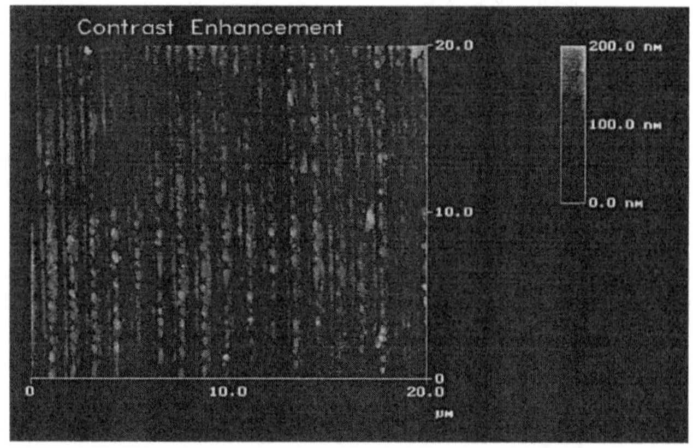

Figure 10. An AFM image that shows the structure of the surface relief grating in a polymer optical preform.

4.2. Effect of Zeroth-Order Diffraction

The existence of the zeroth-order diffraction of a phase mask has very detrimental effects on the quality of optical fiber grating production. Peng et al. [24] reported that the periods of the gratings were not equal to $\Lambda = D/2$ in writing grating in polymer optical preform, when the preform was put directly behind the phase mask, where Λ is the period of the gratings and D is the pitch of the phase mask. Nishii et al. [50] reported a similar situation where the grating period in GeO_2-SiO_2 glasses was the same as that of the phase mask. Both experimental results indicate that the change of the period is not due to the material properties, but instead, is caused by the existence of the zeroth-order diffraction in the phase masks [48, 51–53].

We describe here the experimental study of the effect of the existence of the zeroth-order diffraction. A Bragg grating in a polymer preform was prepared by exposing it to a UV laser beam through a fused silica phase mask, as shown in Figure 9. The mask has a one-dimensional surface relief pattern. Its dimensions were 3×10 mm and the period was 1.0614 μm. The diffraction efficiency of each of the two first-order diffractions was 44% and the zeroth order was suppressed to about 1% for the 248 nm writing wavelength. The sample on which the grating was to be written was cut from a UV-sensitive polymer optical preform. This sample was then exposed to the writing laser beam, thus forming a grating in it with a pattern that corresponds to the interference fringes caused by the diffraction beams.

We examined the structure of the preform grating by illuminating it with an argon laser at a wavelength of 488 nm and observed the resultant interference pattern. The first-order diffraction angle was measured to be about 27.3°, which corresponds to a grating pitch of 1.06 μm, a value close to that of the phase mask period. This result was surprising, because the expected period was nearly doubled. To confirm this result, an atomic force microscope (AFM) was used to characterize the gratings. Figure 10 shows the AFM photograph of the grating written at 248 nm. The grating produced was revealed to be essentially a surface relief type. Its period was 1.06 μm. Figure 11 shows the plot of the AFM cross section of the grating profile. Note that this profile is composed of two sets of oscillations, one deeper than the other, but each oscillation has a common period of about

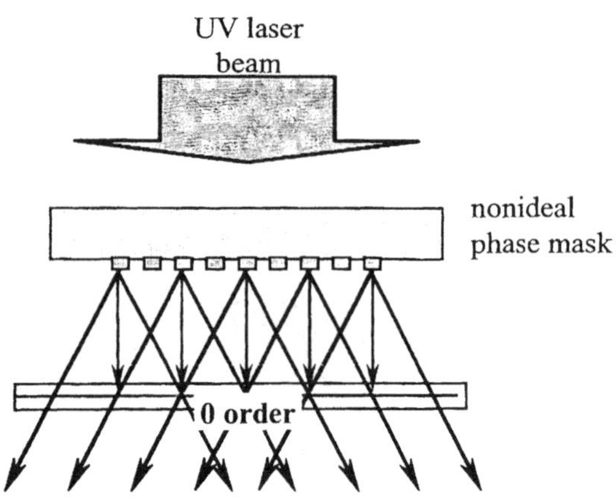

Figure 9. The zeroth-order diffraction from a nonideal phase mask.

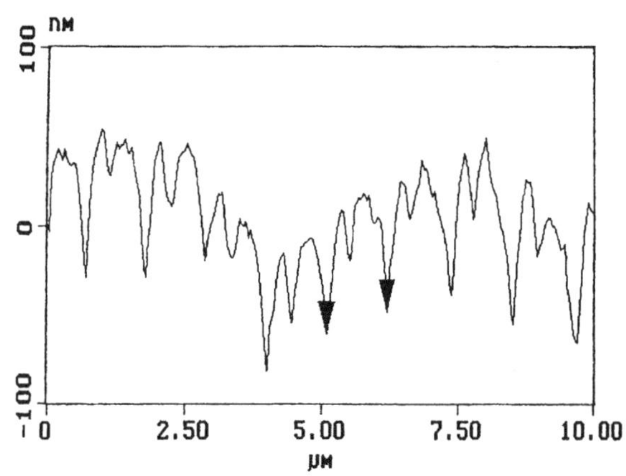

Figure 11. The cross section analysis profile of the grating shown in Figure 10. The vertical axis stands for the depth of the surface relief. Note that due to the AFM limitations and the properties of the grating described in the text, the depth of the relief is much greater than the values shown in this figure.

1.06 μm, as indicated by the two dark arrows shown in the figure. If we denote the deeper one as grating I and the other as grating II, then grating II locates approximately half a period away from grating I. Note here that the relief depth measured by AFM does not reflect its real value due to the limitations of AFM and the grating distortion.

Careful examination of the grating under an optical microscope reveals that grating I is not only created on the surface, but also inside the sample as well. Observations revealed that the gratings have a thickness of 10 μm more than the value shown in Figure 11, but are not uniform in the depth direction. It also was observed that the gratings were divided into layers in this direction and the distance between layers was measured to be about 8.5 μm. Each layer has a grating with the same period of 1.06 μm and there is a shift of about 0.5 μm between layers in the cross sectional direction. Grating II was not observed under the optical microscope due to its low contrast and dimness.

All these results indicate that the grating created by the phase mask technique is not as simple as that predicted by the two beam interference theory. The effects of other diffraction orders need to be consider as well, in addition to the first orders, particularly the zeroth order. To do so, we analyzed the formation of the grating by taking into account the presence of the zeroth order. We adopted the coordinate system shown in Figure 12. The z axis denotes the direction of the illuminating laser beam and the x axis is the direction of the formation of the grating periods. The origin O is chosen in the output plane of the phase mask. The phase change behind the phase mask at position (x, z) with respect to point O is $2\pi S_N/\lambda$ for the Nth-order diffraction, where $S_N = x\sin(\theta_N) + z\cos(\theta_N)$, and θ_N is the diffraction angle of the Nth order with $\sin(\theta_N) = N\lambda/D$, where D is the period of the phase mask.

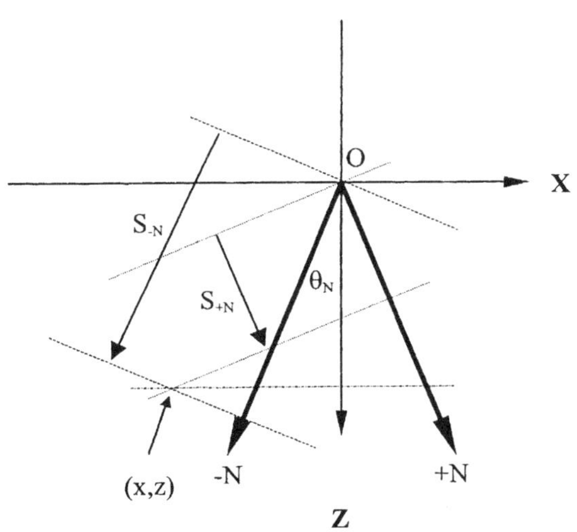

Figure 12. The coordinate system used to simulate three beam interference. X is the direction of the groove array and Z is the direction of the incident laser propagation, which is the same as the normal of the phase mask [53].

The electric field distribution behind the phase mask is

$$E = \sum_{k=-N}^{N} C_k \exp(i2\pi S_k/\lambda) \tag{31}$$

where C_k is the amplitude of the electric field of the kth-order diffraction. Here we consider only the symmetrical diffraction situation where $C_K = C_{-K}$. Then the field becomes

$$E = 2\sum_{k=0}^{N} C_k \cos(2k\pi x/D)\exp(i2\pi z\cos\theta_k/\lambda) \tag{32}$$

Note that in the preceding equation, the zeroth-order amplitude is $2C_0$. The intensity $|E|2$ distribution is

$$I = 4\left\{\sum_{k=0}^{N}[C_k^2\cos^2(2k\pi x/D) + \sum_{j\neq k}[C_kC_j\cos(2k\pi x/D) \times \cos(2j\pi x/D)\cos(2\pi z(\cos\theta_k - \cos\theta_j))/\lambda]\right\} \tag{33}$$

To simplify our discussion, diffraction orders higher than ±1 will be ignored. This neglect is valid in most practical cases because only the zeroth- and first-order diffractions make significant contributions. For the usual interference pattern of the two first-order diffraction beams, the intensity distribution is

$$I_{\pm1} = 4C_1^2\cos^2(2\pi x/D) \tag{34}$$

which demonstrates that the fringe pattern is similar to the structure of the phase mask, that is, the pattern has uniform intensity along both the directions of the zeroth-order beam propagation (z direction) and the pitch D of the phase mask (y direction). The difference is that the pattern has a sinusoidal distribution along the cross section direction (x direction) with a period equal to half of the phase mask period.

If we take the zeroth-order diffraction into account, the distribution changes to

$$I_{0,\pm1} = 4\big[C_1^2\cos^2(2\pi x/D) + C_0^2 + 2C_0C_1\cos(2\pi x/D) \times \cos(2\pi z(1-\sqrt{1-(\lambda/D)^2})/\lambda)\big] \tag{35}$$

Three terms are included in the right-hand side of Eq (35). The first term is the same as Eq. (34), that is, the interference of the two first-order beams. The second term, $4C_0^2$, is a constant, which comes from the zeroth-order diffraction, that reduces the contrast of the fringe pattern of the first term. The third term is due to the interaction between the zeroth-order and the two first-order diffraction beams. It changes along both x and z directions. Along the x direction, this term predicts a sinusoidal distribution in a manner similar to that predicted by Eq. (34) except has a different period, which is exactly equal to the period of the phase mask. Its intensity is dependent on the product of the intensities of the zeroth- and first-order diffractions and also on the position in the z direction. Along the z direction, the distribution is also sinusoidal but with a period

$$z_0 = \lambda/\left(1 - \sqrt{1 - (\lambda/D)^2}\right) \tag{36}$$

To compare the theoretical results from Eq. (35) and (36) with the experimental results presented in the previous section, we set all the values of the parameters in the equations to those used in the experiments,. That is, $\lambda = 248$ nm, $D = 1.0614$ μm, $(2C_0)^2 = 1\%$, and $C_1^2 = 44\%$. In this case, the period of the sinusoidal change in intensity distribution in the z direction is $z_0 = 8.96$ μm, which agrees well with the experimental value of 8.5 μm, with an experimental error of about 5%.

The intensity distributions of the grating in the plane of the phase mask output surface are shown in Figure 13 for different levels of the zeroth-order diffraction. The horizontal axis represents the distance in the cross sectional direction (the x axis in Fig. 12). The curve in Figure 13a, which corresponds to a zeroth-order diffraction level of 1%, consists of two sets of periodic variations, each with the same 1.06 μm period of as that of the phase mask, but with different amplitudes. These are labeled as grating 1 and grating 2 in the figure. The distance between gratings 1 and 2 is equal to $D/2$ (0.53 μm) in the cross sectional direction. This curve bears some resemblance to the AFM curve in Figure 11. Figure 13b–e depicts another four curves with different intensity levels of the zeroth-order diffraction. It can be seen that as the intensity decreases from $(2C_0)^2 = 10\%$ to 0, the interference distribution gradually becomes a uniform, periodic structure with a pitch value of 0.53 μm as shown in Figure 13e where $C_0 = 0$, which corresponds to the usual interference fringe pattern of the two first-order laser beams. Therefore, we can conclude that the emergence of gratings I and II in Figure 11 or gratings 1 and 2 in Figure 13 is due mainly to the intrusion of zeroth-order diffraction.

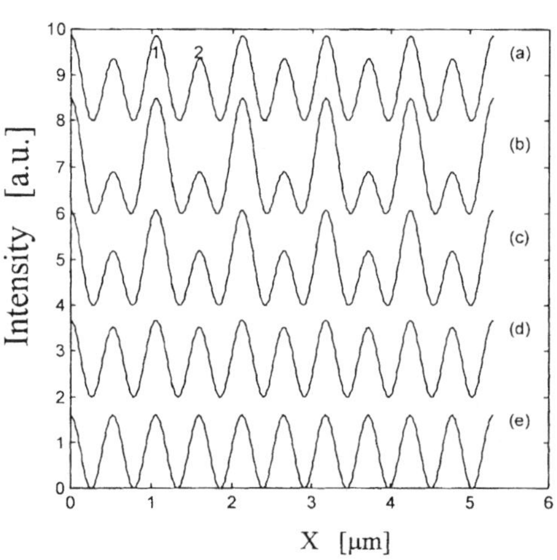

Figure 13. The interference distribution along the x direction at $z = 0$ with different levels of the zeroth-order diffraction $(2C_0)^2$: (a) 1%; (b) 10%; (c) 3%; (d) 0.1%; (e) 0%[53].

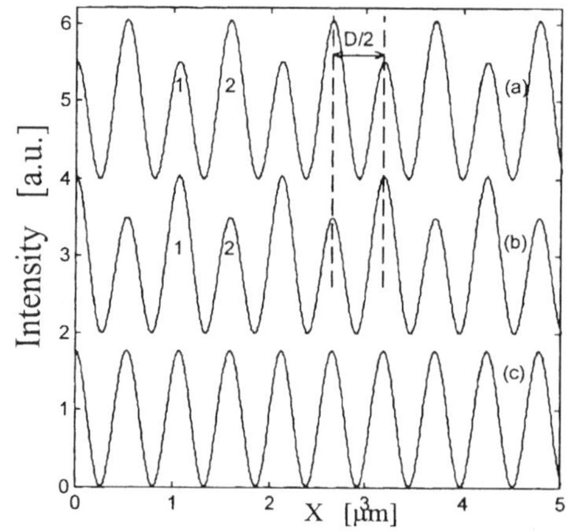

Figure 14. The interference distribution for $(2C_0)^2 = 1\%$ and $C_1^2 = 44\%$ along the x direction at different z positions: (a) $z = z_0/2$ (4.98 μm); (b) $z = 0$; (c) $z = z_0/4$ (32.49 μm) [53].

The extent of the zeroth-order effects on the gratings can be estimated from Eq. (35). At $z = 0$, the intensity peak values of gratings 1 and 2 are $4(C_1 + C_0)^2$ and $4(C_1 - C_0)^2$, respectively. The change in peak intensity from grating 1 to grating 2 is then $4C_1C_0/(C_1 + C_0)^2$. Under our experimental conditions, the change is 27%, whereas for a level of $(2C_0)^2 = 0.1\%$ (the case in Figure 13 d, it is ~10%. Thus, the effects of the zeroth order are substantial even if its intensity is very small.

Figure 14 displays is the calculated distributions of the diffraction intensity for $(2C_0)^2 = 1\%$ and $C_1^2 = 44\%$ at (a) $z = z_0/2$ (=4.98 μm), (b) $z = 0$, and (c) $z = z_0/4$ (=2.49 μm) behind the phase mask. As z varies, the amplitudes of the two pulse sequences change according to Eq. (35). The fringe pattern at $z_0/2$ (as shown in Figure 14 a) has a maximum shift in the cross sectional direction with respect to the fringe pattern in the phase mask output plane ($z = 0$ as shown Figure 14 b). Both curves a and b have the same features as discussed in the previous paragraph and the shift with respect to each other is exactly equal to 0.53 μm, half of the phase mask's period. Note that at $z = 0$, grating 1 has a maximum value, whereas grating 2 has a minimum. On the other hand, when grating 1 becomes a minimum, grating 2 reaches a maximum at $z = z_0/2$. Figure 14c shows the case where grating 1 and grating 2 become equal to form a uniform grating. A comprehensive diffraction distribution can be seen clearly in the three-dimensional graphic in Figure 15.

All the results shown in Figures 13–15 unambiguously predict that the introduction of zeroth-order diffraction by the phase mask into the interference destroys the quality of the Bragg gratings. Its effects are substantial even if the zeroth-order intensity is as small as 0.1%.

In summary, experimental results have shown that zeroth-order diffraction as small as 1% of the total power and first-order diffraction of 44% each result in dramatic effects in the volume structure of the gratings in a polymer optical

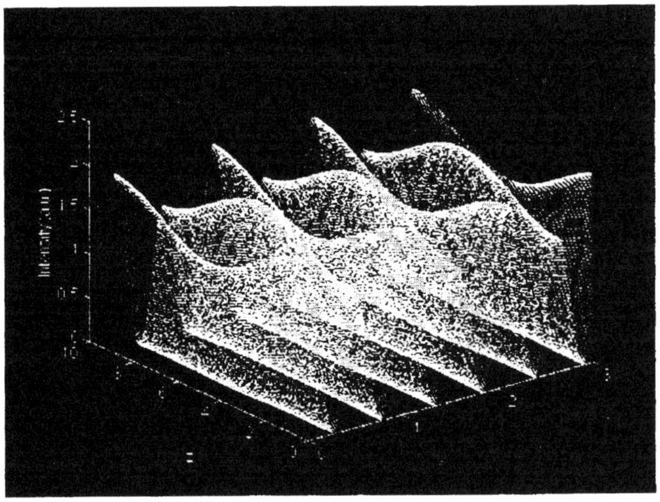

Figure 15. Three-dimensional intensity distribution of the three beam interference [Eq. (35)] with experimental parameters $\lambda = 248$ nm, $D = 1.0614$ μm, $(2C_0)^2 = 1\%$, and $C_1^2 = 44\%$.

preform. Therefore, it is important to eliminate the zero-order interference completely.

4.3. Sagnac Interferometric Scheme

To avoid the interference effect of zeroth-order diffraction during grating writing, we set up a Sagnac interferometer that consists of three prisms to imprint the Bragg grating into the POF as shown Figure 16. As shown in the inset of this figure, the phase mask is placed underneath the fiber and offset so that the UV beam misses the fiber, but is diffracted by the mask. The two first-order diffracted beams form a loop with the three prisms such that they meet and interfere with each other at the fiber core. The zeroth-order beam, which is not in the path of the loop, does not contribute to the interference. The phase has a period of 1.064 μm and is designed for a 248 nm wavelength writing beam. At this wavelength, the zeroth-order diffraction from the mask is less than 1% and the two main orders (± 1) have the greatest power ($\sim$44%). The writing beam is from a frequency-doubled OPO pumped by a frequency-tripled Nd:YAG laser. With a three prism combination, the Sagnac interferometer has the two first-order beams from the mask to form the required interference pattern for the grating. One advantage of this system is that it is possible to select a different writing wavelength away from the 248 nm wavelength without influencing an increased zero diffraction order. The phase mask together with the POF can be transversely shifted by a stepping motor. In this way, a grating can be made with the same length as the mask. A spectrum analyzer is used for on-line monitoring of the grating writing process.

The first POF grating was written on a few-mode POF with a writing wavelength of 325 nm [24]. The reflection spectrum recorded by the spectrum analyzer is shown in Figure 17. Because the fiber has few mode, it has a broadened reflection bandwidth of about 10 nm. The number of peaks

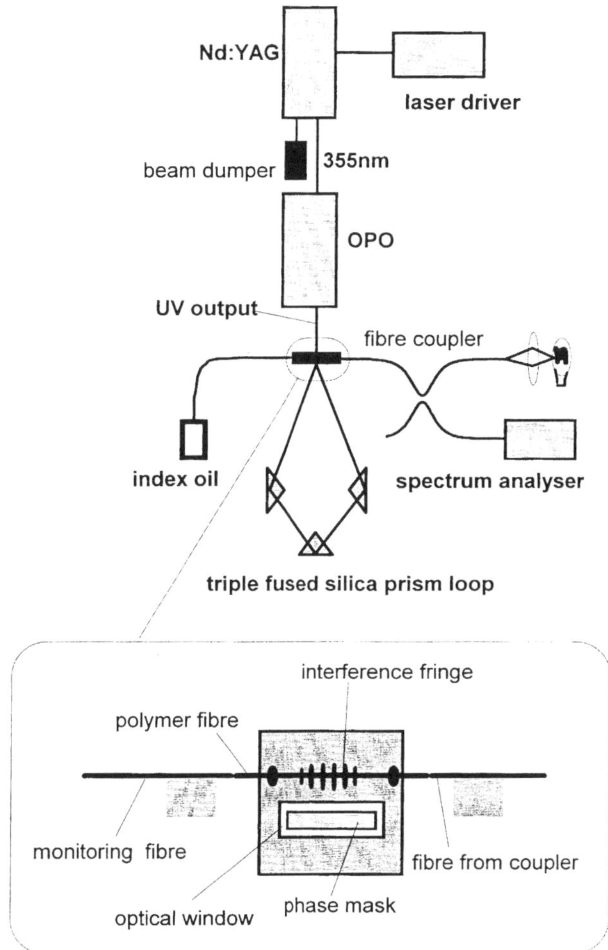

Figure 16. Experimental setup for writing POF grating based on a Sagnac interferometer [54].

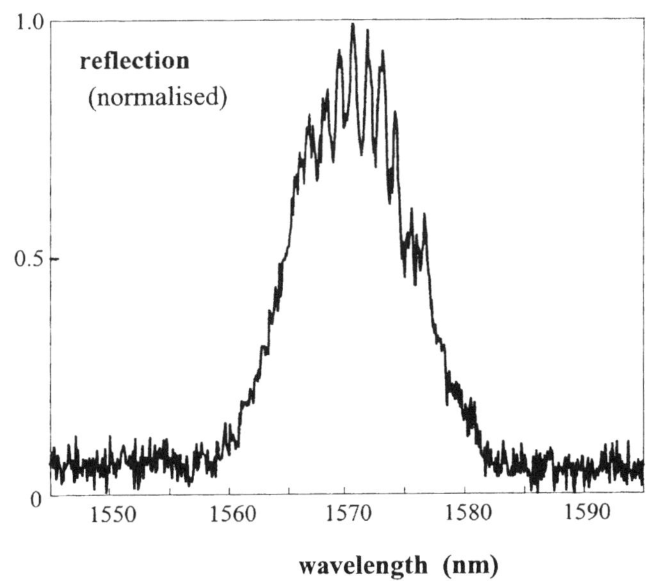

Figure 17. Reflection spectrum of a Bragg grating made in a few-mode POF.

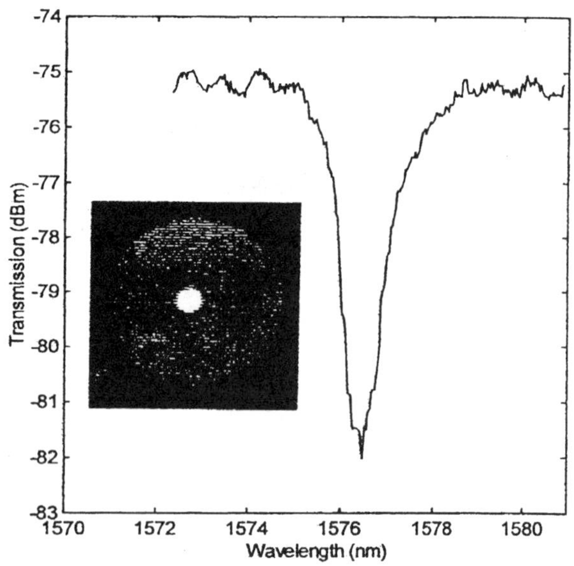

Figure 18. Transmission spectrum of a single-mode POF Bragg grating [54].

in the reflection spectrum is roughly equal to the number of modes in the fiber. Later gratings have been made from a single-mode POF with a reduced bandwidth of ~1 nm. This results in improved reflectivity of the grating to about 80% [54]. Figure 18 shows the transmission spectrum of a single-mode POF Bragg grating. The length of the grating is 1 cm and the bandwidth of the pass band is 0.5 nm.

5. TUNABILITY OF POF BRAGG GRATINGS

The reflection wavelength of a fiber Bragg grating, λ_B, can be expressed in a simple general form from Eq. (8):

$$\lambda_B = \frac{2\Lambda n_{eff}}{N} \tag{37}$$

Here the grating period Λ is determined by the period of the interference pattern (fringe) of the writing light beam. From this expression, we can see that the Bragg wavelength of a grating can be tuned if a particular mechanism is used to introduce changes in the grating period, Λ, and/or the effective index n_{eff}. At least two simple mechanisms can be used for tuning. The first is the mechanical method, which involves applying a longitudinal stress to a fiber Bragg grating so that the grating period is stretched. The second is a thermal method in which the grating is heated up to introduce a change in the grating period. The resulting wavelength shift is due mainly to two factors: a change in Λ due to stress-induced strain and a change in n_{eff} through the photoelastic effect. The stress in the grating arises either from mechanical stretching or from the difference of the thermal expansion coefficients of the constituent fiber components.

In the case of mechanical tuning, the strain is related to the stress σ,

$$\varepsilon = \frac{\Delta L}{L} = \frac{\sigma}{E} \tag{38}$$

where E is the Young's modulus of the polymer. Hence the stress-induced Bragg wavelength shift, $\Delta\lambda_B$, can be expressed easily as

$$\frac{\Delta\lambda_B}{\lambda_B} = (1 - P_e)\varepsilon = \frac{1 - P_e}{E}\sigma \tag{39}$$

where P_e is the effective photoelastic coefficient and ε is the longitudinal strain. Thus the Young's modulus E has a significant influence on the tunability of the Bragg wavelength. Large modulus materials such as silica have less strain sensitivity and thus require larger stress to produce the same strain in comparison to low modulus materials such as PMMA. Note that the Young's modulus of PMMA is about 30 times less than that of silica glass. Moreover, rigid materials usually have low breakdown strain; for silica it is about 1%. This sets a limit on the maximum achievable tuning range. For polymers such as PMMA, because of its great flexibility in material synthesis, however, 10% or more breakdown strain readily could be achieved.

In the case of thermal tuning, a change in the fiber temperature, ΔT, also leads to a shift of Bragg wavelength. This can be expressed by

$$\frac{\Delta\lambda_B}{\lambda_B} = (\alpha + \beta)\Delta T \tag{40}$$

where α is the thermal expansion coefficient and β is the relative thermo-optic coefficient of the fiber material.

5.1. Mechanical Tuning of PMMA POF Bragg Gratings

In the following discussion, we describe the experimental investigations of tuning the POF gratings by means of both mechanical and thermal methods. In the mechanical-tuning experiment, the tuning of Bragg wavelength was realized by a simple mechanical stretching setup (Fig. 19). Before the grating was written, the POF was mounted on a glass slide with both ends glued to it. The slide was then cut in the middle using a diamond saw after the grating was created and the two sides were fixed on two steel blocks connected to two micropositioners. One of the micropositioners was fixed and the other was attached to a translation stage to allow for

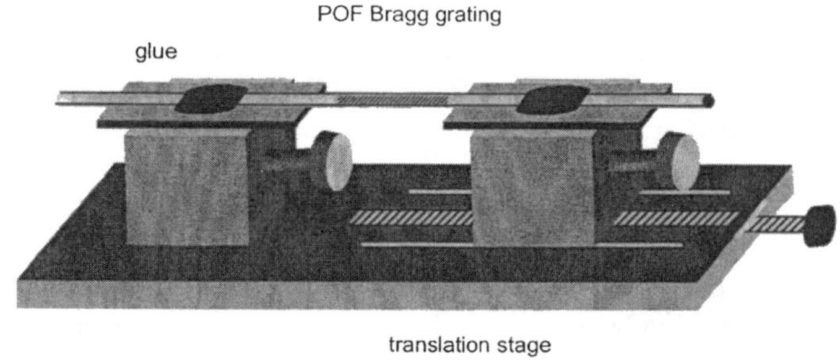

Figure 19. A simple mechanical setup for wavelength tuning of POF grating [54].

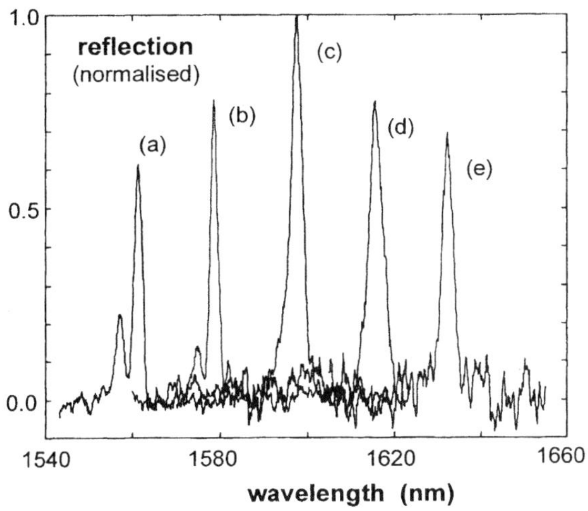

Figure 20. Wavelength tuning of a POF grating by mechanical stretching [54].

stretching. The tensile strain was loaded by rotating the lead screw with a longitudinal displacement of 1 mm per turn.

The reflection spectra for a piece of 25.3 mm POF that contains a 3 mm Bragg grating are shown in Figure 20. Trace (a) shows the spectrum with no tension on the fiber (the original position of the mobile micropositioner), where the resonance wavelength is 1561.12 nm. Traces (b)–(e) depict spectra with strain varying from 1.22 to 4.84%. At this final value, the fiber has been stretched 1.255 mm and the Bragg wavelength has shifted to 1634.22 nm. Thus the total wavelength shift is 73 nm. Whereas the fiber has a cutoff wavelength around 1580 nm, the sidebands in Figure 20a and b are due to the presence of higher order modes in the fiber. Note that the intensity at the Bragg wavelength is not uniform over the shift range, mainly for the following reasons. First of all, the launched white light source has a spectral distribution and the wavelength response of the optical spectrum analyzer used to detect the grating decreases with the wavelength. Second, the multimode fiber coupler employed to launch the grating with a white light source and collect the reflection spectrum has a wavelength-dependent coupling ratio. Finally, as the strain increases, the fiber diameter becomes smaller and this causes the higher order modes to be eliminated [as shown in (a) and (b)]. As a result, the intensity of the fundamental mode increases because of the energy transfer from the higher order modes to the fundamental mode. Then the guided intensity in the fundamental mode should decrease as the core size is further reduced.

The perform that was used to draw the polymer optical fiber was fabricated following the procedure described in [55]. To increase the thermal stability and strength, however, higher temperature was used for core polymerization. A hollow polymer preform filled with core monomer mixture (MMA + EMA + BzMA) was thermally polymerized in an oven. The temperature of the oven was increased gradually from 65 to 85 °C for three days. Note that the temperature used here was still well below the glass transition temperature of the cladding which is usually around 100 °C. Polymer optical fibers were drawn at 270 °C from the preform. The diameter of the fibers is 105 μm with a core size of 7.6 μm. The difference in the refractive index between the core and the cladding is 0.0086, which was measured using the transverse interference method with a microscope after the fibers were drawn. The cutoff wavelength is around 1580 nm. The fiber Bragg grating was prepared using the technique described in the previous section. To increase the intensity of the 325 nm UV laser beams, we added to the setup a pair of cylindrical lenses with focus lengths of 150 and 50 mm, respectively, in front of the phase mask used for grating writing. The beams on the phase mask were reduced from 6 to about 2 mm in the direction perpendicular to the fiber. The efficiently shortened the grating preparation time. For example, the time for saturation of grating reflection decreased from 85 to 15 min. The 3 mm long grating created under higher temperature has a reflection of 50%, which is compatible with that of our previously reported value. This result shows that the higher temperature used here did not degrade the photosensitivity of the POF to the UV light.

The change in Bragg wavelength with strain applied to the fiber is shown in Figure 21 [56]. The circles denote the experimental results and the line is the least-square fitting to the results using $\lambda_B = 1560.61(1 + 0.966\varepsilon)$. The photoelastic constant was deduced to be $p_e = 0.034$. This result shows that the contribution from the photoelastic effect is insignificant on the wavelength tuning and that the wavelength shift is linearly proportional to the strain.

So far a wavelength tuning of 73 nm, the largest range ever reported, in fiber grating has been demonstrated successfully. The Bragg wavelength can be tuned linearly with the tensile strain up to nearly 5%. The result is significant because it tunable range is about 10 times greater than that achieved in a silica optical fiber grating and it would be sufficient to cover the whole bandwidth of erbium-doped fiber amplifiers used in optical WDM communication systems.

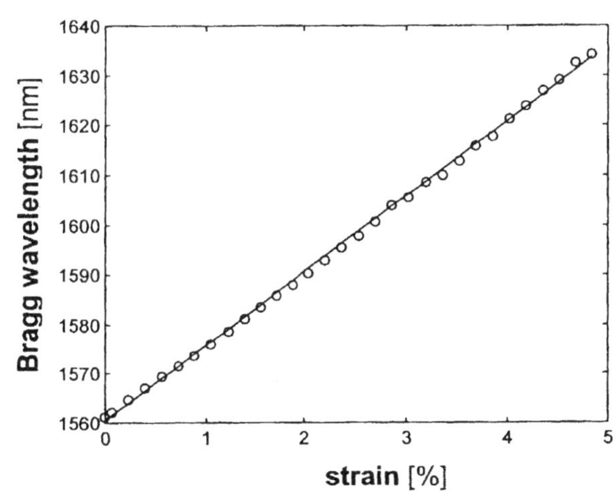

Figure 21. Wavelength tuning of a POF grating versus strain [54].

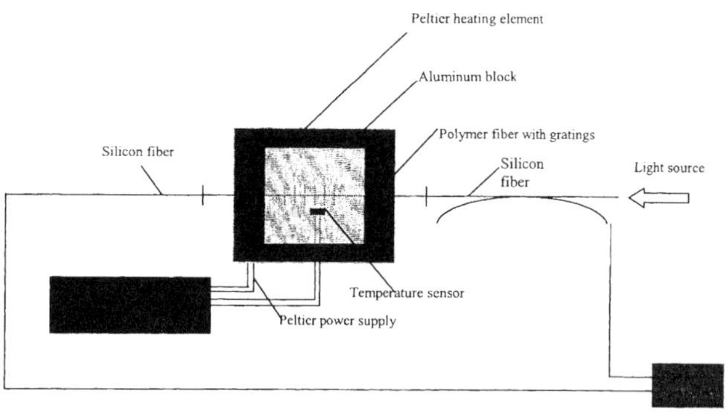

Figure 22. Setup for thermal tuning of a PMMA POF grating.

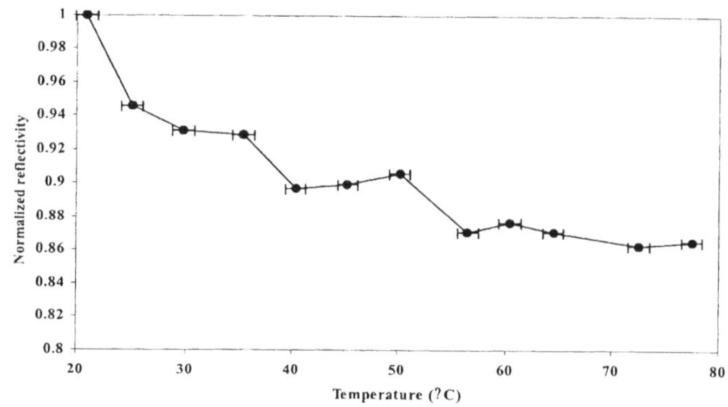

Figure 24. Temperature dependence of a PMMA POF grating.

5.2. Thermal Tuning of PMMA POF Bragg Gratings

To test the thermal tuning of the POF grating, the setup shown in Figure 22 was adopted. A Peltier cell was used as a heat source, the temperature of which is controlled by an electric current. This cell sat on top of an aluminum block, and the POF grating and temperature sensor were inserted between the block and the cell. The POF was quite short and both of its ends were connected to a piece of silica fiber to enable light to be launched into the grating. The transmitted light from the grating was directed to the optical spectrum analyzer (OSA). The accuracy of the temperature measurement was 0.1 °C, but it took about 10 min for the system to reach the desired temperature. When the temperature reached the desired value, the transmission spectrum of the grating was measured using the OSA. A tungsten lamp was used as the light source.

Figure 23 shows the transmission spectrum of the grating obtained from the measurement. The reflectivity at the Bragg wavelength (1578.5 nm) is estimated to be 75% and the spectral width is about 0.5 nm (full width at half maximum). This reflectivity can be improved by (1) increasing the grating length from 3 mm to several centimeters, because the reflectivity of a grating is proportional to the square of the hyperbolic tangent of grating length [Eq 25], and (2) Changing the illumination wavelength from 355 to 325 nm. This is because the photosensitivity effect in the polymer is more pronounced at the shorter wavelength.

Figure 24 shows the change of the reflectivity of the grating as a function of temperature with a step time of 10 min. The change is less than 10%. Included in this figure are the error bars of the measurements due to inaccuracy in determining the temperatures.

Figure 25 shows the shift of Bragg wavelength as a function of temperature of this POF grating. In this figure, two curves are shown. One was obtained when the grating was heated from room temperature to 70 °C, where as the other was obtained when the grating was cooled down. Both curves are not very linear. The error bars due to inaccuracy in determining the temperature are also shown. Note that the heat up curve and the cool down curve more or less coincide with each other. This indicates the absence of hysteresis and shows that thermal means can be used to tune the Bragg wavelength in practice. The tuning range is more than 18 nm over a temperature variation of 50 °C. This is much larger than the corresponding range of silica fiber grating, which is less than 1 nm.

The thermal stability of the grating in polymer fiber is further investigated by maintaining the grating at an elevated temperature and monitoring its reflectivity as a function of time. Figure 26 shows the change of reflectivity as a function of time when the grating was maintained separately at 30, 50, and 65 °C. Although these plots bear a

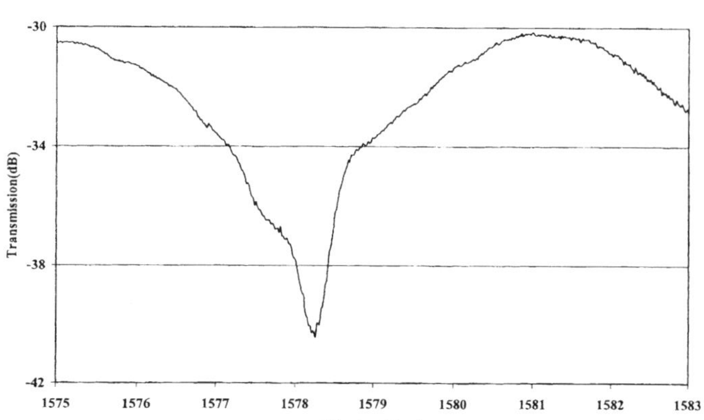

Figure 23. Transmission spectrum of a single-mode PMMA POF grating.

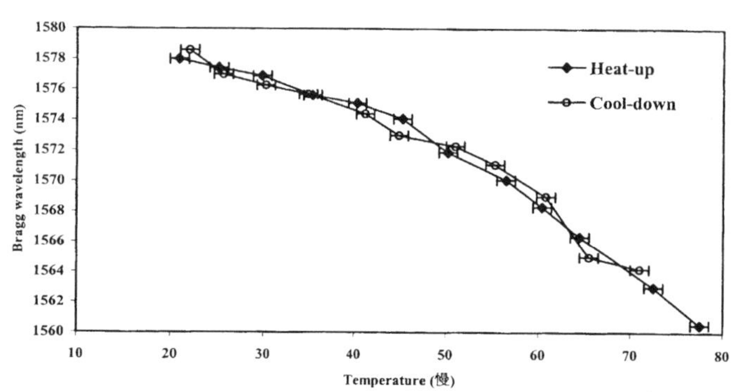

Figure 25. Temperature dependence of Bragg wavelength in a PMMA POF grating.

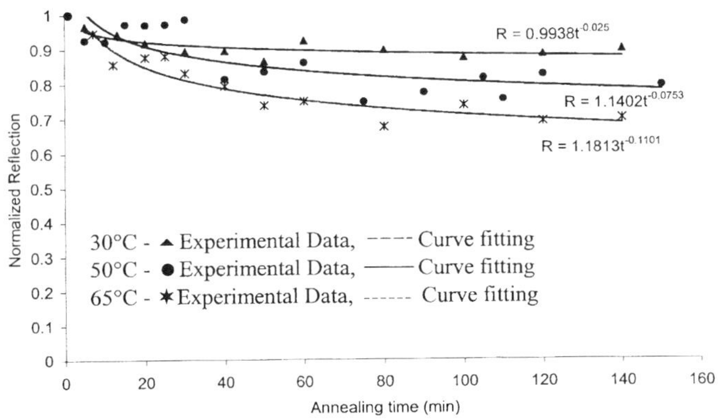

Figure 26. Thermal stability of a PMMA POF grating.

remarkable resemblance to those of silica fiber gratings [57, 58], we found they do not fit the Erdogan model [57] for non-hydrogen-loaded gratings; neither do they fit the Baker model [58] for hydrogen-loaded gratings. In the end, we resigned to the simple curve fitting

$$\text{normalized reflection } R = At^{-a} \tag{41}$$

where A and α are shown in Figure 26. There is a notable difference in thermal stability between polymer fiber gratings and silica fiber gratings in that, for the polymer case, there is no dramatic decrease of the reflectivity when t is small, whereas in the case of glass gratings, the decrease is very pronounced in hydrogen-loaded and non-hydrogen-loaded gratings. The fact that these plots do not fit both Erdogan and Baker models may be due to the different photosensitivity mechanisms: for polymer, it is chiefly due to photopolymerization [57, 59]. Just as in the case of silica fiber gratings, the long term stability of polymer fiber grating requires burning in at an elevated temperature for a short time.

6. CYTOP POF GRATINGS AND THERMAL TUNING

The CYTOP fiber was developed by Keio University and Asahi Glass Company in Japan. It has excellent chemical, thermal, electrical, and surface properties similar to conventional fluoropolymers such as Teflon. In addition, it is characterized by its high optical transparency. This is because the CYTOP fiber is composed of perfluorinated polymer consisting of C-C, C-F, and C-O bonds. These bonds have absorption wavelengths at 7.7–10.0 μm compared to that of C-H bond, 3.3–3.5 μm, the composition PMMA fiber. Hence at visible wavelengths, the CYTOP fiber has much lower intrinsic loss than the PMMA fiber. By optimal design of the index profile, researchers at Keio University managed to fabricate CYTOP-based graded-index optical fiber that has loss down to 40 dB/km even in the near infrared region [59]. It also has been shown that the theoretical attenuation limit of CYTOP-based fiber is similar to that of silica-based fiber (~0.3 dB/km at 1550 nm) [61]. Furthermore, multimode fiber made of this material was shown to have a larger bandwidth than both the PMMA-based graded-index fiber and GeO_2-SiO_2-based multimode fiber. With these advantages, the novel CYTOP graded-index fiber is very promising for high-speed data communication and telecommunication.

More recently, the photosensitivities of this perfluorinated CYTOP polymer material and POFs made from CYTOP have been investigated jointly by researchers at the University of New South Wales, Keio University, and Asahi Glass Company. Considerable photosensitivity of the perfluorinated polymer has been revealed [61].

6.1. CYTOP POF Gratings

Bragg gratings can be written readily in PMMA fibers. However, this is not the case for CYTOP fibers. Apparently, the photosensitivity effect in CYTOP fiber is less than that in PMMA fiber. The photosensitivity effect in both PMMA and perfluorinated polymer is initiated by the irradiation of the UV beam. This causes an increase in the polymer density because of the cross linking or the polymerization of the residual monomer; this is also called the photopolymerization of the polymer. As a result, the refractive index increases. Our experimental results show that the modulation index of the CYTOP grating is only 0.6 that of PMMA grating. This means that the change of polymer density in the CYTOP grating is only 0.6 that of the PMMA grating. This in turn means that the CYTOP polymer is denser than PMMA. In fact, it can be shown that the theoretical density of PMMA is 1.19 g/cm^3 and that of perfluorinated polymer is 2.03 g/cm^3. The corresponding density ratio is therefore 0.58. This figure agrees very well with 0.6, the experimental value. This explains the difficulty of writing gratings in CYTOP fibers. Instead, we found that gratings can be written when the fiber is squashed into a slab. Our study concentrated on such a structure.

The slabs were made by squashing the fiber between two heated clean metallic plates. We tested two kinds of slabs in our experiment. The first one was made from the CYTOP fiber with PMMA cladding and it was 150 μm thick. Then we removed the PMMA cladding and squashed the remaining pure CYTOP core to fabricate another slab with a thickness of 100 μm. Bragg gratings were then written on these slabs using the simple phase mask technique as shown in Figure 27. The sample was directly exposed with diffracted beams from a silica phase mask under UV illumination. The UV writing beam at 355 nm wavelength is the output from a frequency-tripled Nd: YAG laser in pulse mode at a repetition rate of 10 Hz and pulse energy of 280 mJ. Because the phase mask, which has a period of 1.061 μm, is designed to operate at a wavelength of 248 nm, it has a significant zeroth-order diffraction at the 355 nm wavelength. As a result, this doubles the period of the interference fringe and, therefore, the period of grating written (1.061 μm) [53]. The diffraction pattern can be directly observed using a He-Ne laser in

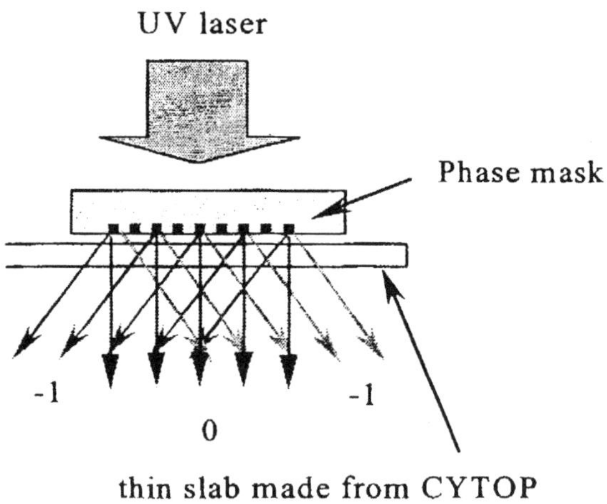

Figure 27. Grating writing on a CYTOP slab [56].

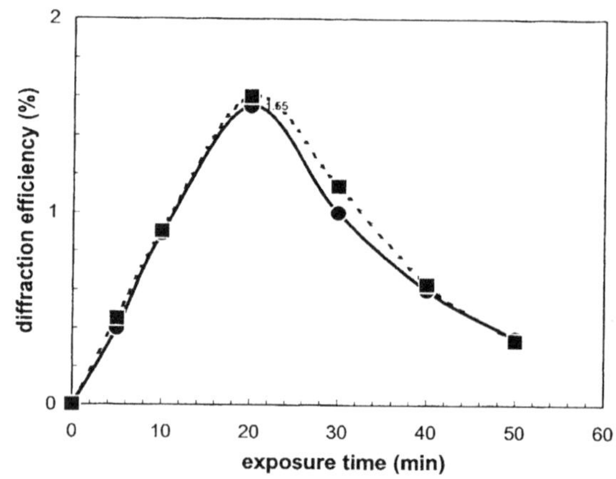

Figure 29. Grating diffraction efficiency versus UV exposure time with the PMMA jacket removed (solid line) and with index matching liquid added (dotted line.)

an arrangement like that shown in Figure 28. The diffraction angle, θ, of the first order was measured to be 36° in the experiment. The grating period can be calculated by taking the grating diffraction equation as

$$\Lambda = \frac{m\lambda}{\sin\theta} \tag{42}$$

where m is the diffraction order, Λ is the grating period, and λ is the incident light wavelength. The period of the CYTOP slab grating is calculated to be 1.07 μm, which agrees well with the prediction of period doubling in the presence of zero diffraction. By measuring the powers in the first orders (P_{-1}, P_{+1}) and in the zero order (P_0), the diffraction efficiency η can be calculated from

$$\eta = \frac{P_{-1} + P_{+1}}{P_{-1} + P_0 + P_{+1}} \tag{43}$$

The diffraction efficiency of several gratings formed by varying the UV exposure time was measured and the results are shown in Figure 29 (solid line).

To ascertain whether the grating written in a CYTOP sample is of a surface relief type or of a bulk type, diffraction efficiencies also were measured when a refractive index matching oil was spread onto the slab surfaces. The differences in diffraction efficiency were compared to determine whether the diffraction contribution resulted from a surface relief grating, a bulk grating, or a combination of both. In a surface relief type grating, the diffraction disappears when index matching oil is spread. The experimental results with index matching oil are included in Figure 29 (dashed line). It is clear that the diffraction efficiency changes little and that the gratings written are of bulk type, that is, they are the result of the change of refractive index, not due to material ablation. This proves the existence of the photosensitivity of CYTOP. Observe from this figure that the efficiency increases as the exposure time increases up to 20 min and that a maximum diffraction efficiency was obtained. Note that the underlying mechanisms associated with photosensitivity require further investigation.

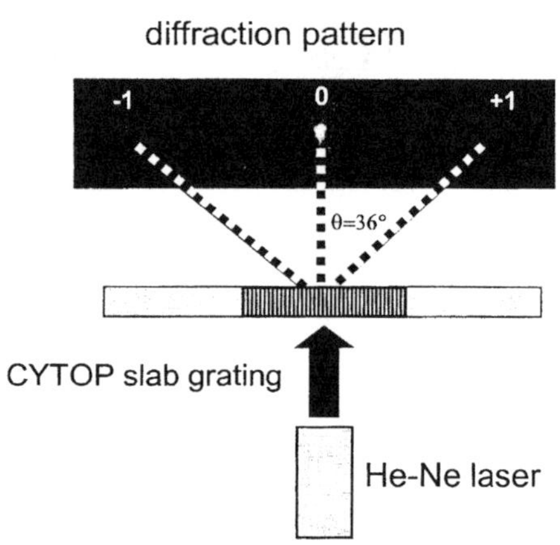

Figure 28. Arrangement for grating testing, including a diffraction pattern from a CYTOP grating [56].

6.2. Thermal Tuning of CYTOP POF Gratings

To test the thermal tuning of CYTOP fiber grating, we used the setup shown in Figure 30. The fiber grating was immersed in a transparent bath of water and the bath was placed in a glass container that also had a thermal heater. A He-Ne laser beam was shown laterally onto the grating from underneath. The angle of the first-order diffraction was detected at a screen above the grating. This angle was measured as the water temperature, which was controlled by the thermal heater, increased Figure 31 shows the change of diffraction angle as a function of temperature. We can convert the change of the diffraction angle into the change of Bragg wavelength of the fiber grating. The result is shown in Figure 32. It can be seen from this figure that the Bragg wavelength of the grating can be shifted from 1410 to 1419 nm over a temperature change of about 60 °C. Furthermore, the tuning is linear and independent of the direction of the temperature change. This should be compared with the tunability of PMMA fiber grating. In Section 5, we reported that over the same tuning temperature range, the Bragg wavelength shifts over 18 nm, that is, double that of the CYTOP

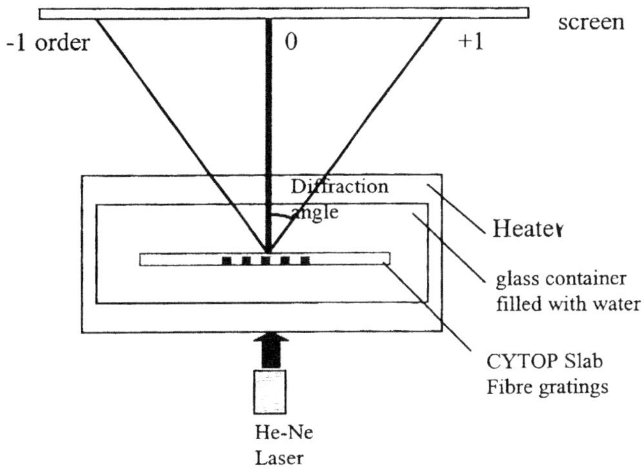

Figure 30. Thermal tuning of CYTOP fiber grating [56].

fiber grating. However, the tuning curve shown in Figure 25 is not as linear as that shown in Figure 32.

6.3. Thermal Stability of CYTOP POF Gratings

We also measured the grating diffraction efficiency of the Cytop fiber grating as the temperature changed from 20 to 80 °C. The result is shown in Figure 33. It can be seen that the efficiency practically remained unchanged. This is in marked contrast to the behavior of PMMA fiber gratings that are shown in Figure 24. In this case, the reflectivity has dropped from the normalized value of 1 to 0.88.

Further tests of the grating diffraction efficiency of Cytop fiber at elevated temperatures over a long period of time also were carried out. Figure 34 shows the result. It can be seen that the efficiency practically remains unchanged over nearly 200 min at elevated temperatures of either 45 or 60 °C. This behavior shows that Cytop fiber grating is even more stable than glass fiber grating. In the latter, the reflectivity suffers a sharp drop in the first few minutes before it stabilizes to a constant value.

Baker et al. [58] showed that the decay characteristic of the normalized reflectivity $\Re$ follows the Erdogan power law

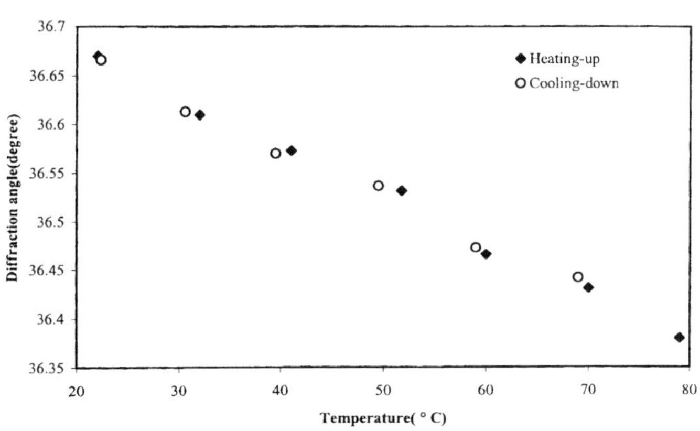

Figure 31. Change of diffraction angle with temperature [56].

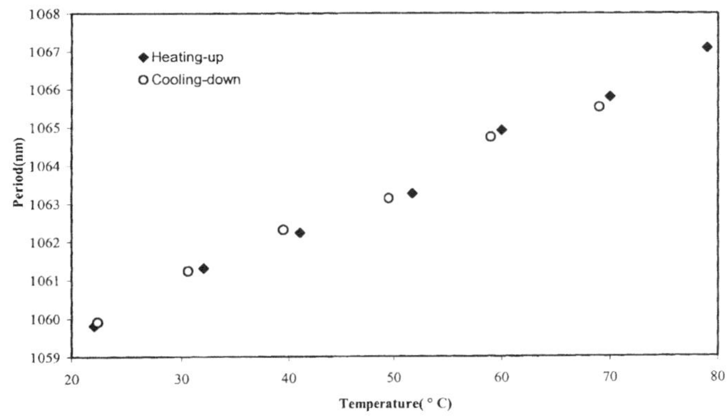

Figure 32. Change of grating period as a function of thermal tuning.

of time t for non-hydrogen-loaded glass fiber,

$$\Re = \frac{1}{1 + At^{\alpha}} \tag{44}$$

where the parameters A and α have been evaluated at absolute temperature T for boron and germanium codoped fiber,

$$A = 0.192 \times 10^{-3} \exp(13.1 \times 10^{-3} T)$$
$$\alpha = T/2941$$

whereas for germanium-doped fiber,

$$A = 1.86 \times 10^{-3} \exp(7.64 \times 10^{-3} T)$$
$$\alpha = T/5250$$

However, for hydrogen-loaded glass fiber, $\Re$ follows a different characteristic [58],

$$\Re = 1 - K \log\left(\frac{t}{\pi}\right) \quad \text{for } t > \tau \tag{45}$$
$$= 1 \qquad t \leq \tau$$

where $K = 0.093$ and $\tau = A\exp(E_a/RT)$, where $E_a =$ 105 kJ/mol and $A = 9 \times 10^{-15}$ min.

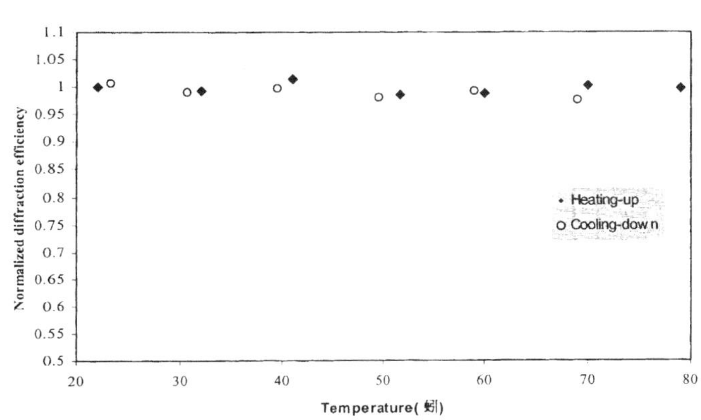

Figure 33. Temperature dependence of the diffraction efficiency of a CYTOP fiber grating.

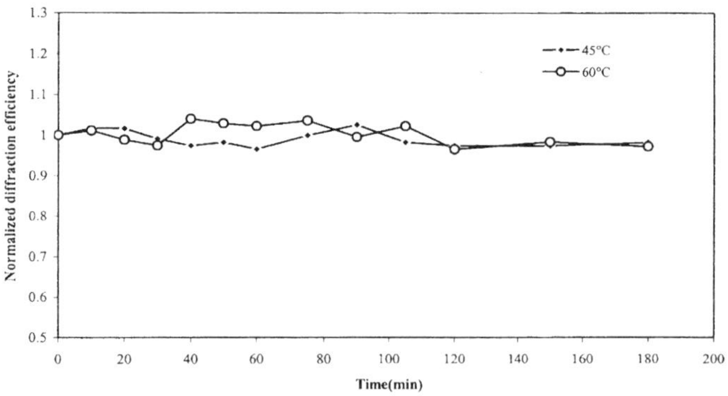

Figure 34. Thermal stability of the diffraction efficiency of a CYTOP fiber grating.

For the CYTOP fiber grating, we can simply represent

$$\Re = 1 \tag{46}$$

in view of Figure 34. On the other hand, the thermal stability of the grating reflection of PMMA fiber at an elevated temperature is shown in Figure 26. Comparing Figures 34 and 26, we can see that the CYTOP fiber grating is much more thermally stable than PMMA fiber grating.

7. APPLICATIONS

The development of POF gratings is still in its infancy. The unique characteristics that lead to potential applications in optical systems are their mechanical and thermal tunabilities. However, before POFs find practical applications, other grating qualities must be studied, such as repeatability, mechanical strength, fatigue strength, and, above all, life expectancy. Another important problem is low to join it to standard silica glass optical fibers in the system. Fusion splicing is out of the question, because of the vast difference in melting temperatures. The V-groove technique may be one solution, but for single-mode fiber connection, this inevitably will lead to large joint loss and, hence, large insertion loss of the grating. The temperature stability of the joint also depends on the temperature characteristics of the glue in addition to the thermal stability of the grating. Inspite of all these problems that we believe will be solved in the long run, there are several potential applications of POF gratings where glass fiber gratings cannot compete. These applications are described in the following subsections.

7.1. Tunable Add–Drop Multiplexer–Demultiplexer

The optical add–drop multiplexer is an important device needed in wavelength division multiplexing (WDM) communication and sensing systems. Its function is to drop or divert a particular wavelength from the trunk fiber to a certain location and to add the same wavelength from that location to the trunk fiber. Figure 35 shows two common versions of

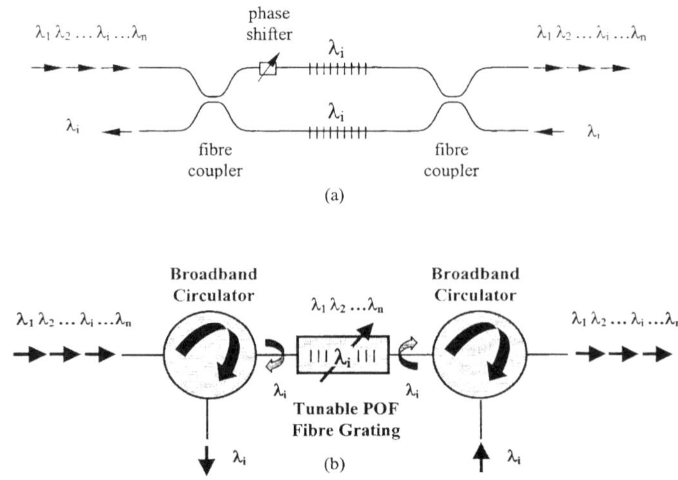

Figure 35. Fiber grating application in an optical add-drop multiplexer: (a) in a Mach-Zehnder interferometric configuration; (b) in a configuration using a pair of optical circulators. A tunable or reconfigurable OADM can be constructed by using a tunable fiber grating.

OADM. In Figure 35a, it is in the form of a Mach–Zehnder interferometer in which both arms are written with an identical grating, that drops the wavelength λ_i via the lower arm of the left-hand side 3 dB coupler. Similarly, the same wavelength can be added to the output wavelength stream via the lower port of the right-hand side 3 dB coupler. In the standard version of this OADM, the gratings are usually in glass fiber or silica waveguide form, and are fixed at a Bragg wavelength λ_i so that the device can operate only for that wavelength. Now, if they waveguides are replaced by polymer fiber gratings or polymer waveguide gratings, they can be tuned mechanically or thermally to operate at any desired wavelength.

In Figure 35b, the OADM is in the form of a combination of a fiber grating connected at either end by a broad-band circulator. The incoming wavelength stream enters the left port of the first circulator. Inside the circulator, this stream circulates clockwise, and leaves the right port and meets the fiber grating. Because the grating is designed to reflect wavelength λ_i, this wavelength reenters the same port and circulates clockwise to the bottom port. This wavelength λ_i is then dropped out of the port. The remaining wavelengths in the stream pass through the grating with no significant attenuation and enter the left port of the second circulator. These wavelengths circulate clockwise in the circulator and then leave the right port. A new wavelength λ_i is added to the outgoing stream by being sent into the second circulator through the bottom port. It circulates clockwise and reaches the left port within this circulator. It then leaves this port, and is met and reflected back to the port by the fiber grating. On reentering the port, λ_i joins the remaining wavelength stream and appears as one of the output wavelengths at the left port. Now if the grating is fixed and with no possibility of tuning, only one wavelength can be added or dropped. However, polymer optical fiber grating affords a wide tuning range through either mechanical or thermal means. Hence,

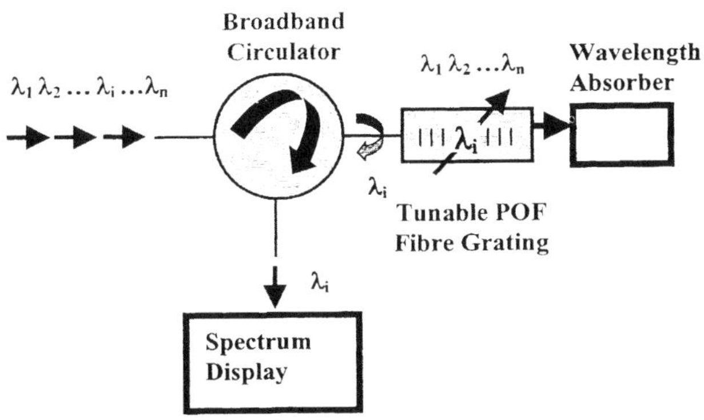

Figure 36. A simple optical spectrum analyzer.

replacement of the glass fiber grating with POF will enable the OADM to add or drop any wavelength desired.

7.2. Simple Optical Spectrum Analyzer

If we remove the second broad-band circulator in the OADM shown in Figure 35b, we achieve the basic structure of an optical spectrum analyzer as shown in Figure 36. The second circulator is now replaced by a wavelength absorber that simply can be a piece of black velvet attached to the end of the fiber Bragg grating. The purpose of this absorber is to prevent light from reflecting back to the input end of the grating, thus interfering with the signal delivered to the spectrum display unit. Instead of the absorber, an optical fiber isolator may be inserted between the first circulator and the fiber grating. Of course, we need to calibrate the tuning of the POF grating to produce a relationship between the amounts of tuning and wavelength shift. Mechanical tuning is preferred in this situation because it is faster than thermal tuning. This simple OSA can be incorporated into a personal computer so that control and data collection can be effected using software.

7.3. Adjustable Chromatic Dispersion Compensator

When an optical pulse travels within a standard optical fiber, its width tends to broaden in proportion to the distance traveled, because different wavelength components in the pulse travel with different velocities and, hence, have different time delay. This phenomenon is called chromatic dispersion of the fiber. Mathematically, the dispersion is equal to the first derivative of the time delay with respect to wavelength. As a result, the dispersion effectively reduces the information carrying capacity of the fiber. Many techniques have been suggested to compress the pulse at the end of the fiber back to its original width (Fig. 37). One of these approaches is to use a properly designed fiber grating. Figure 38a shows the delay characteristics of the fiber and of the grating, whereas Figure 38b shows the corresponding dispersion characteristics of the fiber and of the grating required to compensate for the fiber dispersion. The idea is that those wavelengths that

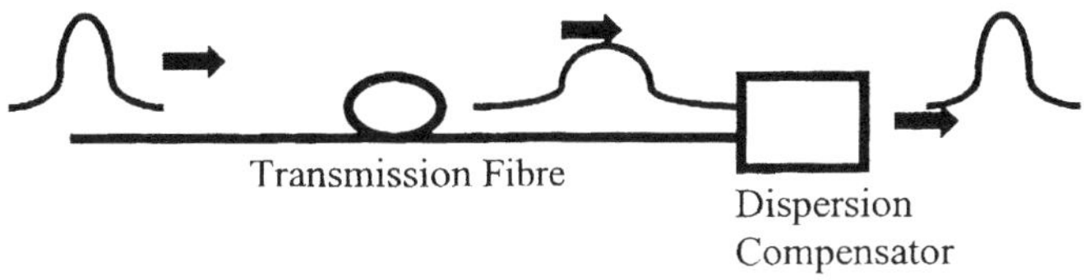

Figure 37. Optical fiber dispersion compensation.

suffer large delays in the transmission fiber will be speeded up by the grating, where as those that suffer least delays will be slowed down by the grating so that, on the whole, all the wavelength components in the pulse will arrive at the output at the same time. However, if the granting dispersion is not tunable, the compensation can be effective only for a particular state of the fiber. In practical situations such as undersea optical cables, the fiber dispersion can change due to water pressure and stresses. It is, therefore, desirable to have a fiber grating that has a tunable dispersion. This goal is achieved by the use of POF gratings where the tuning can be effected either mechanically or thermally.

7.4. Individual Interrogation in a WDM Sensor Array

A large number of wavelengths are being used in WDM systems for fiber sensing. This technique allows many thousands of sensor heads to be incorporated into a single length of optical fiber. The generic structure of a WDM sensor array is shown in Figure 39. It consists of a laser array that delivers N light beams with wavelengths λ_1, λ_2, λ_3, ..., λ_N and a WDM multiplexer that combines these beams into a single fiber connected to one of the ports of the circulator. The circulator then delivers all the beams to the sensor array via another port. The sensor array consists mainly of a single

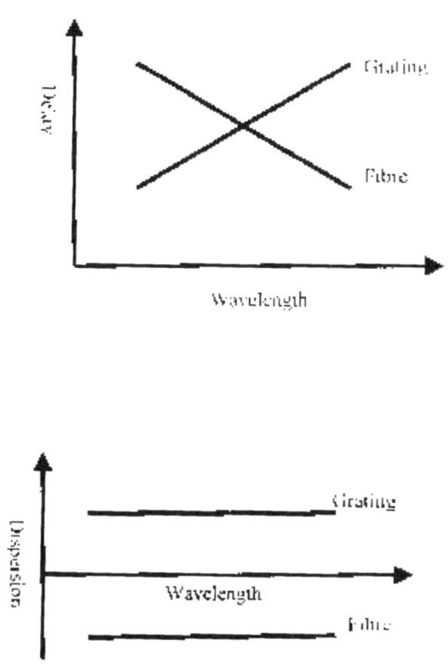

Figure 38. Dispersion compensation by means of fiber grating: (a) delay characteristics of fiber and grating; (b) grating dispersion characteristic required to compensate for fiber dispersion.

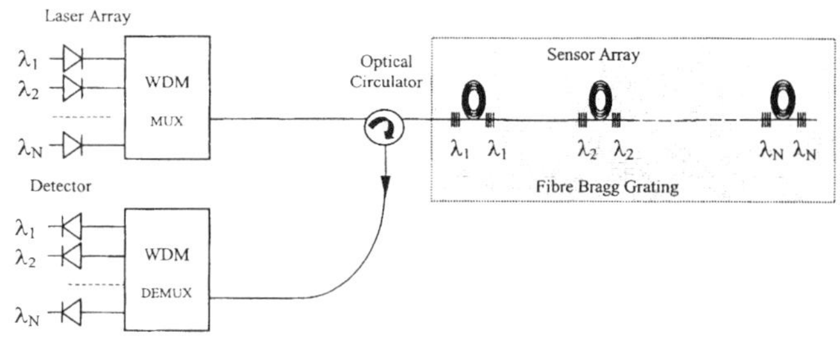

Figure 39. Fiber grating application in a WDM sensor array.

fiber that has pairs of gratings written at regular intervals. The fiber between the grating pair is coiled to form a sensor head. Each pair has a different Bragg wavelength. Hence, each sensor head responds to a particular wavelength. The return signals from all the wavelengths are directed to the WDM demultiplexer via the third port of the circulator.

Sometimes it is more convenient to interrogate the signal individually from an arbitrary wavelength instead of demultiplexing all the signals at once as shown in Figure 39. In this case, the WDM demultiplexer can be replaced by the tunable OADM described in Section 7.1.

8. FINAL REMARKS

Optical fiber Bragg gratings form the backbone of many devices for communication, sensing, and optical signal processing. However glass fiber gratings, in spite of many superior qualities, suffer an important disadvantage, that is, limited wavelength tunability. This is where POF gratings come to the rescue. The latter have a tuning range at least 10 times greater than the former. For this reason, new tunable optical devices can be developed. The examples quoted in Section 7 are only some obvious ones. Projections for the future envision new devices that make full use of the tunability of POF gratings.

The gratings described in this chapter are the simplest kind, that is, gratings that have constant pitch. More advanced versions, such as chirped gratings and apodized gratings, will be developed later and surely will lead to the development of more interesting devices.

The writing of gratings in POFs depends on their photosensitivity effect. We have described an extensive experimental investigation into this effect in POFs—both PMMA fiber and CYTOP fiber. Although we found that PMMA is about twice as sensitive as CYTOP, the real mechanisms of the change of refractive index due to exposure of the fiber to UV light are not yet clear. It is interesting, however, to find that gratings written in CYTOP fibers are not only more thermally stable than those in PMMA fibers, but they are also more stable than the gratings in glass fibers. Again, the reason for this will not be clear until the actual mechanism of the photosensitivity effect in CYTOP fibers is revealed.

In any case, CYTOP fiber is an interesting and important development, not only because it is thermally stable, but because its attenuation is very low: the latest reported measured figure is 13 dB/km. More importantly, the theoretical loss figure of this fiber is as low as 0.3 dB/km, a figure similar to that of glass fiber.

The development of POF gratings is still in a preliminary stage. A fairly long list of important issues remain to be addressed, including physical property issues (such as long term thermal, optical, chemical, and mechanical stabilities, attenuation, and dispersion), material issues (photoinduced processes and underlying principles, photosensitivity enhancement, elasticity tailoring, synthesis techniques, etc.), and grating fabrication issues (optimal writing wavelength, optimal exposure intensity and time, and desirable fabrication scheme) as well as component techniques (practical tuning mechanism, packaging, connection, etc.). All these topics are challenging and important to increase our understanding and to enable the development of POF photonic components for practical applications.

ACKNOWLEDGMENTS

This research was supported by the Australian Research Council and the Australian Photonic Cooperative Research Centre. The authors acknowledge contributions to the work on photosensitivity and Bragg grating in polymer optical fibers from our colleagues T. Whitbread, Z. Xiong, B. Wu, H. Liu, H. Hewa-Gamage, and L. Luo. The work on the photosensitivity of CYTOP POFs was carried out in collaboration with Y. Koike of Keio University and Y. Watanabe of Asahi Glass Company.

REFERENCES

1. K. O. Hill, Y. Fuji, D. C. Johnson, and B. S. Kawasaki, *Appl. Phy. Lett.* 32, 647 (1978).
2. K. Enner, R. I. Laming, and M. N. Zervas, *J. Lightwave Technol.* 16, 807 (1998).
3. P. Leisching, H. Bock, A. Richter, D. Stoll, and G. Fischer, *Electron. Lett.* 35, 591 (1999).
4. A. Inoue, T. Iwashima, T. Enomoto, S. Ishikawa, and H. Kanamori, *IEICE Trans. Electron.* E81-C, 1209 (1998).
5. W. H. Loh, B. N. Samson, L. Dong, G. J. Cowle, and K. Hsu, *J. Lightwave Technol.* 16, 114 (1998).
6. M. Ibsen, M. K. Durkin, M. N. Zervas, A. B. Grudinin, R. I. Laming, *IEEE Photon. Tech. Lett.* 12, 498 (1998).
7. S. L. Tsao, J. Wu, and B. C. Yeh, *Proc. SPIE.* 3746, 438 (1999); Cavaleiro, F. M. Araujo, L. A. Ferreira, J. L. Santos, and F. Farahi, *IEEE Photo. Tech. Lett.*, 11, 1635 (1999).
8. C. S. Hong, C. Y. Ryu, B. Y. Koo, C. G. Kim, and S. H. Yun, *Proc. SPIE*, 3988, 371 (2000).
9. N. Takahashi, K. Tetsumura, K. Imamura, and S. Takahashi, *Proc. SPIE*, 3541, 18 (1999).
10. M. V. Dewernter, P. Ferdinand, E. Bocherens, R. Carbone, H. Berange, S. Bourasseau, M. Dupont, and D. Balageas, *J. Intelligent Mater. Syst. Struct.* 9, 785 (1999).
11. K. O. Hill, "Proceedings of the Conference on Lasers and Electro-Optics," Anaheim, CA, 1994, pp. 241–242.
12. B. Malo, F. Bilodeau, D. C. Johnson, and I. M. Skinner, *Proc. SPIE*, 1590, 83 (1993).

13. J. Albert, B. Malo, F. Bilodeau, D. C. Johnson, K. O. Hill, Y. Hibino, and M. Kawachi, *Opt. Lett.* 19, 387 (1994); K. Beckley and O. Herman, "Proc. of CLEO'97," Piscataway, NJ, 1997.
14. M. C. Cardakli, S. Lee, A. E. Willner, V. Grubsky, D. Starodubov, and J. Feinberg, *IEEE Photon Technol Lett.* 12, 552 (2000); N. Ghani, S. Dixit, and T. S. Wang, IEEE Commun. Mag. 38, 72 (2000).
15. M. D. Todd, C. C. Chang, G. A. Johnson, S. T. Vohra, J. W. Pate and R. L. Idriss. Proceedings of the International Modal Analysis Conference IMAC, 1999, vol. 2, pp. 1719–1725.
16. J. L. Cruz, A. Dieez, M. V. Andres, A. Segura, B. Ortega, L. Dong. *Electron. Lett.* 33, 235 (1994); S. Jin, R. P. Espindola, H. Mavoori, T. A. Strasser, and J. J. DeMaro. *Electron. Lett.* 34, 2158 (1998).
17. H. Okayama, Y. Ozeki, and T. Kunii. *Electron. Lett.* 33, 881 (1997).
18. S. Y. Kim, S. B. Lee, S. W. Won, S. S. Choi, and J. Jeong. *Electron. Lett.* 34, 104 (1998).
19. T. Eftimov, M. C. Farries, S. Huang, N. Duricic, D. Grobnic, B. Keyworth, and J. Obhi, "European Conference on Optical Communication (ECOC'98)" vol. 1, pp. 127–128. IEEE, Piscataway, NJ, 1998.
20. Y. Huang, K. Liu, W. Guo, and X. Ren, *J. Beijing Univ. Post Telecom.* 22, 11 (1999).
21. E. R. Lyons and H. P. Lee, *IEEE Photon. Technol. Lett.* 11, 1626 (1999).
22. L. E. Adams, H. Mavoori, S. Jin, and R. P. Espindola, "Technical Digest of OFC'99", 1999, Vol. 3.
23. P. L. Chu and G. D. Peng, *Proc. SPIE* 3470, 120 (1998).
24. G. D. Peng, Z. Xiong, and P. L. Chu, *Opt. Fibre Technol.* 242 (1999).
25. Z. Xiong, G. D. Peng, B. Wu, and P. L. Chu, *IEEE Photon. Technol. Lett.* 11, 352 (1999).
26. P. L. Chu, G. D. Peng, and Z. Xiong. "Proceedings of the 8th International POF Conference," Chiba, Japan, July 1999, pp. 242–245.
27. H. M. Schleinitz, "Proceedings of the International Wire and Cable Symposium," 1977, p. 352
28. T. Kaino, M. Fujiki, and K. Jinguji. *Rev. Electron. Commun. Lab.* 32, 478 (1984).
29. G. Giaretta, W. White, M. Wegmueller, and R. V. Yelamaty, "Technical Digest of OFC'99," 1999, Vol. 3.
30. Y. Koike and T. Ishigure, *IEICE Trans. Electron.* E82-B, 1287 (1999).
31. M. G. Kuzyk, U. C. Paek, and C. W. Dirk, *Appl. Phys. Lett.* 59(8), 902 (1991).
32. H. Kanbara, M. Asobe, K. Kubodera, and T. Kaino, *Appl. Phys. Lett.* 61, 2290 (1992).
33. Z. Xiong, G. D. Peng, and P. L. Chu, *Opt. Eng.* 39(3), 624 (2000).
34. A. Tagaya, S. Teramoto, T. Yamamoto, K. Fujii, E. Nihei, K. Sasaki, and Y. Koike, *IEEE J. Quantum Electron.* Vol. 31(12), 2215 (1995).
35. A. Tagaya, Y. Koike, E. Nihei, S. Teramoto, K. Fujii, T. Yamamoto, and K. Sasaki, *Appl. Opt.* 34, 988 (1995).
36. G. D. Peng, P. L. Chu, Z. Xiong, T. Whitbread and R. P. Chaplin, *Opt. Commun.* 129, 353 (1996).
37. G. D. Peng, P. L. Chu, Z. Xiong, T. Whitbread, and R. P. Chaplin, *IEEE J. Lightwave Technol.* 14, 2215 (1996).
38. D. J. Welker, J. Tostenrude, D. W. Garvey, B. K. Canfield, 2nd M. G. Kuzyk, *Opt. Lett.* 23, 1826 (1998).
39. A. Maslyukov, S. Sokolov, M. Kaivola, K. Nyholm, and S. Popov, *Appl. Opt.* 34, 1516 (1995).
40. A. Costela, F. Florido, I. Garcia-Moreno, R. Duchowicz, F. Amat-Guerri, J. M. Figuera, and R. Sastre, *Appl. Phys. B* 60, 383 (1995).
41. T. Kobayyashi, K. Kuriki, N. Imai, T. Tamura, K. Sasaki, Y. Koike, and Y. Okamoto, *Proc. SPIE* 3623, 206 (1999).
42. A. Yariv, "Optical Electronics in Modern Communications," Chap. 13. Oxford University Press, New York, (1997).
43. A. Ankiewicz, Z. Wang, and G. D. Peng, *Opt. Commun.* 156, 27 (1998).
44. R. Kashyap, Opt. Fibre Technol. 1, 17 (1994).
45. W. J. Tomlinson, I. P. Kaminow, E. A. Chanderross, R. L. Fork, and W. T. Silfvast, *Appl. Phys. Lett.* 16, 486 (1970).
46. I. P. Kaminov, H. P. Weber, and E. A. Chandeross, *Appl. Phys. Lett.* 18, 497 (1971).
47. G. Meltz, W. W. Morey, and W. H. Glenn, *Opt. Lett.* 14, 823 (1989).
48. K. O. Hill, B. Malo, F. Bilodeau, D. C. Johnson, and J. Albert, *Appl. Phys. Lett.* 62, 1035 (1993).
49. K. O. Hill and G. Melz, *J. Lightwave Technol.* 15, 1263 (1997).
50. J. Nishii, H. Yamanaka, H. Hosono, and H. Kawazoe, *Opt. Lett.* 21, 1360 (1996).
51. J. D. Prohaska, E. Snitze, and J. Winthrop, *Appl. Opt.* 31, 3896 (1994).
52. Z. S. Hegedus, *Appl. Opt.* 36, 247 (1997).
53. Z. Xiong, G. D. Peng, B. Wu, and P. L. Chu, *J. Lightwave Technol.* 17, 2361 (1999).
54. Z. Xiong, G. D. Peng, B. Wu, and P. L. Chu, "Proceedings of the 24th Australian Conference on Optical Fibre Technology," Sydney, July 1999, pp. 135–138.
55. G. D. Peng, P. L. Chu, L. Xia, and R. A. Chaplin, *J. IREE Australia* 15(3), 289 (1995).
56. H. Y. Liu, G. D. Peng, P. L. Chu, Y. Koike, and Y. Watanabe, "Proceedings of the 25th Australian Conference of Optical Fibre Technology," Canberra, 2000, pp. 48–50.
57. T. Erdogan, V. Mizrahi, P. J. Lemaire, and D. Monroe, *J. Appl. Phys.* 76, 73 (1994).
58. S. R. Baker, H. N. Rourke, V. Baker, and D. Goodchild, *IEEE J. Lightwave Technol.* 15, 1470 (1997).
59. M. J. Bowden, E. A. Chandross, and I. P. Kaminow, *Appl. Opt.* 13, 112 (1974).
60. N. Yoshihara, "Proceedings of the Polymer Optical Fiber Conference," September 1997, paper Tub-2.
61. N. Tanio and Y. Koike, *J. Polymer* 32, 43 (2000).
62. H. Y. Liu, G. D. Peng, P. L. Chu, Y. Koike and Y. Watanabe, Electronics Lett., 37, 277 (2001).

CHAPTER 5

Polymer-Based Integrated Optics Devices

Alfred Driessen, Arne Leinse, Sami Musa, René M. de Ridder, and Sulur
MESA+ Institute, Lightwave Devices Group, University of Twente, Faculties of Electrical Engineering and Applied Physics P.O. Box 217, 7500 AE Enschede, The Netherlands

CONTENTS

1. INTRODUCTION

From a system point of view, a book on polymer optical fibers demands a section where the discussion does not deal with the fiber itself, but with the modules connected by the fiber. It is, of course, not possible to review the complete field of micro- and integrated optics. Instead we concentrate on polymeric integrated optics devices that are specifically appropriate for use in connection with polymeric fibers or that enable a desired functionality thanks to the properties of a specific polymer.

Figure 1 gives a schematic example of a multiwavelength communication network—a star network—for distances that can be bridged by polymeric fibers, that is, in the range from

ISBN: 1-58883-012-8
Copyright © 2004 by American Scientific Publishers
All rights of reproduction in any form reserved.

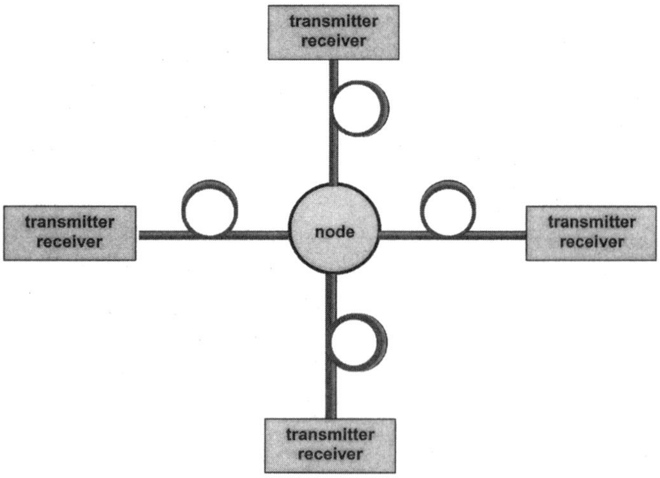

Figure 1. Schematic view of a multiwavelength communication star network.

approximately 1 m to 1 km. The continuously increasing demand for bandwidth probably is best solved by an optical network where the only electro-optic conversion steps are at the transmitter, where the optical signal is generated, and at the receiver, where it will be detected. The intermediate node(s) allows for routing and wavelength switching, all in the optical domain. A single transceiver, consisting of various transmitters and receivers (see Fig. 2) for each wavelength, already has quite complex modules with fiber–chip couplers, (de-)multiplexers, modulators, light sources, and light detectors.

In long-distance core networks, all components are commercially available, featuring presently up to 40 Gbit/s transmission speed for a single wavelength. The transmission medium employed is the silica-based single-mode fiber, because of its extremely low optical absorption. Economic issues are less relevant because the costs are shared by a very large number of users. The most important factor that determines the technology used to realize optical components is performance. High-speed modulators, for example, have been demonstrated in polymer technology that allow data transmission in excess of 100 GHz [1], a speed that cannot be obtained by alternative technologies. In addition, thermo-optic switching and tuning are most efficiently done in polymer technology because of the large thermo-optic coefficients of polymers.

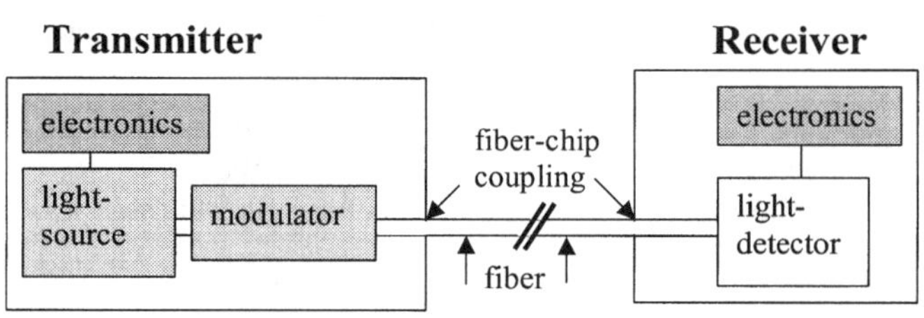

Figure 2. Schematic view of a communication link with a transmitter and a receiver.

In short-distance networks, local area networks (LANs), and the last kilometer to the computer—the access network, the situation is completely different. Polymer fibers or other multimode fibers are probably more adequate because of easier handling and connecting technology. Cost is now of primary importance because the number of customers making use of a transceiver can be as low as one. New technologies have to be employed that combine high reliability and high performance with user-friendly design and low cost. Polymer-based integrated optics devices could be a technological answer to these challenges, because they are based on potentially low-cost materials and can be produced by mass-production methods such as molding. The amazing success of the standard compact disk with submicrometer precision stamping illustrates the capacity of precision polymer technology. A massive introduction of optical technologies in short-distance communication is expected to take place in the coming decade that will result in optical modules becoming mass-produced consumer products. Much research has to be done in the fields of materials, new device principles, and improved mass-production technology to arrive at the desired products.

A very optimistic view on the potential of polymeric integrated optics devices prevailed some 10–15 years ago. Because the succeeding steps were probably more cumbersome than expected, enthusiasm dropped and research activities diminished. The last few years have seen a revival of industrial and academic research in the field of integrated optics, in general, and also in the field of polymer technology. Pushed by the rapid increase in bandwidth-consuming services on the Internet, fiber to the home is expected to become available in the second decade of this century. Before that, high performance, low-cost integrated optics modules for the mass market have to be developed. Based on this observation, we expect strong development of the technology in the coming years.

This chapter presents a review of polymer-based integrated optic devices with a strong bias on work carried out at the Lightwave Device Group of the MESA$^+$ Institute at the University of Twente. In our experience, only an integral approach can lead to innovative devices. Therefore, in this chapter we not only treat materials and technology aspects, but also discuss design options and system demands. Several excellent reviews are available, for example, Diemeer et al. [2], Eldada and Shacklette [3], and Bosshard et al. [4], but we hope to supply complementary information and insight on the subject that are useful also.

After a brief introduction, in Section 2 we summarize materials and technological aspects of polymeric devices. Emphasis is given to the realization of multimode polymeric waveguides based on commercially available products. In the next section we consider devices for optical telecommunication, in general, single-mode devices in accordance with the transmission medium—the standard single-mode fiber. In the fourth section we discuss devices for short-distance networks based on multimode fibers. As a consequence, the integrated optics modules have to be based on multimode planar waveguides, that is, a nearly completely different approach in comparison with the single-mode devices in the core network. The following section discusses active devices where polymers are the best choice because of their specific performance, for example, electro-optic switching. In the sixth section we give a summary and conclusions. The chapter ends with a list of references, which were not necessarily selected only by importance and merit, but also by availability and personal preferences.

2. POLYMER MATERIALS AND TECHNOLOGY

2.1. Introduction

Polymer waveguide technology has a great potential for economic mass production of complex planar photonic circuits that comply with the severe requirements imposed by applications in communication systems. The low-cost prospect arises from the availability of a wide range of cheap optical polymers and the simplicity of fabricating waveguides from them. A significant subset of optical polymer materials has shown excellent optical, chemical, and mechanical characteristics, that are very attractive for applications in long- and short-distance data communication. For instance, most of these polymers are transparent in the wavelength range from 400 to 2000 nm, and losses of polymer waveguides can be as low as 0.1 dB/cm in the three telecommunications windows around 850, 1300, and 1550 nm. The available optical polymer materials have a refractive index in the range from 1.3 to 1.7, with the possibility of tailoring and fine-tuning the index of some of these materials. Polymers also can be used for active devices for routing, switching, and even high-speed modulation of optical data, because all polymers have large thermo-optic coefficients and, in addition, a variety of polymers have shown good electro-optic properties. A large number of optical polymers have proven to be thermally, chemically, and mechanically stable, thus fulfilling the heavy requirements needed for operation under harsh environmental conditions.

The most appealing characteristic of polymer waveguide technology is the simplicity and flexibility of waveguide fabrication methods. Polymer thin films can be deposited in a wide thickness range by spin or dip coating using relatively simple equipment. A variety of channel waveguide fabrication methods exists, ranging from existing microtechnology techniques such as etching, to mass production methods developed especially for polymers, including molding and laser delineation.

Polymer waveguide technology for communications applications must satisfy severe requirements with regard to both the polymer materials as well as the waveguide fabrication methods. In this subsection, we, therefore, present an overview of recent advances in optical polymer materials and waveguide fabrication technology. We discuss properties that are relevant for telecommunications as well as short-distance applications.

2.2. Materials Properties of Optical Polymers

For a polymer to be selected as a material for waveguide technology, it has to satisfy the following set of requirements:

- Low optical losses (not more than 0.1 dB/cm) in the communication spectral windows around 800, 1300, or 1500 nm
- Low wavelength dispersion
- Low birefringence
- Low polarization-dependent losses
- Thermally stable optical and mechanical properties
- Resistant to humidity
- Good mechanical properties such as flexibility and toughness
- Low cost

In the literature a large number of polymers have been used to fabricate waveguides. In Table 1 of their review paper, Eldada and Shacklette [3] listed the key properties of optical polymers developed worldwide, with an emphasis on single-mode waveguides. At the University of Twente, Sulur [5] made a compilation of polymers for multimode waveguides; see Table 1[6–16]. The following discussion details the properties of optical polymers in accordance with the requirements just given.

2.2.1. Optical Losses

Optical losses in polymers are mainly caused by absorption and scattering. In general, optical polymers are transparent in the visible-wavelength region, and are strongly absorbing in ultraviolet (UV) and throughout the far infrared (IR) regions. The general absorption characteristics of optical polymers are represented in Figure 3 [17], which presents the transmittance spectra of several optical polymers. The spectra show that these polymers are totally opaque in the UV and the IR beyond $\lambda = 2100$ nm. There are additional absorption bands centered around the wavelengths 1150, 1350, and 1675 nm. The strong absorption in the UV region is caused by electronic transitions, whereas high absorption loss in the IR region is due to vibrational excitation of molecular bonds. The higher harmonics vibration of the C—H bond (C—H bond overtone) also causes high loss in polymers in the visible region. For example, in poly(methyl methacrylate) (PMMA), the sixth harmonic vibration of the C—H bond appears at a wavelength of 627 nm and contributes to the absorp- -tion loss of 0.427 dB/m [18]. Absorption bands caused by vibration transitions and their overtones can be shifted to the longer wavelength regions outside the telecommunications spectral windows by replacing the lighter atoms such as hydrogen in C—H bonds with heavier ones such as deuterium and halogens. The wavelength shift takes place because the transition energy is inversely proportional to the square root of the mass of the atoms in the bond; hence, by increasing the mass, the transition energy decreases. This method has

Table 1. Overview of optical polymers used for multimode waveguides

Core layer		Cladding layer		Insertion loss		
Polymer	Index	Polymer	Index	α (dB/cm)	λ (nm)	Ref.
PMMA	1.49	PFMA	1.39	1.4	680	6
		CYTOP	1.34	<0.1	675	7
		Silicon resin	1.41			
Mixture of urethane		Mixture of epoxy		0.7	1,300	8
diacrylate, tripropylene		dicrylate, propoxylated		0.5	1,000	
glycol diacrylate,		glycerol triacrylate,	1.49	0.5	633	
propoxylated	1.55	1-hydroxycyclohexyl				
glycerol triacrylate		phenyl ketone				
Photolime gel	1.52	Glass	1.50	0.5	632.8	9
polymer				1.1	632.8	
Norland 63	1.56	Electrolite 4481	1.48	0.4–0.6		10
Fluorinated polymide	1.535–1.551	Silicon dioxide	1.48	0.6	830	11
d-PMMA	1.49	UV-cured epoxy resin	controllable	0.02	830	12
Silicon resin	Controllable	Silicon resin	controllable	0.2	1,300	13
				0.5	1,550	
SAN	1.562–1.572	PMMA	1.49	0.2	1,300	14
				0.5	1,550	
d-PFMA	Controllable	UV-cured epoxy resin	controllable	0.02	680,830	15
				0.07	1,310	
Polyguide^{TM}	1.485–1.51	Polyguide^{TM}	1.485–1.51	0.08	800	16
				0.35	1,300	
				1.6	1,550	

been used successfully by many workers to reduce polymer absorption losses in the infrared region [3, 15, 19].

Contaminants trapped in a polymer film (such as water or metals) can cause extra absorption losses. Water absorbed in a polymer is responsible for increasing loss through the vibration of O—H bond [20]. Metal contaminants, such as cobalt, chromium, and magnesium, in a polymer cause absorption due to electronic transition within the metal ions.

Scattering losses can be categorized as extrinsic and intrinsic [21]. Intrinsic losses are due to Rayleigh scattering, which is a wavelength-dependent mechanism (inversely proportional to λ^4). Extrinsic losses are caused by irregularities in the films such as thickness and index inhomogeneity, side-wall roughness, and voids. Wall roughness and surface irregularities are the prominent causes of scattering loss. Their influence can be reduced by choosing an appropriate fabrication technique or applying roughness-smoothing methods [6].

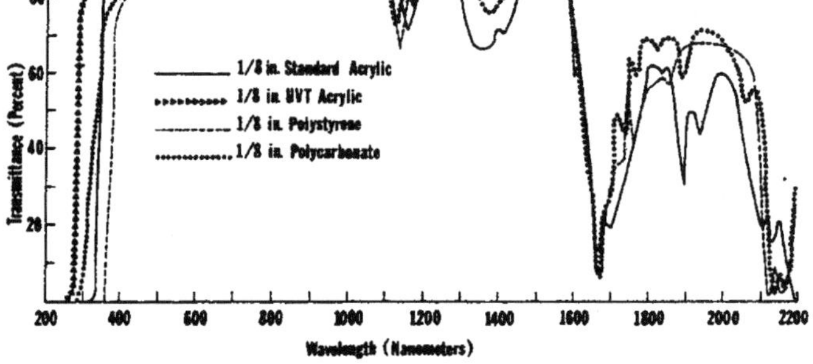

Figure 3. Spectral transmittance for some optical polymers. Reprinted with permission from R. R. Shannon and J. C. Wyant, "Applied Optics and Optical Engineering," Vol. II. Copyright © 1979, Academic Press, New York.

2.2.2. Refractive Index

Optical polymer materials can exist in a broad refractive-index range from $n = 1.3$ to 1.7. Note that no optical polymer material exists that has a refractive index as high as that offered by inorganic materials like Si ($n = 3.5$) or Si_3N_4 ($n = 2$) that could allow the design of ultra-compact photonic structures like photonic bandgap materials do. The refractive indices of a variety of polymer materials can be tailored and precisely controlled to suit a specific design purpose. This additional characteristic rarely is found in other waveguide technologies (with the exception of SiO_xN_y technology [22]). There are many different methods for tailoring the refractive index of a polymer. These methods have to be worked out carefully, because mixing solutions of polymers with different indices of refraction is often not possible or results in layers that have poor quality after spin coating. The following list gives several examples of commonly used methods.

- The refractive index of d-PFMA [deuterated poly (fluoromethacrylate)] can be controlled by changing the fluorine content [15]. An increase in the fluorine content causes a decrease in the refractive index of d-PFMA. In this way, the refractive index can be controlled in the range from 1.363 to 1.484 at a wavelength of 1300 nm with an accuracy of better than 0.001.

- The refractive index of thermosetting silicone resin is controlled by changing the composition ratio of phenyl and methyl substituents, resulting in a refractive index in the range of 1.395–1.543 at 1300 nm [13].
- The refractive index of fluorinated polyimide can be controlled precisely by varying the 6FDA/TFDB content [6FDA: 2,2-bis (3,4-dicarboxyphenyl) hexafluoropropane and TFDB: (2,2′-bis (trifluoromethyl)-4, 4′-diaminobiphenyl] [23]. The refractive index of fluorinated polyimide can be controlled easily between 1.520 and 1.620 for transverse electric (TE) light polarization [between 1.512 and 1.517 for transverse magnetic (TM) polarization] by changing the 6FDA/TFDB ratio.
- For acrylic polymers, the refractive index can be tailored by changing the ratio of high-index and low-index resins. An index range from 1.494 to 1.549 can be obtained with an accuracy of 0.002 [24]. The acrylic polymer is synthesized from high-index epoxy diacrylate and low-index aliphatic urethane diacrylate as the resins, propoxylated glycerol triacrylate as the diluent, and phenyl ketone as the photoinitiator.

2.2.3. Refractive Index Dispersion

Many communication applications require that the refractive index of the waveguide material be wavelength-independent, that is, $dn/d\lambda = 0$ or at least is negligibly small. Figure 4 compares the wavelength dependence of the refractive index of a number of polymer thin films to that of SiO_2, including polycarbonate, polystyrene, polyurethane, PMMA, UV15 (UV-curable resin), and MasterSil (UV15 and MasterSil are commercially available from Masterbond [25]). These polymer layers have been realized and characterized at the University of Twente by Sulur [5]. The wavelength dispersion in all these polymers is slightly higher than that of SiO_2, but much lower than the dispersion in semiconductors. Similar wavelength dependence of the refractive index also has been obtained in the acrylic optical polymers developed by Allied Signal [26].

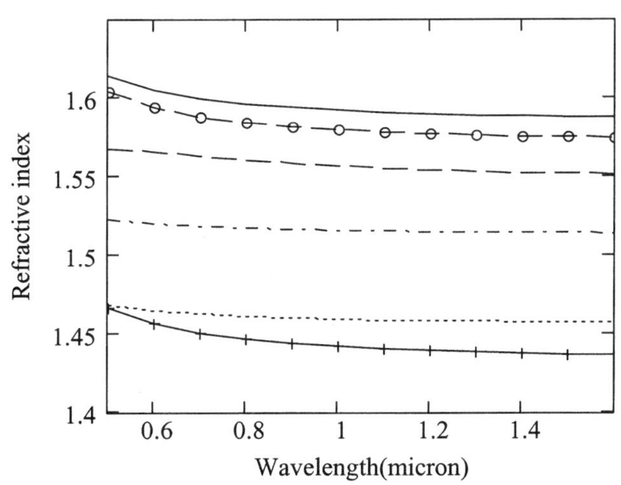

Figure 4. Refractive index as a function of wavelength for polycarbonate (solid line), polystyrene (dashed line with open circles), polyurethane (dash line), UV15 (dash-dot line), MasterSil (solid line with crosses), and SiO_2 (dot line).

2.2.4. Birefringence

Birefringence, which is the difference between the refractive indices of a material at two different polarizations, for example, TE and TM polarizations, is an undesirable effect, because it can cause polarization mode dispersion and lead to unacceptable power penalties for transmission. In polymer films, birefringence may notably increase during the film deposition process due to the orientation of polymer molecules. Several mechanisms, such as stress, the spin deposition process, and polymerization, have been identified as the main causes of birefringence [3, 13, 27]. Stress originates mainly from the difference in thermal expansion coefficient of the polymer film and that of the substrate [27]. In spin-coated films, birefringence is higher for heavy molecular weight polymers than for polymers of lighter molecular weight. For instance, the birefringence of silicone resin (less than 5×10^{-4}) is less than that of polysiloxane (8×10^{-3}) [13], which has a molecular weight that is 1 order of magnitude higher. The reason is that heavier molecules are more susceptible to orientation by spinning than lighter ones. Birefringence also could be caused by the orientation of the monomers during the polymerization process. However, cross-linked halogenated polymers have shown very low birefringence (10^{-6}) [3].

2.2.5. Electro-optic Properties

Nonlinear optical polymers have interesting electro-optic properties. These effects are often described in terms of the nonlinear susceptibilities $\chi^{(2)}(-\omega, \omega, 0)$ and $\chi^{(3)}(-\omega, \omega, 0, 0)$, where ω is the angular frequency of the optical field. We concentrate here on the well-known ($\chi^{(2)}$-based) electro-optic Pockels effect, which describes a linear relationship between the refractive index and an externally applied electric field. This effect has been used extensively for modulators and switches in crystalline inorganic materials, especially lithium niobate. In polymers, the linear electro-optic effect is based on electro-optic molecules or groups that, for a macroscopic effect, need to be aligned in a process called poling. Although this adds some complexity compared to the crystalline materials, which usually do not need poling, it also adds design flexibility by allowing selective poling. An important advantage of suitable polymer materials is that their permittivity does not vary strongly over a very broad frequency range from zero up to visible light frequencies. This property facilitates the design of high-frequency devices where the propagation speed of the driving radio frequency (rf) or microwave signals should be matched to that of the optical wave. An overview of polymer electro-optic devices can be found in Steier et al. [28], Teng [29], de Ridder et al. [30], and Section 5.

An important issue to be dealt with in electro-optic devices is the birefringence of the electro-optic material that arises from the alignment of its molecules. In applications where the polarization of the input signal can be controlled, this is not a problem, but in other cases, for example, when input signals are randomly polarized with large fluctuations in time that are generated by passing through a long length of standard telecom fiber, it is quite a challenge to obtain the desired polarization-independent behavior.

A special type of material that also must be mentioned here is electro-optic organic crystals (for an overview, see Bosshard et al. [4]). These materials do not need an additional poling treatment and have very high electro-optic coefficients (more than 100 pm/V). Examples of these materials are DAST and Mero-2 [31, 32]. The rest of this chapter deals with polymers only.

2.2.6. Thermo-optic Properties

Materials that have a large thermo-optic coefficient (*dn/dT*) have a great potential for use in active device applications like switches and tunable filters. Polymers have a relatively high thermo-optic coefficient (1 order of magnitude larger than that of silica, but with opposite sign) [2, 33]. The thermo-optic coefficient of polymers is negative, which means that the refractive index of the polymer decreases as the temperature increases. The thermo-optical coefficients of some polymers and, for comparison, SiO_2 are shown in Table 2 [2, 33–35]. The use of waveguides that employ polymers and silica layers simultaneously can lead to the design of athermal waveguides, which allow critical functions, for example, wavelength multiplexing, without the need for active thermal stabilization of the device [35].

2.2.7. Thermal Stability and Other Aging Properties

Optical and mechanical properties of polymers may change as a function of temperature. For instance, when heated to high temperatures, optical polymers may undergo undesirable transformation, including chain scissions, development of strength-destroying types of cross linking, and formation of low molecular weight products. Also, upon thermal aging, optical polymers may develop strong absorption bands in the UV region that may tail off to the visible part of the spectrum. Therefore, it is important to have optical polymers that have thermally stable optical and mechanical properties, so they are compatible with high process temperatures and can stand operation in harsh environmental conditions. The thermal stability of polymers is determined to a great extent by the polymer structure. A variety of methods can be used to improve the thermal stability of polymers, such as introduction of polar and bulky substituents into the polymer, use of catalysts that can induce polymerization of thermally resistant polymers, and utilization of the cross-linking property of some polymers. Optical polymers that are thermally stable up to very high temperatures have been reported in the literature. Thermal gravimetric measurements performed on a number of optical polymers, including polyimides, deuterated polysiloxane, and nonhalogenated acrylic polymers, have shown that these polymers are thermally stable up to 400 °C [3, 36]. Optical loss measurements on fluorinated polyimides, deuterated polysiloxane, and acrylic polymers have shown that the losses remain almost unchanged up to temperatures of 360, 200, and 290 °C, respectively [36]. Optical loss measurements carried out as function of time and temperature on cross-linked acrylic optical polymers developed by Allied Signal have predicted that the loss increase will remain within 0.1 dB/cm if these polymers where heated at 100 °C for 65 years [3].

For the long-term stability of optical polymers, consideration also should be given to the role of oxygen in combination with high optical intensities like those found in optical waveguides. Even if the wavelength of the light is such that no direct single photon absorption is possible, there can be long-term degradation effects, especially when oxygen is present. Detailed experiments on conjugated oligomers films by Lüer et al. [37] showed that there is a correlation between oxygen, irradiation of light, and mobile charge carrier formation.

Table 2. Thermo-optic coefficient of selected polymers in comparison with SiO_2.[a]

Material	SiO_2	PMMA	PS	PC	SAN
$(dn/dT) \times 10^{-4}$ C^{-1}	0.101	−1.05	−1.4	−1.07	−1.1

[a]PS, polystyrene; PC, polycarbonate; SAN, styrene acrylonitrile.

2.2.8. Humidity Resistance

Water absorption can lead to a drastic degradation of the optical quality of a polymer film. If a polymer film incorporates some water molecules, optical losses in the infrared will increase due to O—H bond vibration transitions. The amount of water absorbed in a film is directly related to the voids in the film. The humidity resistance of a certain polymer film can be tested by measuring its optical losses while the film is being cycled between very dry and humid conditions. Various polymers, including polysiloxane, polyimides, and Allied Signal acrylic polymers, have shown excellent humidity resistance. No significant increase in optical losses is found in waveguides made of these polymers that were subjected to humidity resistance tests for time periods ranging from 200 to 1000 [3, 27]. In general, highly cross-linked polymers are resistant to humidity, because they are tightly bound in networks and hence have a low level of voids.

2.2.9. Mechanical Stability

In certain communication applications, polymer waveguides will be subjected to various kinds of external forces that may cause them to deform or break. In such cases, the polymer materials should have a sufficient degree of toughness and flexibility. The mechanical properties of polymers are determined largely by their structure and their molecular weight. A number of research groups in the world have succeeded in developing optical polymers that have high levels of toughness of flexibility [2, 3].

2.3. Fabrication of Polymer Optical Waveguides

In the foregoing text, the materials properties of optical polymers were discussed. The next step is the realization of optical waveguides based on these polymers, which includes the deposition of polymer films and their structuring. So as to end up with the desired waveguide pattern. Fabrication methods used to produce polymer waveguides have to comply with the following set of general demands:

- Optical losses introduced by fabrication, such as that caused by scattering on rough side walls, have to be kept to a minimum.
- The method has to be simple and reproducible.
- The fabrication process should be suitable for low-cost mass production.
- High precision, sometimes well below 1 μm, is required.

There are at least three fundamental categories of polymer channel waveguide fabrication techniques, namely, (1) optical lithography combined with etching, (2) molding, and (3) photolithographic delineation. Important aspects of these three approaches are presented next.

2.3.1. Optical Lithography Combined with Etching

The most common approach to realizing optical waveguides is deposition of a layer stack that has the desired optical properties, and structuring it by optical lithography and subsequent etching. The advantage of this approach is versatility, because all kinds of deposition methods and etching processes can be applied. The number of processing steps, in most cases more than 20 and some of them very critical, makes this method relatively expensive and less suitable for low-cost mass production.

The deposition of a polymer layer stack is far from trivial. The first problem is adhesion of the undercladding (see Fig. 5) to the substrate and adhesion between subsequent polymer layers. Then the deposition process, in many cases

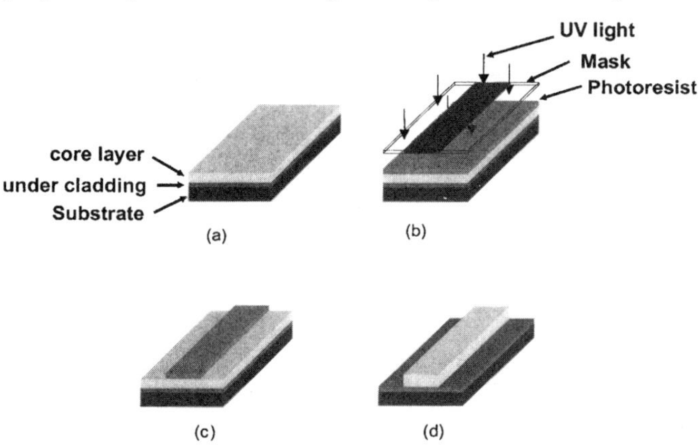

Figure 5. Fabrication of a polymer waveguide using lithography and etching: (a) deposition of the waveguiding layer stack with undercladding and core layer; (b) photolithography process of the resist layer; (c) formation of the waveguide pattern by photolithography; (d) realization of the channel waveguide by etching.

spin coating, should result in homogeneous, crack-free layers that fully cover the entire chip area. Between subsequent deposition stages, the layers should be cured to avoid partial dissolution of existing layers by the freshly deposited layer. Sulur [5] extensively studied deposition properties of commercially available polymers for thick multimode waveguides. His results are presented in Table 3.

The structure of the waveguide core can be defined by removing unwanted parts of the core layer using an etching process. In such a process, first a core layer is deposited by spin or dip coating on a lower index cladding layer. After a photolithography process using a photoresist, a waveguide pattern is formed. The channel waveguide is then realized by etching; see Figure 5.

Dry or wet etching can be used to structure polymer waveguides. Among the dry etching methods, reactive ion etching (RIE) is the most efficient and widely used, because it has good selectivity, little undercut, and high productivity. During the RIE process, a plasma discharge produces ions that directly or indirectly enhance the etch rates of selected material. Generally lower pressures and higher ion energies are required compared to plasma etching. The pressures used in RIE are 1–100 mtorr [38] and the frequency of the rf is 2.45 or 13.56 MHz. The etch rate, the selectivity ratio between etching of the polymer layer and the mask, and the precision of the microfabrication are controlled by the gas pressure and gas flow rate during the etching process.

Pure oxygen plasma is most commonly used to etch polymers [6, 7, 39]. Certain gases can be added to oxygen in the gas feed to enhance the concentration of highly reactive atomic oxygen free radicals that increase the etching rate. One common additive is tetrafluoromethane (CF_4). In addition to enhancing the oxygen atom concentration, CF_4 is a source of atomic fluorine that participates in surface chemical reactions

RIE has an inherent problem, namely, it creates side-wall roughness and changes the cross-sectional waveguide shape. The polymer debris that is formed by the etching process may be redeposited on the side walls where the ion bombardment is much weaker, thus causing roughness. The roughness can be minimized by blowing with an inert gas, by heat treatment, or by dipping in a solvent [7]. Blowing with N_2 physically removes polymer films and particles that are attached to the side walls. Heat treatment is done at a temperature above or near the glass transition temperature to relax the residual stress of the core and to flatten the core surface. Dipping for a short period in a polymer solvent after etching results in removal of small particles and smoothes the surface. RIE has been used successfully to fabricate both single and multimode polymer waveguides that have very low losses; for an overview, see Table 1 and [3].

Because photoresist cannot withstand long etching times, it is necessary to use a two-stage masking process if thick (>10 μm) core layers have to be etched. In such a process, first a nonerodable masking material (e.g., Al, SiO_2, Ti, Ni, or Cr) is deposited on the core layer. Then this

Table 3. Adhesion and coverage properties of commercially available optical polymers for application in multimode waveguides (rows, upper layer; columns, lower layer).[a]

Polymer	Epoxy Resin	PMMA	PC	PU	PS	MasterSil 773	UV15
Epoxy resin	A+ C+	A+ C+				A− C−	
PMMA		A+ C+	A+ C+ Cracks	A+ C+ Cracks	A+ C+ Cracks	A− C−	A+ C+
PC		A+ C+ Cracks	A+ C+			A− C−	
PU		A+ C+ Cracks		A+ C+		A− C−	
PS		A+ C+ Cracks			A+ Air bubbles	A− C−	
Master Sil 773		A− C+		A− C+	A− C+ Cracks	A+ C+	
UV15		A+ C+					A+ C+

[a]PMMA, polymethyl methacrylate; PC, polycarbonate; PU, polyurethane; PS, polystyrene. Adhesion is measured by pulling-off the layer after scratching. Easy pulling-off means poor adhesion. Good coverage means the surface is completely covered by the new layer, A+, good adhesion; A−, poor adhesion; C+, good coverage; C−, poor coverage.

high resistance material is patterned using conventional photolithography and etching, and finally the core material is etched through the resulting nonerodable mask. The mask layer can be removed later after etching. For illustration, in Figure 6 a scanning electron microscope (SEM) picture is given of a highly multimode polyurethane waveguide after etching of the core layer (64 × 120 μm), Figure 7 shows the cross section of polymer waveguides that have cores (40 × 20 μm) made of UV15 [25] and PMMA cladding layers.

2.3.2. Molding Techniques

The term "molding" categorizes a number of low-cost replication techniques, including casting, compression molding, injection molding, soft embossing, hot embossing, and LIGA (Lithoraphie Galvanoformung und Abformung) [3, 9, 14, 40]. The choice of method is chiefly determined by the nature of the polymer involved, for example, the curing method required and initial raw form—powder, pellets, or sheet.

As an illustrative example of a molding technique, the fabrication steps involved in compression molding as given in [9] are described briefly. Compression molding, like all other molding techniques, requires a mold. In this example, the mold was created by fabricating the desired structure in a mechanically polished brass substrate using a diamond turning machine and replicating this structure in nickel using

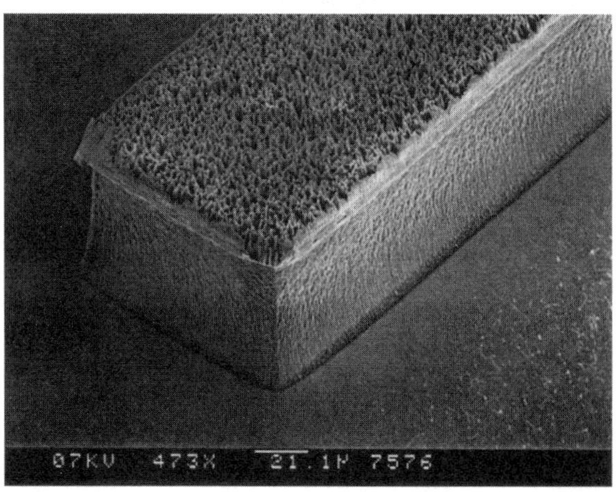

Figure 6. SEM picture [5] of a multimode polyurethane waveguide structured by RIE etching using a nonerodable Al mask. The waveguide dimensions are 64 × 120 μm

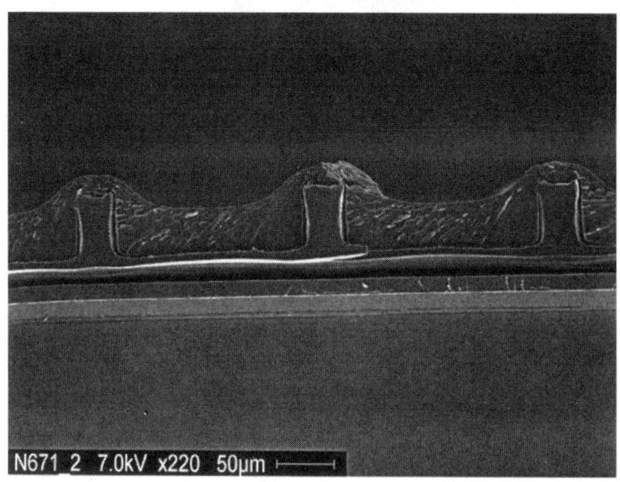

Figure 7. SEM picture of a cross section of multimode polymer waveguide structured by RIE etching using a nonerodable Al mask. The core (dimension 40 × 20 μm) is made of UV15 [25]; the upper and lower cladding is PMMA layers.

electroforming. The compression molding process starts by depositing a film of polymer, from which the structure is going to be made, onto a substrate. The thickness of the deposited film should be equal to the height of the intended structure. Then both the mold and the polymer film are heated to the polymer T_g temperature. After heating, the two parts are pressed together. The polymer film, which is softened by heat, takes exactly the shape of the mold. The final step involves hardening the created structure by heat treatment or UV curing.

Molding is the most suitable technique for mass production. A structure can be fabricated in large numbers using only one mold. Fabrication of optical waveguides with extremely low losses have been demonstrated using various molding techniques; see [9, 14]. However, the high accuracy that is required for telecommunication applications is difficult to obtain using molding processes. Therefore, molding techniques are better suited for fabrication of multimode waveguides used in short-distance data communication, where fabrication tolerances are more relaxed compared to those required in telecommunications.

2.3.3. Photolithographic Delineation

Polymer waveguides can be realized by externally induced dopant diffusion. There are two types, namely, photolocking and selective polymerization. In both, a monomer is doped with a certain photosensitive dopant. The doped monomer is spun onto a substrate and the layer is exposed under a mask for UV light. The dopants are locked into the host polymer during the photo-polymerization reaction. When the polymer material is heated after removal of the mask, the exposed regions are polymerized, whereas out-diffusion of the dopants can occur in the unexposed regions. In photolocking, the core regions are exposed and the dopants in the cladding are out-diffused, which lowers the refractive index, because the dopants have higher refractive indices than the host polymer. PMMA has been doped with benzophenone and coumarin to form waveguides by this technique. In selective polymerization, a lower refractive index material is out-diffused from the core region, leaving a high-index waveguide. This method offers flexibility in creation of stable core layers by tailoring the molecular properties of the host or the dopant. It provides smooth top surfaces and side walls of the core layer. PMMA, polystyrene, and polyimide have been used for waveguide applications by utilizing this method [16].

A related method is called photolithography delineation [14, 41] or contact printing lithography [3] because it is similar to negative-tone photoresist. A photosensitive polymerization initiator is added to a monomer to provide a means of photochemically initiated polymerization. The monomer is spun onto a substrate and exposure is performed using a mask. The exposed region is polymerized and the unexposed region remains monomeric. The pattern is developed by conventional wet etch of the unreacted material with a standard organic solvent such as methanol.

Internal polymerization-driven diffusion reactions in the absence of dopants is a method of polymer waveguide fabrication that utilizes diffusion of monomers during the polymerization process. The monomer layer is exposed by using a mask for a certain dose of light with a wavelength of 350–400 nm. During exposure, the light initializes polymerization in the waveguide region and monomers begin to diffuse to that region. Due to this mass transfer, this region has a higher refractive index. The refractive index of the guiding layer can be controlled by the total dose of light, diffusion time, diffusion temperature, and presence of oxygen. This process was used to produce Polyguide™ (DuPont) polymers [16].

Without using a mask, waveguides can be fabricated by laser writing. This technique utilizes a controllable focused UV laser beam and UV-curable polymers [29, 42] or cross-linked polymers [43]. The UV-curable polymer is a photo polymer that is deposited by spin coating and cured with UV light exposure. The cross-linked PMMA becomes highly soluble in the developer solution after exposure. A computer controls the movement of the focused UV light beam to pattern the waveguide structure. After exposure, the pattern is developed with a standard organic solvent such as acetone. In the case of UV-curable polymers, the unexposed region will be removed, leaving a channel waveguide pattern; for cross-linked polymers, the exposed region is soluble and can be dissolved by wet etching.

3. POLYMER-BASED WAVEGUIDE DEVICES FOR OPTICAL TELECOMMUNICATION

3.1. Introduction

Optical telecommunications has been the main application field driving the development of integrated optics devices. Devices of interest in this field include power splitters, switches, modulators, wavelength multiplexers/demultiplexers, optical cross-connects, dispersion compensators, gain equalizers, light sources, and optical amplifiers. Most of these functions can be realized in polymer material systems in principle, and several have been. In this section, the discussion is limited to passive devices (including thermo-optically actuated ones). Active devices that involve nonlinear electromagnetic material properties like the electro-optic effect are discussed in Section 5.

A very important requirement for telecommunications devices has been low insertion loss, although it is mitigated by the availability of optical amplifiers. One of the strong points of polymer technology is the ability to obtain a wide range of refractive-index contrasts and layer thicknesses, thus allowing the design of waveguides that have modal field distributions that are well matched to standard single-mode telecom fibers. In addition, the waveguide losses (absorption and scattering) can be quite low, so that overall insertion loss compares favourably to that obtained in most other materials systems. Due to the relatively large thermo-optical coefficient of polymers, it is possible to realize active devices that require low switching power.

In principle, polymer layer structures can be deposited on many different substrates of arbitrary size, although in many cases the lithographic fabrication steps needed to define the lateral waveguide structures, restrict the substrate choice to be compatible with standard microelectronics processing. There are, however, fabrication technologies available, like molding, that avoid critical and costly lithographic and etching processes, and hence offer the prospect of low-cost integrated-optic components.

In the remainder of this section first we give an overview of general design issues with regard to polymer-based monomode optical waveguide devices. Then we briefly describe the most important device classes for application in optical telecommunications and present an overview of realized devices with an emphasis on more recent results.

3.2. Design of Polymer-Based Devices

Refractive-index contrast is one of the most important design parameters for channel waveguides. Increasing the index contrast generally leads to increased losses due to surface scattering and less efficient fiber–chip coupling. In addition, the dimensions of single-mode waveguiding structures generally need to be reduced, which leads to more severe requirements on processing accuracy. On the other hand, increased contrast decreases the radiation losses at waveguide bends and, in most cases, also decreases the overall losses because the optical pathlength in compact high-contrast structures is greatly reduced. Some types of devices, like (MMIs) can be better realized in high-contrast systems. Also, small structures may be modulated with better power efficiency and higher speed than larger low-contrast structures. Hence, as always, the designer's task is to find a satisfactory trade-off between conflicting requirements.

3.2.1. Boundary Conditions: Refractive Index, Layer Thickness, and Lateral Dimensions

Freedom in the design of integrated optic circuits is conditioned by the available technology. A key parameter in designing optical waveguides is the refractive index n and, even more important, the attainable refractive index contrast. In polymer systems the range of available refractive index values is generally limited. The absolute minimum is $n = 1$ (for air), but for polymer cladding materials the value $n = 1.33$ [for poly(tetrafluoroethylene-co-hexafluoropropylene)] is a realistic minimum. The maximum known polymer index is $n = 1.7$ for poly(pentabromophenylmethacrylate) [44]. As explained in Section 2, materials that have different indices cannot be arbitrarily combined due to process and functional compatibility requirements. In some material systems, the available refractive indices may be limited to fixed values for core and cladding layers. In other systems, there is considerable flexibility, for example, in many polymer systems a continuum of refractive index values can be made available by mixing different components [13, 15, 24, 27, 45]. Although this design freedom offers an attractive flexibility, especially in process development, it implies the risk of poor reproducibility and standardization. In practice, single-mode planar polymer waveguide systems have been demonstrated with index contrasts in the range of 0.2% up to a few percent.

The next important design parameter is the layer thickness, which, in practice, does not impose severe restrictions because in most polymer systems it can be varied within a wide range from a few tens of nanometers to hundreds of micrometers.

The attainable lateral dimensions are determined mostly by the available lithographic processes. The limits are the same as for standard microelectronic processing (about 150 nm at the turn of the century). The minimum feature size that can be attained by photochemical shaping (photobleaching or cross linking) of the polymers themselves is ultimately limited to approximately half the processing wavelength for thin layers and about the layer thickness for thick layers. To realize the potential for fabricating low-cost devices, the minimum feature size should not be chosen too small; a value on the order of a micrometer is reasonable. Such sizes are also well attainable using low-cost mass-production techniques like casting, molding, or embossing.

3.2.2. Monomodal Behavior and Polarization Independence

The operation of almost all functional devices is based on interferometry. Also, optical fibers for telecommunications are generally monomode. To ensure correct operation of interferometric devices and to enable efficient coupling to monomode fibers, it is essential that the basic building block, the channel waveguide, is designed to be single mode. This is a very common requirement in integrated optics, for which standard solutions exist, leading to restrictions on waveguide dimensions that depend on a given refractive-index contrast. The task becomes somewhat less trivial if a large modal-field size (e.g., for efficient fiber coupling) must be realized in a high-index-contrast system. Solutions are based on having higher order modes radiate away into the slab region, which is either part of a ridge waveguide [46, 47] or a high-index cladding region of a rib waveguide [48].

Note that the term "monomode' is actually somewhat misleading, because in most waveguides at least two orthogonal modes, denoted (quasi) TE and (quasi) TM, are guided. This is also a desired behavior that enables accommodation of input signals that have arbitrary (and unpredictably time-varying) polarization that arises from standard optical fibers. To process both polarizations equally, which is an important requirement in optical communications, both the effective refractive index and the waveguide attenuation should be independent of polarization (low birefringence and low polarization-dependent loss, respectively). The birefringence of optical waveguides is caused by both the material properties and geometrical effects. If a polymer is completely amorphous, a zero material birefringence is expected. However, during deposition or polymerization, some ordering of the polymer molecules may be induced. Additionally, strain induced by the vastly different thermal expansion coefficients of polymer layers and substrates like glass or silicon can cause an appreciable amount of birefringence. On the other hand, the geometrical birefringence can be designed to be zero, or even to have such a value that (partially) compensates for the material birefringence [49]. Published birefringence values range from smaller than 10^{-6} to a few percent [3].

Several potential causes of polarization-dependent loss should be taken into account. The polymer material itself can show a polarization-dependent absorption due to deposition- or polarization-induced molecular ordering. The waveguide structure can be such that (quasi) TE and TM modal fields have differently extending evanescent field tails, which may be attenuated in the substrate or electrode structures. Moreover, TM modes couple in a fundamentally different way to metallic films than TE modes do. Also, even if a straight waveguide channel were completely polarization-independent, any waveguide bend would introduce some asymmetry, causing TE and TM modes to behave differently. Polarization-dependent scattering losses can be caused by the processes used to shape the waveguide core (e.g., etching), which may lead to "vertical" walls that have a different surface roughness than the "horizontal" interfaces. Finally, as a consequence of birefringence, switches and other active devices may have different operating characteristics for both polarizations, which again translate into polarization-dependent loss.

3.2.3. Trade-off: Fiber Coupling Efficiency versus Bend Radius

To incorporate them into a system, optical devices for telecommunications should be connected to optical fibers (generally by providing them with fiber "pigtails"). Efficient coupling to a fiber calls for the modal field in a waveguide to have approximately circular symmetry and a relatively large diameter of about 9.5 μm. These requirements lead many designers to choose a rib-type channel waveguides with low-index contrast of 0.2–0.4% and that have a relatively large 6×6 or 8×8 μm^2 square core cross section (e.g., [27, 50]). Although this approach can lead to very low insertion loss, it is certainly not favorable for obtaining compact devices because the low-index contrast does not allow for a small bending radius [51]. However, a higher index-contrast compact waveguide technology does not necessarily lead to strongly increased insertion loss. First, in a compact device, propagation distances are smaller, so the overall propagation loss also can be smaller. In addition, the coupling efficiency can be improved by mode-field adapters (also known as spot-size converters). At the fiber side this can be done, for example, by using a short section of "small-core", fiber, which has a larger index contrast and hence a more strongly confined modal field than standard fibers, that can be adiabatically spliced to a standard telecom fiber [52]. In the polymer device, the modal field can be locally expanded, for example, by tapering down the waveguide width [53].

An interesting approach to increasing the integration density is to use a vertical integration scheme by stacking several layers of waveguide patterns [54].

3.2.4. Tolerances and Reproducibility

In any fabrication process there is a limit to the accuracy with which devices can be made. For integrated optics devices the most important features are dimensions (thickness of layers, and width and depth of structures made in these layers) and refractive index. The position of structures with respect to other structures that are part of the same device also can be critical. Increasing fabrication accuracy usually implies disproportionately increased costs. Moreover, the material properties are not constant in time, for example, due to temperature changes or aging; this is especially important with polymers. For these reasons, designs should be made robust with respect to changes in material properties and deviations from the nominal dimensions. Devices based on interferometric interaction, like directional couplers or Mach–Zehnder interferometer switches, are very sensitive to small deviations, so they are not the first choice in polymer-based device design unless some trimming or feedback mechanism can be incorporated.

3.2.5. Losses

The requirements on optical losses in integrated optics devices for telecommunications are demanding. The increasing complexity of the required optical functions tends to offset the increased loss budget offered by the availability of optical amplifiers. For each device, a fiber-to-fiber insertion loss of at most a few decibels can be tolerated. In addition to the intrinsic functional loss (e.g., the splitting loss of a power splitter), several other loss mechanisms should be taken into account.

A first type of loss depends mainly on the choice of materials and processing methods [21, 43]. As discussed in Section 2.2.1, the waveguiding materials have strongly wavelength-dependent optical absorption. Scattering losses due to inhomogeneity of the waveguiding layers and surface roughness at the interfaces between materials (e.g., caused by etching the waveguiding channels) are strongly process-dependent. Generally the surface scattering effects are dominant; they vary approximately as the square of the ratio of roughness to wavelength and as the index contrast to the power 2.5 [6]. Waveguide designers cannot do much to reduce this kind of loss except choose a different technology. Then, losses can occur due to limited confinement of the optical field within the waveguiding channel: the interaction of the evanescent field tails with a substrate or with a nearby electrode structure may cause a strong attenuation. Here, of course, the designer should provide for sufficiently thick buffer layers. Finally, as discussed in Section 3.2.3, radiation losses occur in sharp bends and there are losses associated with the fiber–chip coupling. The latter is caused by the mismatch (in both size and symmetry) of the modal field distributions in the fiber and the planar polymer waveguide, and by alignment errors of the fibers with respect to the chip. Depending on design and alignment accuracy, this coupling loss may range from below 0.05 to over 0.1 dB per interface (e.g. [15, 55]).

3.2.6. *Thermo-optic Interaction*

The thermo-optic effect, which describes the temperature dependence of the refractive index, has been used to provide an actuating mechanism in many "passive" materials like silica and polymers. A good overview of the physics of the thermo-optic effect in polymers is given in [2]. In transparent polymers, the thermo-optic coefficient $\partial n/\partial T$ is about 10'–4" K^{-1} at room temperature (see Table 2), which is an order of magnitude larger than in most inorganic materials. Although this strong effect is favorable for actuating devices that use moderate power (well below 100 mW per single functional element), its drawback is that environmental temperature changes may affect device operation. Therefore, the design should aim at a mode of operation where the differential temperature is more relevant than the absolute temperature. In this way, overall thermal drift has little effect. Unfortunately this plan cannot always be realized (e.g., in interferometric wavelength filtering devices that use branches of different length), so the temperature should be well regulated in such cases. Thermal drift can be reduced by applying a compensating expansion effect of the substrate [3]. Since high temperature may affect the physical and chemical stability of polymer structures, designers should observe strict thermal limits, depending on the particular material (e.g., well below 100 °C for PMMA-like materials [15] or up to over 300 °C polyimides [27]).

Macroscopic thermal processes tend to be slow. However, in integrated optic polymer devices this low speed is mitigated by the relatively small volume involved and the fact that a substrate such as silicion or aluminum may act as a very good heat sink. The switching time of devices made on such substrates can be well below 1 ms, which is sufficient for may applications in optical telecom, such as protection switches and tunable filters. If an application demands much higher speed, the electro-optic effect (see Sect. 5) may be an option.

3.2.7. *Design Tools for Polymer-Based Single-Mode Devices*

All the standard design tools of integrated optics, ranging from quasianalytic tools such as effective index method and coupled-mode theory to numerical mode-solvers and beam-propagation methods (BPM) [56], can be useful for designing polymer devices. In addition, thermal and stress modeling tools, often based on finite-element methods, are needed, especially for thermo-optic devices. For the purpose of thoroughly optimizing a device design, these tools need to be integrated with a programmable layout editor, for example, the commercially available C2V software suite [57]. In addition to such general-purpose tools, more specific programs may be needed to model some of the fabrication processes, for example, those related to photopolymerization.

3.3. Devices

3.3.1. *Elementary Passive Waveguide Structures*

Polymer waveguide technology has been investigated since the 1970s [58]. For some of the early work, we refer to the review articles by Keil (PMMA at the Berlin Heinrich Hertz Institute) [59], Booth (acrylate-based Polyguide™ technology at DuPont) [16, 55], Diemeer (BeamBox™ technology at Akzo Nobel, now JDS Uniphase) [2, 60], and Eldada and Shacklette (acrylate at Allied Signal) [3]. Several kinds of polymer waveguides have been investigated at NTT: deuterated poly(fluoromethacrylate), which has low birefringence (5×10^{-6}) but also low thermal stability [15], deuterated polysiloxane[36] and fluorinated polyimide [27], both of which have high thermal stability but also high birefringence (on the order of 10^{-2}) and polarization-dependent loss (0.4 dB/cm), and deuterated silicone resin which offers a compromise because it has both reasonably low birefringence ($<5 \times 10^{-4}$) and acceptable thermal stability [13].

Optical channel waveguides can be formed in polymers using several methods, for example, photoinduced polymerization combined with wet chemical etching [24, 45], photobleaching [61], plasma etching [62, 63], direct electron-beam writing [64], mechanical replication techniques like embossing or molding [65], and etching trenches in silicon or silica that are then filled with polymer to yield inverted ridge waveguides by virtue of the planarizing properties of polymers [47]. An interesting property of many polymers is that after fabrication of waveguide structures, the refractive index can be changed locally with UV radiation, for example, for trimming to the exact splitting ratio of a directional coupler [65]. On the other hand, without proper encapsulation, this mechanism affects long-term stability.

Planar waveguide devices must be fiber pig-tailed to be of any practical use. Alignment of the fiber with respect to the waveguide is a critical issue. Active alignment, which involves measuring the insertion loss during the pig-tailing process, can position the fiber with sub micrometer accuracy and, hence, achieve negligible misalignment loss, but it is an expensive technique, that is not well suited for mass production. Passive alignment relies on accurately dimensioned and positioned fiber guiding grooves in the substrate of the optical circuit. To achieve at reasonable cost the high-precision positioning of the grooves with respect to the waveguides, a self-aligned process, which defines the positions of waveguides and grooves in a single step, is most desirable [66].

3.3.2. *Thermo-optic Switches*

The first reports on planar thermo-optic polymer devices appeared at the end of the 1980s [16, 67]. The main structures used to switch light in thermo-optic devices are Mach–Zehnder interferometers (MZI), switched directional couplers (SDC), and digital optical switches (DOS), as shown in Figure 8. Although MZI and SDC devices can be designed to operate at considerably lower switching power (up to a factor >10), the DOS structure is generally preferred due to its monotonous switching behavior. In a system context, one of the most important properties of a switch is its isolation: the attenuation of the power at an output port in the "off" state with respect to the power at an output port in the "on" state. A lack of isolation at the switch level may cause different kinds of cross talk at the system level. The isolation provided by thermo-optic polymer switches is not perfect. For MZI and SDC-type switches, even small errors in waveguide dimensions or refractive index (especially in the coupling regions) can significantly reduce the attainable isolation to values below 20 dB. Moreover, because their switching characteristics are not monotonous, the applied heater power should be accurately controlled. Even then, maximum isolation may be obtained only for a single polarization (due to waveguide birefringence) and a limited wavelength range (because they are interferometric devices, the couplers are wavelength-dependent devices). Conversely, a DOS has relatively liberal tolerances on birefringence, fabrication accuracy, and driving conditions. Its monotonous transfer function enables polarization and wavelength-independent switching by somewhat overdriving the switch.

Some typical DOS examples are found in [33, 47, 48, 60, 68]. An interesting variant is a DOS that has three output channels (Fig. 9) that can be operated as either a 1×3 switch [69] or a 1×2 switch with increased isolation [70].

A hybrid integration technique that combines the very low-loss silica waveguide technique with the strong thermo-optic effect in polymers employs a variant of the switched

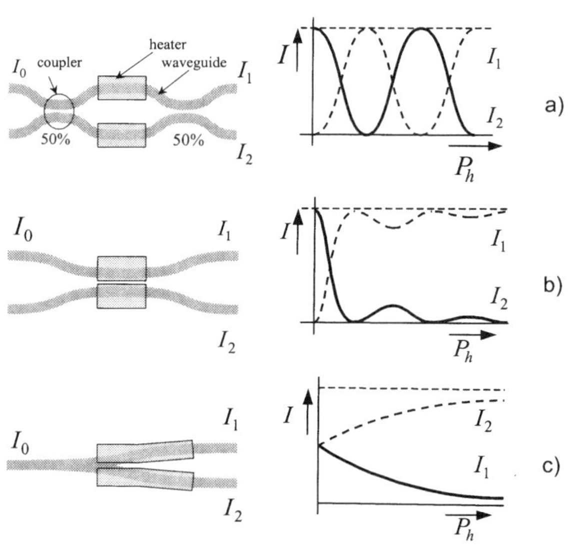

Figure 8. Important thermo-optic switching structures: (a) Mach-Zehnder interferometer; (b) switched directional coupler; and (c) digital optical switch. The left-hand column shows a schematic top view of the waveguide and heater structures; the right-hand column gives the corresponding idealized transfer functions of optical output powers I versus the heater power P_h (only one heater driven).

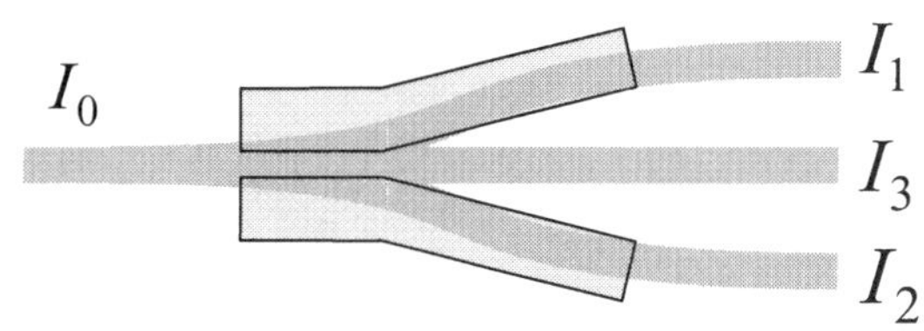

Figure 9. Schematic view of a 1×3 digital optical switch ("Φ" or "W" junction) a DOS with one input and three output ports. If the middle output guide is not used, a 1×2 switch functionality with increased isolation is obtained.

directional coupler, where coupling takes place vertically between a silica waveguide and a polymer waveguide stacked on top [71]. This approach might have the additional advantage of increasing integration density.

3.3.3. Switch Matrices

Elementary switches can be combined to form a fabric of higher functionality. The isolation provided by a single DOS can be larger than 25 dB. A much higher isolation can be obtained by cascading switches. On the other hand, switches can be combined to form an $n \times m$ matrix that connects n inputs to m possible output ports. The choice of the matrix architecture, like crossbar, dilated Benes, or recursive tree, in combination with the properties of the basic switching element determines the overall insertion loss, cross talk, power consumption, and size properties. For DOS-based polymer matrix switches, the recursive tree structure seems to be optimum. Using its BeamBox™ technology, Akzo Nobel together with the University of Twente realized an 8×8 matrix that had an insertion loss of about 10 dB and a typical isolation of 30 dB [72]. A group at the Technical University of Berlin demonstrated a 4×4 matrix that used oversized BCB (Benzo Cyclo Butene) polymer ridge waveguides [47], where the isolation was enhanced to better than 30 dB (42 dB average) by clever operation of otherwise unused switch elements and optimizing waveguide crossings [73].

For comparison, we mention the recently demonstrated 16×16 thermo-optic switch matrix realized in NTT's (Nippon Telephone & Telegraph) highly developed silica technology that features a 6.6 dB insertion loss and 53 dB isolation [74].

3.3.4. Wavelength Selective Devices

Wavelength-division multiplexing (WDM) is currently the most important technology in optical communications. Several polymer-based devices for WDM applications have been demonstrated, such as optical filters [75], demultiplexers [13, 50, 76–81], and add–drop (de)multiplexers [82–86]. Especially the thermo-optic properties of polymers are attractive here because they provide for a method of tuning or switching wavelength channels, or obtaining athermal behavior with no need for thermal stabilization [81].

We briefly describe an example of an optical cross-connect (OXC) that was realized in a slightly modified version of JDS Uniphase BeamBox™ technology at the University of Twente [86]. A schematic drawing of the OXC is shown in Figure 10a. It consists of two interconnected four-channel optical add–drop multiplexers (OADMs). The four wavelengths were chosen around a central wavelength of 1555.75 nm (192.7 THz) according to the ITU (International Telecommunication Union) standard, with 400 GHz (3.2 nm) channel spacing. Each OADM consists of the following components shown in Figure 10a: (1) one $16 \times 6\ mm^2$ 8×8 phased array that is used in a fold-back configuration [87] as a 1×4 demultiplexer plus a 4×1 multiplexer; (2) four add–drop switches S_i, each consisting of two cascaded 1×2 DOS-type switches. Direct cross talk from add (A_i) to drop (D_i) channels is minimized by using an add-after-drop configuration. The four output wavelengths λ_i of the demultiplexer can be individually switched (by switch d_i) either to individual drop waveguides D_i or to the input switch t_i of the corresponding channel of the multiplexer. Each such input switch t_i selects between the through signal and an add waveguide A_i.

To fit the device on a 100 mm Si substrate, the index contrast of the polymer system was increased to a value of 1%

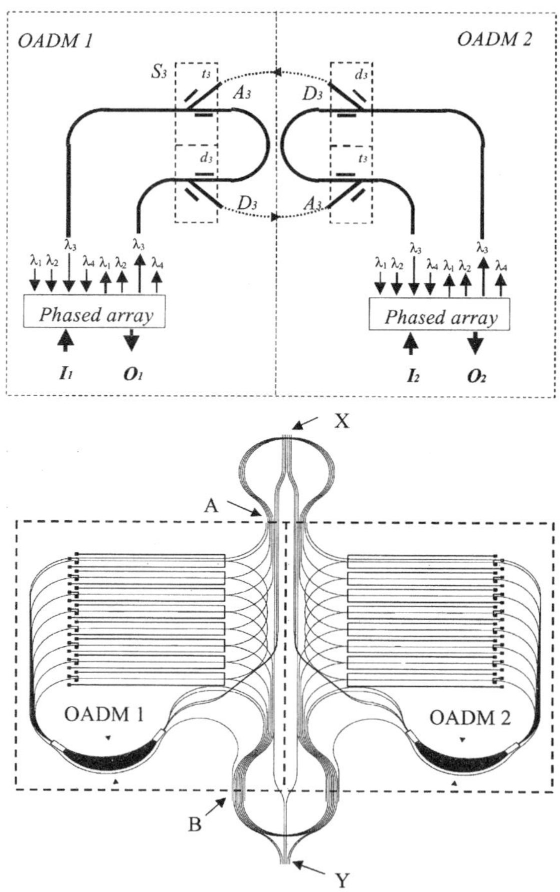

Figure 10. Optical cross-connect that consists of two optical add-drop multiplexers (OADMs). Each add-drop multiplexer consists of a single 8×8 phased-array demultiplexer which is used first to demultiplex four wavelengths and then, in fold-back configuration, to multiplex the same four wavelengths; (a) functional schematic; (b) layout. Dicing along the dotted lines separates the OADMs. Input and output channels of the OXC are available at positions X and Y. Input and add channels of OADM 1 are at A; output and drop channels are found at B.

[48]. Minimum bend radius is 6 mm and waveguide crossing angles are not smaller than 30°, corresponding to a measured maximum loss of 0.05 dB/90° and 0.1 dB per crossing, respectively. The mask layout, shown in Figure 10b, defines an OXC as two interconnected OADMs, which can be separated easily by dicing along the dotted lines. The small black triangles at both sides of the phasars are alignment marks for dicing a slot in which a half-wave plate is inserted to reduce polarization sensitivity [88]. The cross talk of the OADMs was below −29 dB and the polarization-induced wavelength shift was 0.03 nm.

3.4. Outlook

Good performance for many of the functions for modern optical telecommunications has been demonstrated with both passive and thermo-optically actuated polymer-based single-mode waveguide devices. Strong points of the technology are the large thermo-optic coefficient, which leads to low switching power requirements, and efficient fiber–chip coupling. The potential low cost of this technology, which was promised in the past, has not yet materialized, mainly because the cost of the present device generation is dominated by fiber–chip coupling and packaging. The cost advantage may become significant as soon as the complexity of the integrated optic circuits has grown sufficiently. Although, in principle, the technology allows for virtually unlimited substrate size and hence very large-scale integration, most demonstrated devices have been produced using conventional lithographic techniques, which constrain the circuit size to that of current wafer sizes used in micro electronics fabrication. In fact, this is even a disadvantage of current polymer technology because its relatively low-index contrast limits the attainable integration density compared to high-contrast semiconductor technology. In addition, other low(e.g., silica [74]) to medium-contrast (e.g., silicon oxynitride (SiON) [89, 90]) technologies present strong competition for polymer technology. A breakthrough in increasing circuit complexity in polymer waveguide technology may come via application of large-scale replication techniques like embossing or molding, although a chicken-and-egg like problem remains how to fabricate the required large-area master templates. Other promising approaches include hybrid integration with other waveguide technologies [91, 71], vertical integration using several waveguide-containing layer stacks [54], or even photonic crystal structures [92].

4. MULTIMODE INTEGRATED OPTICAL COMPONENTS FOR SHORT-DISTANCE COMMUNICATION

4.1. Introduction

Multimode fiber technology and, increasingly, also multimode planar waveguide devices are largely employed in short-distance communication applications such as LAN and interconnects. The revival of the use of multimode optical technologies in this area is driven and stimulated by the need for higher bit rates and the prospective low cost of these technologies. Multimode optical technologies have the advantage of being potentially low cost in comparison with single-mode solutions, but their major drawback stems from the limited bandwidth-length product which is due to modal dispersion. As mentioned in Section 1, cost is probably the most important issue in short-distance communications. Several factors contribute to the low-cost prospects of optical multimode technologies:

- The large core of the multimode fiber (typical diameter $\geq 50\ \mu$m) allows for relaxed alignment tolerances for coupling to an optical multimode chip or to other fiber segments. In this way, expensive alignment tools and procedures that are needed in the case of single-mode fiber technology can be avoided.
- In general, multimode integrated optical components require less fabrication precision as compared to single-mode components and, hence, can be fabricated cheaply (for example, by molding technologies).
- A wide range of cheap polymers that have excellent optical, mechanical, and chemical properties (see Section 2), and readily can be fabricated into waveguides are commercially available.
- Cheap light sources are available in the wavelength range of 600–900 nm, where silica as well as polymer multimode fibers show relatively low absorption. In addition, these light sources can be easily aligned to multimode fibers or planar waveguides.

As briefly indicated in Section 1, broad-band communication becomes really attractive only when the optical functionality is not restricted to point-to-point transport of the data, but also includes routing, switching, wavelength-division multiplexing, and other advanced data handling. Complex optical multimode networks require multimode optical components that are capable of performing these functions. Integrated optical devices appear to be most suited for these applications in comparison with multimode fiber and bulk optics because of their robustness and the possibility to achieve low cost and dense integration. As a consequence, considerable effort is being devoted to the development of multimode integrated optical devices in their most challenging issue of device design and fabrication technology. A review of these contributions and a highlight of the state-of-the-art in development of multimode integrated optics follows.

Polymer-based waveguide technology has emerged as the most suitable candidate for the fabrication of large-cross-section multimode integrated optical waveguides. Other alternatives, including plasma-enhanced chemical vapor deposition SiO_2 [93] and ion exchange in glass [94] were considered earlier, but polymer waveguide technology with the prospect of low-cost fabrication techniques appears to be the most cost effective. Much work has been done to optimize multimode waveguide fabrication technologies using promising polymers [3, 15, 21, 95]. Spin and dip coating have been used to deposit thick polymer layers. Large-cross-section channel waveguides have been fabricated using reactive ion etching (RIE), photopatterning, molding, and hot and soft embossing [3, 5, 15, 96]. An overview of optical polymers used in multimode technology, their properties, and the various methods used for the fabrication of multimode waveguides was given in Section 2.

Several multimode integrated optical elements have been successfully realized in polymer waveguide technology [97–99], including complete transceiver modules that comprise light sources, power splitters, (de)multiplexers, and detectors [100]. However, the performance of these components has not reached the maturity of their single-mode counterparts. The large number of optical modes in multimode components poses extra challenges for the design when compared to single-mode components, because key parameters, such as the phase, effective refractive index, and polarization, are no longer well defined.

In the following discussion, we give an overview of multimode integrated optical components. We focus on issues related to the design and device performance; materials aspects already were dealt with in Section 2. In Section 4.2, the challenges of modeling multimode integrated optical components are treated. Coupling light into multimode waveguides is presented in Section 4.3, followed by a discussion of multimode bent waveguides in Section 4.4. Thereafter, an overview of important multimode components that have been designed and fabricated so far is given. Section 4.5 deals with multimode power splitters and star couplers, and Section 4.6 concerns the important field of multimode wavelength-division (de)multiplexers.

4.2. Modeling Tools

Realistic modeling is important for device analysis, for understanding device performance, and, finally, for the design. Because of the large size and the consequent large number of modes in multimode structures, it is difficult to find suitable modeling tools. These tools, mostly developed for single-mode or free space structures, can be applied with only varying degrees of success in highly multimodal waveguides. An evaluation of these methods is given in the following paragraphs.

A direct method for modeling the propagation of light in a multimode structure is to evaluate the superposition of the guided modes in the structure. However, because the number of modes in these structures is very large, ranging from hundreds to thousands of modes, there is a numerical difficulty in accurately determining the fields and the propagation constants of all the modes. Moreover, tracing all these modes along the structure is cumbersome due to the rapid variation in the phase dependence of each mode. Therefore, this method cannot be used.

Beam-propagation method techniques [56], which completely disregard modal considerations, emerge as a sound alternative. Indeed, two-dimensional and even the more complex three-dimensional BPM have been successfully applied to model highly multimodal structures [93, 98, 101]. The disadvantage of these methods is that they need a fine calculation grid (5–10 points per unit of length: the wavelength) and small calculation steps to obtain accurate results. With the large waveguide cross sections (squares and rectangles with side lengths of tens of micrometers) and long structures ranging from millimeters to a few centimeters) these methods need large computer memory and become time consuming.

Ray optics has been historically used to model multimode waveguides. Initially, ray optics started with simple models to simulate the excitation and propagation of light in multimode waveguides. Later it was developed to incorporate more sophisticated wavelength-dependent effects such as absorption and bending loss. Ray tracing methods have been used to model all kinds of multimode components, including bent waveguides, Y junctions, star couplers, and grating-based wavelength-division multiplexers [97, 102]. Analytical models have been derived to calculate the output of multimode Y junctions, couplers, and bends. In fact, ray optics is the only practical method that is capable of calculating bend loss in multimode bent waveguides [103]. Ray optics has accuracy closer to that of the wave theory as long as the waveguide is highly multimodal [103]. Ray tracing techniques have the drawback that they cannot handle wave-related phenomena such as diffraction and interference.

4.3. Coupling Light into Multimode Planar Waveguides

Although coupling light into multimode planar waveguides is much simpler than in the case of the single mode, there are some considerations that have to be taken seriously into account so as to obtain optimum coupling of the light in the most cost effective way. Light can be coupled from a plastic optical fiber (core diameter typically 200–1000 μm), from a glass fiber (typical diameter 62.5 μm or 50 μm), or directly from a light source. Among the many light sources that can be used for short-distance applications, vertical cavity surface emitting lasers (VCSELs) are very attractive. Efficient coupling of light from a VCSEL into multimode waveguides using a 45° mirror machined into the edge of the waveguide has been demonstrated by different researchers [100, 104].

Coupling light from a multimode glass fiber into a multimode planar waveguide is somewhat simpler than in the case of coupling from a plastic fiber with a much larger diameter. Optimum coupling can be obtained when the overlap between the areas of the rectangular planar guide and the fiber core is maximized, and, simultaneously, the numerical aperture is set equal. In this case, the advantage of polymer waveguide technology is very clear, because waveguides as thick as 100 μm can be easily made. Coupling from a large diameter plastic optical fiber requires more effort. Of course, polymer waveguides with cross sections that are compatible with plastic optical fiber can be realized, and, indeed, waveguides with a cross section of $250 \times 200\ \mu\text{m}^2$ have been demonstrated [105]. However, large cross-sectional waveguides result in a very large device, which may exclude the possibility of dense integration.

As cost is a bigger issue than optical losses in data communication applications, less efficient coupling methods can be accepted to keep the device size sufficiently small. Accepting the loss of part of the intensity, light can be coupled from a plastic fiber to a multimode waveguide with much smaller cross section (e.g., $50 \times 50\ \mu\text{m}^2$) using a lens or a horizontally tapered waveguide. The principle of coupling from a very thick plastic optical fiber (200–1000 μm diameter) to a $50 \times 50\ \mu\text{m}^2$ rectangular waveguide using a lens is schematically shown in Figure 11. The lens is used to collect the light from the plastic fiber and focuses it on a small spot that dimensionally matches that of the rectangular waveguide. Maximum coupling efficiency in this system can be obtained if the numerical aperture of the planar waveguide NA_w is matched to that of the fiber NA_f according to

$$\text{NA}_w = \frac{d_f}{h_w}\text{NA}_f \qquad (1)$$

where d_f is the diameter of the fiber and h_w is the height of the planar waveguide. Equation (1) demonstrates that if small waveguide dimensions are required, the numerical aperture of the waveguide has to be very large. In this way, the lowest limit of the waveguide dimensions, at which maximum coupling from the fiber is still obtainable, is determined by the maximum available refractive-index contrast. The maximum refractive-index contrast offered by optical polymer materials that are extensively used to fabricate multimode waveguides is, in general, not sufficiently large to give efficient light coupling into a relatively compact waveguide (e.g., with dimensions $50 \times 50\ \mu\text{m}^2$).

4.4. Bent Multimode Waveguides

Bent waveguides are essential elements in planar optical multimode circuits. Therefore, the bending losses have to be known for a successful design. In addition to the losses that

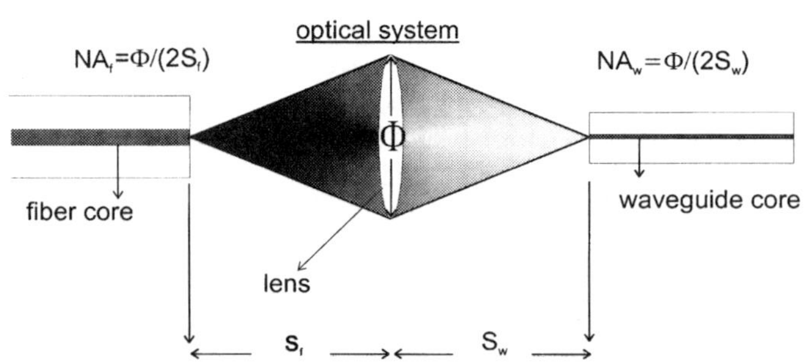

Figure 11. Schematic view of fiber-chip coupling from a plastic optical fiber to a multimode rectangular polymer waveguide using a lens. S_f and S_w are the object and the image length of the lens, respectively; The relationships between the NAs of the waveguide and the core, the diameter of the lens (Φ), and S_f and S_w are given for the case of optimum coupling efficiency.

occur in straight waveguides, bends suffer radiation losses at the outer curved interface, a kind of optical tunneling. In the ray picture this can be explained by the increased angle between the ray and the normal due to the curved outer interface. After a few reflections, this angle eventually can exceed the critical angle for total reflection, resulting in partial transmission. Modeling approaches used in the case of single-mode waveguides such as bend mode-solvers cannot be used for multimode bend structures. A geometrical optics approach is the most suitable alternative. Using ray optics concepts, an analytical expression for the power transmitted through a bent structure can be derived [106] as

$$T = \frac{P_{\text{out}}}{P_{\text{in}}} = 1 - \frac{1}{2}K - \frac{1}{6}K^2 - \frac{1}{8}K^3 - \frac{1}{8}K^4 + O(K^5) \tag{2}$$

where $K = \frac{\text{NA}^2 R}{w}$, w is the width of the channel, R is the bend radius, NA is the numerical aperture, and P_{in} and P_{out} denote input and output power, respectively. Equation (2) has shown excellent agreement with results of two- and three-dimensional ray tracing modeling performed on bent waveguides with different bend radii [106]. Therefore, this rather simple analytical expression can be used to predict losses in bent waveguides with good accuracy. A comprehensive theoretical study on losses in bent multimode waveguides can be found in [103].

4.5. Multimode Power Splitters and Couplers

Multimode power splitters and couplers are key components in optical short-distance networks. They distribute an incoming signal to N output ports and convert the simple point-to-point connection to a tree- or starlike network. Power splitters and couplers have been used in optical interconnection systems in the back plane [107], in on-board chip-to-chip connections [108–110], and in local area networks [99].

In recent years, there has been a considerable research effort directed toward the development of suitable design and fabrication methods for these power splitters [97–99, 101, 111]. Different designs of $1 \times N$ and $N \times N$ power splitters based on BPM and on ray tracing modeling have been realized in polymer waveguide technology.

An overview of $1 \times N$ and $N \times N$ couplers that have been produced so far follows.

4.5.1. 1 × N Power Splitters

4.5.1.1. Y Junctions Y junctions are the simplest form of multimode power splitters. They consist of an input waveguide that is split into two output branches. When $1 \times N$ splitters with output ports $N > 2$ are desired, the Y junctions can be cascaded. Two basic designs of Y junctions exist, namely, spatial and angular power splitters, as presented in Figure 12. In the spatial splitter, the output branches can be two S bends or two straight channels that subtend an angle α between

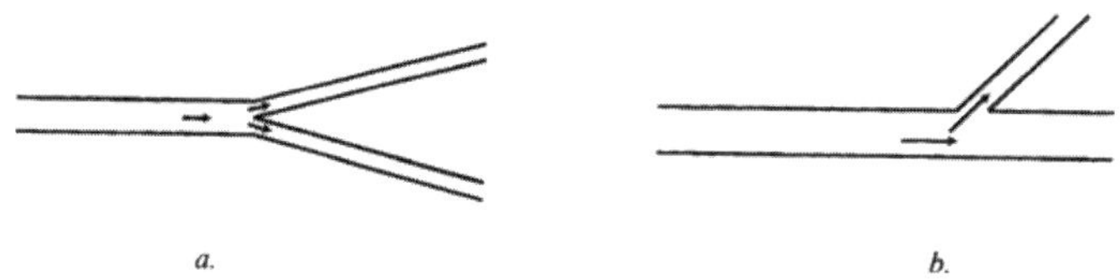

Figure 12. Schematic top view of a Y-junction-based splitter. (a) spatial Y junction; (b) angular Y junction.

them (see Fig. 13). The splitting ratio in a spatial splitter is equal to the ratio of the width of the output channels. When a spatial Y junction is used as a combiner, the widths of the output branches have to be at most equal to half the width of the input waveguide. If the widths of the output branches are larger than half the width of the input channel, a taper is needed to match the width of the output branches to that of the input waveguide. Introduction of a taper in a combiner structure causes losses and, hence, should be avoided. In the angular splitter, part of the solid-angle aperture of the input channel is coupled into the output branch (see Fig. 12b), and thus extreme splitting ratios can be obtained.

Results of extensive modeling using two-dimensional ray tracing and experimental measurements have shown that multimode Y junctions have excellent performance as 50:50 power splitters and combiners [111]. Note that these results were obtained with input fields that had uniform distributions. The performance of a multimode Y junction can be drastically hampered by nonuniform illumination of the input channel. This can be a critical practical problem and a serious drawback, for example, when light is coupled from a light source directly into the Y junction.

4.5.1.2. MMI-Based* 1 × *N Splitters The principle of these power splitters is based on the self-imaging effect, which takes place in multimode waveguides due to multimode interference. Understanding of the self-imaging phenomenon and the modeling of the devices based on it can be best obtained by mode propagation analysis. Simon and Ulrich [112] applied this method to study the self-imaging properties of waveguides that were highly multimodal in the lateral as well as in the transverse directions. They showed that the self-imaging conditions for highly multimodal waveguides are analogous to the self-imaging conditions in the widely studied case of couplers that are multimodal only in the lateral direction [113] if p^2 is set to be an even integer, where p is the aspect ratio of the coupler section, that is, the width divided by the height. The propagation distance, L_{MMI} at

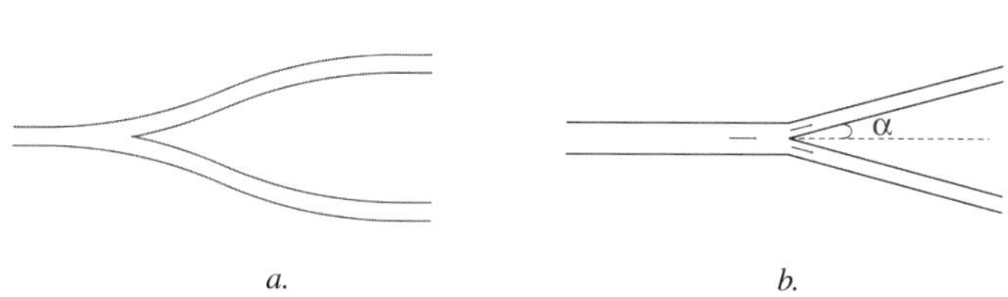

Figure 13. Two different designs for a spatial Y junction.

which multiple images of the input field can occur along the propagation direction can then be given by

$$L_{\mathrm{MMI}} = \frac{M n_c w^2}{N\lambda} \tag{3}$$

where M ($M = 1$ for the first image plane) and N are integers that have no common devisor, N is the number of the output ports, n_c is the refractive index of the core layer, and w is the width of the coupler section. As an example, for a 1×2 coupler ($N = 2$), Eq. (3) predicts a coupler length $L_{\mathrm{MMI}} = 8.88$ mm for a MMI section width of $w = 100\ \mu$m. This width was chosen to provide sufficient isolation between the two multimode output waveguides, which each have a large cross section of $40 \times 40\ \mu m^2$, to match the 50 μm diameter multimode fiber. The refractive index of the waveguide core was chosen as $n_c = 1.51$, as determined by the available material. The design wavelength was 0.85 μm, in the spectral region suitable for short-distance communications applications. The predicted device length is relatively large compared to that of a Y junction, and the possibility of further integration of the coupler with other components on a chip may be excluded, especially if the coupler is designed for a large number of output ports, resulting in a much larger width.

Musa et al. [99] showed that the device length can be substantially reduced when the MMI section is tapered. By introducing a taper, the coupler length L_{MMI} can be scaled down to

$$L_{\mathrm{MMI}} = \frac{M n_c w_0^2}{N\lambda\chi} \tag{4}$$

where χ is a geometric factor that depends on the initial width w_0 of the coupler section and the minimum width w_1 in the middle of the MMI section (see Fig. 14). Using Eq. (4), the 1×2 coupler length in the preceding example can be reduced to 2.84 mm, which is comparable to the length of the Y-junction-based splitter. Note that, according to Eqs. (3) and (4), when a power divider with a larger number of output ports is considered, the length increase introduced by the larger device width is reduced by a factor $1/N$. This is different for the case of Y-junction-based splitters, where, to obtain the larger number of output ports, the Y junctions have to be cascaded, resulting in a much longer device length.

This class of devices cannot be modeled using ray tracing because it is impossible to handle interference phenomena. Therefore, a three-dimensional BPM program was used [57]. Figure 14b presents a top view as well as cross-sectional slices that show the field distribution in the coupler as obtained by three-dimensional BPM simulations. The top view of the field distribution is similar to that obtained for multimode couplers that are single mode in the transverse direction [113]. BPM results have shown that these splitters exhibit very low excess loss and good power uniformity between the output channels, even for very short devices [99].

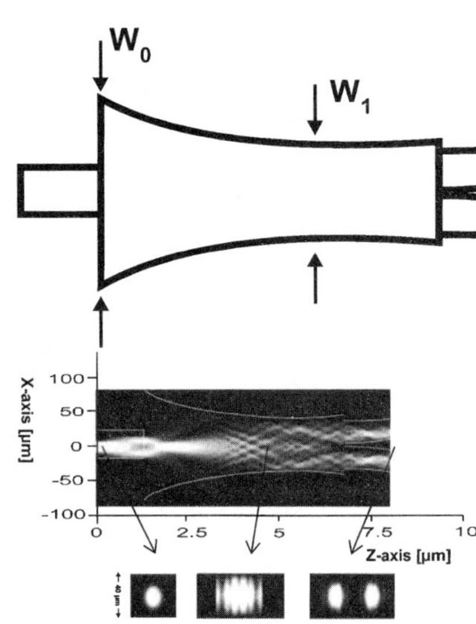

Figure 14. A 1×2 tappered splitter based on multimode interference for multimode input and output waveguides: (a) top view of basic layout; (b) top view of the field distribution obtained by a three-dimensional BPM. The calculated field at three cross sections (input port, MMI section, and output ports) also is given [99].

4.5.2. *N* x *N* Splitters

Multimode star couplers (i.e. $N \times N$ splitters) require a much more sophisticated design than $1 \times N$ multimode splitters do. They should have small size, uniform power distribution over the output guides for light launched from each of the input channels, and low excess loss. In the literature a number of designs that have acceptable performance can be found: symmetric and asymmetric tree star couplers [97], mixer rod star couplers [97], and MMI-based tapered couplers [101]. In the following paragraphs we present these different approaches and discuss their performance.

4.5.2.1. Symmetric and Asymmetric Tree Star Couplers

The input part of a 4×4 symmetrical star coupler is shown in Figure 15. The four inputs are combined in pairs by symmetrical Y junctions, which, in turn, are cascaded down to the central mixing rod. The output part is the mirror image of the input part. The design of the Y junctions is based on S bends of the type shown in Figure 13a. The asymmetrical star coupler design is basically similar to that of symmetric star couplers; the only difference is that the Y junctions are

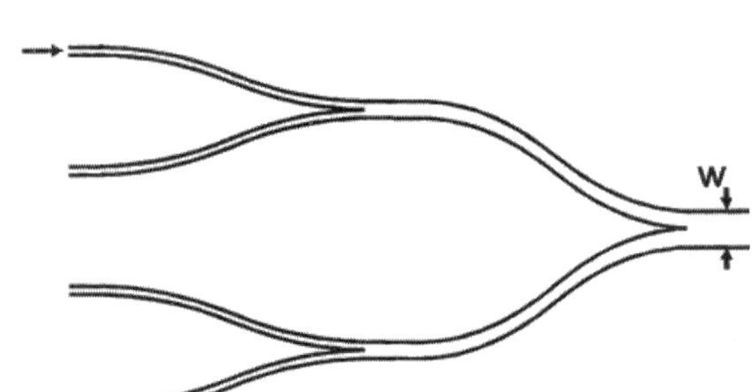

Figure 15. Basic layout of the input part of a 4×4 symmetric Y-junction star coupler for multimode planar waveguides.

asymmetrical. The use of asymmetric Y junctions reduces the number of S bends and, hence, makes the device shorter.

The design and fabrication of 8 × 8 symmetric and asymmetric tree star couplers were reported by Israel et al. [97]. In their design, each access waveguide had a cross section of 45 × 45 μm^2 and a NA between 0.2 and 0.3 to ensure an efficient coupling to the standard multimode fiber. The radii of curvature of the S bend were set to R = 16 and 21 mm, which resulted in device lengths of 38 and 42 mm for symmetric star couplers, and 23 and 30 mm for asymmetric star couplers, respectively. The design was done using ray tracing methods.

The 8 × 8 symmetric and asymmetric star couplers were fabricated using DuPont polymer waveguide technology [16]. Measurements on symmetric star couplers have shown low excess losses (average 0.16 dB) for $R = 25$ mm and small power imbalance of about 0.5 dB. For the asymmetric star coupler, the excess losses were determined to be below 1.5 dB and the power imbalance was ~1 dB. Symmetric couplers have better performance in terms of power imbalance and excess losses, but have the disadvantage of longer length than asymmetric star couplers. The measurement results are in good agreement with those of the modeling.

4.5.2.2. Rod Mixer Star Coupler A typical $N \times N$ rod mixer star coupler is shown in Figure 16. A fan of access guides is connected to a fan of output guides through a mixer region as shown in the figure. An 8 × 8 rod mixer star coupler has been designed and fabricated [97]. The device was modeled using ray tracing, which gave results that are in good agreement with the experimentally determined values. The obtained excess losses and imbalance, 2.57 and 1.42 dB, respectively, are high compared to the values obtained for symmetrical and asymmetrical 8 × 8 star couplers.

An alternative design for a rod mixer star coupler was proposed by Ilic et al. [98], in which the mixer region is truncated so that the distance between the outer input or output access guides the distance between the outer input or the inner ones (see Fig. 17). In this way, all access guides can enter or leave the mixer region at the same angle; in addition no reflecting areas between the individual access guides remain. As a result, power imbalance and insertion loss are improved when compared to a conventional rod mixer coupler.

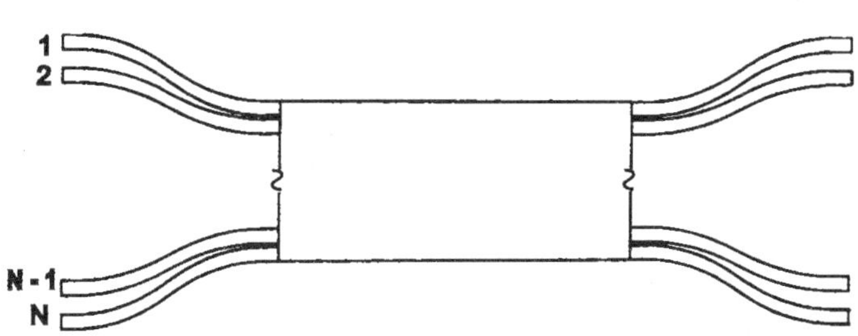

Figure 16. Layout of an $N \times N$ rod mixer star coupler for multimode planar waveguides.

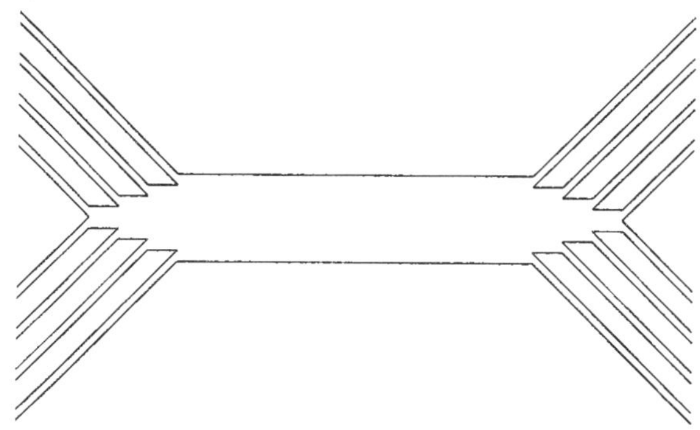

Figure 17. Basic layout of a truncated 8 × 8 star coupler for multimode waveguides. Reprinted with permission from I. Ilic et al. *IEICE Trans. Commun.* E-80B, 135 (1997). Copyright © 1997.

4.5.2.3. Multimode Interference-Based Star Couplers The multimode interference $N \times N$ star coupler utilizes the self-imaging effect and waveguide taper similar to the $1 \times N$ splitters in section 4.5.1.2. Figure 18 shows a basic layout of a 4 × 4 star coupler. It differs from the $N \times N$ rod mixer star coupler presented in the previous subsection in that the coupler section is tapered and all the dimensions, including thickness, width, and length, are chosen carefully in accordance with the design rules in section 4.5.1.2. so as to obtain self-imaging.

Musa et al. [101] designed a 4 × 4 star coupler based on this approach by using three-dimensional BPM calculations. The imaging properties of such devices are shown in Figure 19, which presents a top view of the intensity distribution. The intensity pattern is similar to that observed in other MMI devices that are single mode in the vertical direction [113]. The intensity is seen to be focused at the entrance of each of the four output ports. This property, a result of the self-imaging effect, improves the obtainable power imbalance and excess losses compared to other rod mixer multimode star couplers already discussed [97, 98].

The performance of the device can be expressed in terms of the excess loss, power imbalance, and length. Figure 20 shows the simulation results of the excess losses (open circles) and the power imbalance (dark circles) as a function

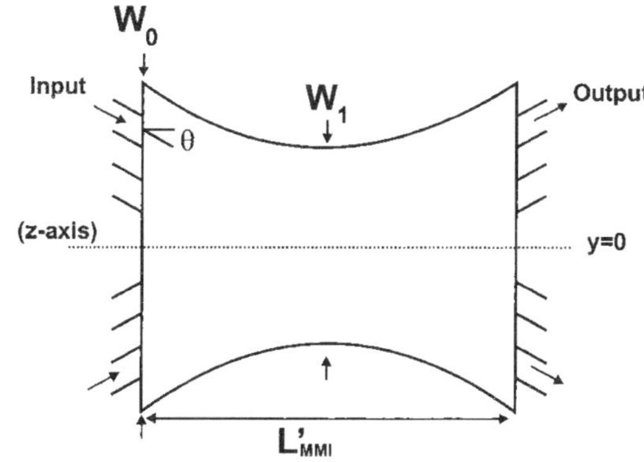

Figure 18. Top view of 4 × 4 multimode tapered star coupler for multimode waveguides based on the self-imaging effect.

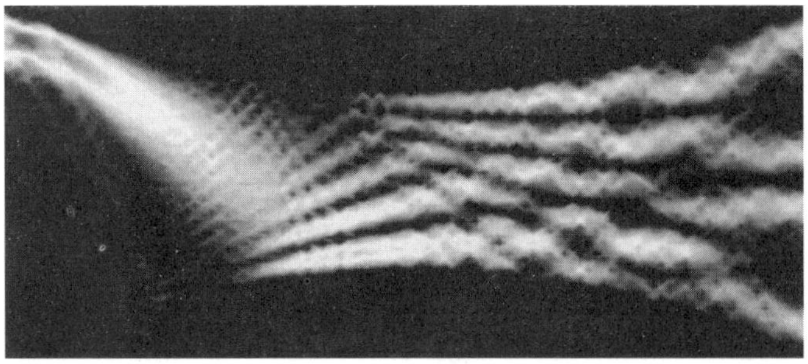

Figure 19. Top view of the intensity distribution of an 4×4 multimode interference tapered star coupler for multimode waveguides obtained by three-dimensional-BPM calculations [101].

of coupler length. Notice that the excess loss increases as the coupler length decreases. Excess loss of approximately 0.5 dB is obtained for a coupler length of 29.7 mm, which is about 50% shorter than the straight coupler. The increase in excess loss with the decrease in coupler length is caused by the loss of power in the higher order modes that become leaky as the coupler width narrows. The lowest power imbalance, approximately 0.3 dB, is obtained for coupler lengths that satisfy the self-imaging condition given in Section 4.5.1.2 (see circled data points in the figure). This observation confirms the validity of the mode propagation analysis given in the previous section. Further reduction of the device length results, however, in increasingly larger values for excess loss and imbalance.

4.6. Multimode Wavelength-Division (De)Multiplexers

Multimode optical fiber is predominant in many building and campus backbone links in LANs. Although multimode fiber has the advantage of relaxed alignment tolerances compared to single-mode fiber, its bandwidth is much narrower. The limit in the bandwidth of the multimode fiber is set by the modal dispersion. A 62.5 μm diameter multimode fiber, for example, has a bandwidth length product of about

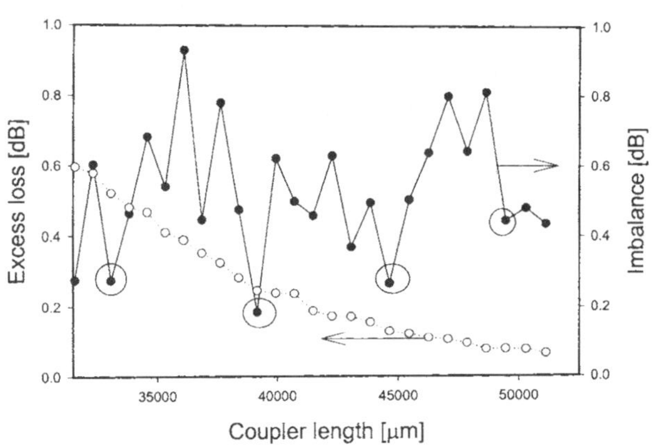

Figure 20. Calculated excess loss and power imbalance as a function of coupler length for a 4×4 tapered star coupler obtained by three-dimensional-BPM calculations [101].

160–200 MHz × km at a wavelength of 850 nm, where low-cost VCSELs are commercially available. With the advent of bandwidth-demanding applications in LANs such as the 10 Gbit/s ethernet, the limited bandwidth of multimode fiber becomes a serious problem.

Several solutions that have been proposed to improve the bandwidth–distance product of multimode fiber include the following approaches:

- Using a wavelength window around 1300 nm, because it is known that around this wavelength, the bit rate–distance product of multimode fibers is three times higher than at $\lambda = 850$ nm. However, lasers that operate at 1300 nm have much higher costs than VCSELs at 850 nm. The feasibility of this alternative will increase when VCSELs that operate around 1300 nm (which are still in the research and development stage) become available comercially.
- A sound and feasible alternative is to use graded-index instead of step-index fibers. The modal dispersion in graded-index fiber is much lower than in step-index counterparts. For instance, graded-index plastic optical fiber has shown a bit rate–distance product of 1.1 GHz × km at 1300 nm [114].
- Increasing the bandwidth of the multimode fiber by using restricted launching conditions, that is, by directing a narrow laser beam to a specific position in the facet of a multimode fiber, the bandwidth has been shown to increase substantially [115]. The transmission of an 11 Gbit/s signal over 1 km of multimode fiber has been demonstrated [115]. The disadvantage of this method is that it is very sensitive to misalignment which thus undermines the advantage of using multimode fiber.
- Adopting a multiple logic scheme to transmit the signal instead of conventional binary logic. With multiple logic, higher bit rate can be squeezed out of the signal. However, this scheme requires a very high signal-to-noise ratio at the receiver side.
- By using WDM schemes, several different wavelengths, each containing a separate signal, can be sent simultaneously through a fiber. In this way, the fiber bandwidth is multiplied by the number of wavelength channels. WDM is the most attractive scheme for expanding the bandwidth of multimode fiber. Moreover, it can even be used in combination with other methods.

Despite the promising characteristics of WDM schemes, multimode fiber-compatible WDM devices pose serious challenges to the designer because of the competition between spectral and modal dispersion. The difficulty increases further when the low-cost and compact size requirements for short-distance applications are imposed. A number of multimode wavelength division (de)multiplexers have been suggested and demonstrated using mainly dielectric filters or gratings. Here, we present on overview of this activity with special attention to the most recent developments.

4.6.1. Grating-Based WDM (De)Multiplexers

In the past, a number of grating-based multimode (de)multiplexers were developed. These devices were fabricated in a variety of planar waveguide technologies, including ion-exchange, silicon, and glass. Different grating types have been employed, including straight gratings with planar focusing lens, Rowland-type gratings [116], and chirped gratings [117]. Liu and Chen [118] demonstrated an interesting micro-optical component that functions as both a wavelength-division (de)multiplexer and a $1 \times N$ power splitter. In this device, a set of stacked holographic gratings is used to both couple light vertically from a multimode fiber into a glass slab, and to (de)multiplex it. Another set of gratings is used to couple the signals vertically out again to multimode fibers. A number of methods, such as holography, lithography, mechanical ruling, and chemical etching [118, 119], have been used to realize these gratings. Most devices of this sort have shown good characteristics such as low cross talk and insertion loss, robustness, and compact size. However, the precise, submicrometer technologies needed to fabricate these gratings again make cost effectiveness an issue.

4.6.2. SpectraLAN Project

SpectraLAN is a Hewlett–Packard project to produce multimode transceiver modules for LANs. It represents the most serious work done so far to realize low-cost multimode (de)multiplexers. A complete multiplexing transceiver module, including multimode multiplexing and demultiplexing devices, has been produced using polymer waveguide technology.

The multiplexer is a 4×1 Y-junction-based combiner that couples light from four different VCSELs into the output waveguide by using a 45° mirror cut into the end of the waveguide. The output channel is connected to a multimode fiber. To reduce the insertion loss, the output waveguides were made wider than the input channels. The obtained insertion loss for this combiner is 2.1 dB, which is much lower than in single-mode Y-junction-based combiners (6 dB) [100]. However, the insertion loss in the multimode Y-junction-based combiners is dependent on the excitation conditions.

The demultiplexer consists of a 1×4 splitter, similar to the combiner, with tiny dielectric filters at each output channel. Each filter transmits a signal with a certain wavelength that is detected by a photodiode at the end of the output waveguide. Filtering the signal after splitting leaves the device with an inherent loss of 6 dB. The simple demultiplexer design is compact and can be fabricated at very low cost.

Figure 21 shows the transmission spectrum of a 1×4 splitter demultiplexer with wavelength channels centered at 820, 835, 850, and 865 nm. The device shows an insertion loss of about 8.5 dB per channel. The loss increases further with increasing number of output channels, making the device unsuitable for a high number of wavelength channels.

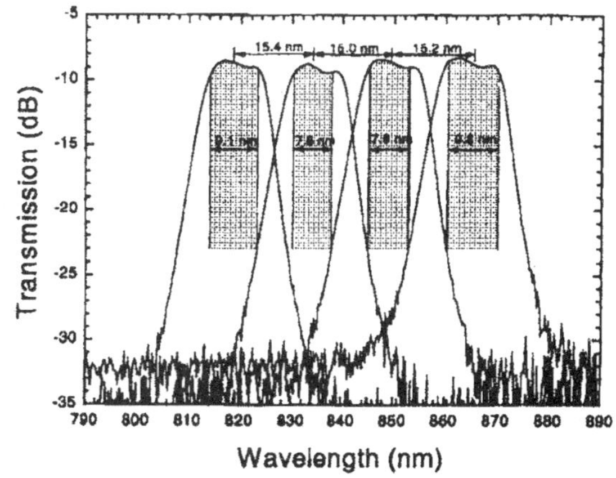

Figure 21. Transmission spectrum of the 1×4 splitter demultiplexer of the SpectraLAN system. Reprinted with permission © 1998, IEEE from L. Aronson et al. *IEEE Photon. Technol. Lett.* 10, 1489 (1998). Copyright © 1998, IEEE, New York.

To overcome the problem of the inherent 6 dB loss in the splitter demultiplexer, an alternative zigzag demultiplexer was developed [120]. In this design, dielectric pass-band filters and gold-coated mirrors are placed successively at each turn of the zigzag. Light propagating in the zigzag structure is reflected at each turn of the zigzag alternatively by the dielectric pass-band filters and the gold-coated mirrors. Only light that has a wavelength within the pass-band of the first dielectric filter is transmitted to the output waveguide. Wavelengths that are outside the pass-band filter are reflected back into the zigzag structure and eventually get reflected by a gold mirror toward the next filter. The light continues to propagate into the zigzag structure through successive reflections and transmissions until the entire signal is filtered out. The transmitted light is detected with GaAs detectors placed immediately after the filters. The insertion loss in this design is determined only by the material losses and the filters' transmission losses, and not caused by the device design as in the case of the splitter demultiplexer. Results of transmission measurements obtained from a 1×4 zigzag demultiplexer showed an insertion loss of 2.1 dB, which is much lower than in the case of the splitter demultiplexer.

5. ACTIVE DEVICES BASED ON NONLINEAR OPTICAL POLYMERS

5.1. Introduction

Many basic concepts for planar waveguide devices were introduced in the preceding sections. The few active devices discussed so far were based on the thermo-optic effect of the polymers and had rather limited speed. For faster operation, other material effects should be applied. In Section 2.2.5, optical nonlinear polymers were introduced, where the index of refraction is a function of electrical field $n = n(E)$, or a function of light intensity $n = n(I)$, the Pockels and the Kerr effect, respectively. Especially the dc Kerr effect, which is based on third-order nonlinearity, has an enormous technological potential, because it allows realization of the optical "transistor," that is, a device that has high speed all-optical functions like switching and amplification, and serves as a logical gate. Certain polymers have a relatively high third-order nonlinearity [121], but the intensity needed for switching is still too high for practical applications. For more details on third-order polymer-based devices, see the work of Stegeman [122, 123], the overview by Bosshard [124], and Driessen et al. [125].

In the following subsections, we concentrate on planar waveguide devices that make use of second-order nonlinear polymers for high-speed modulation and switching. In the first section, we deal with the relevant aspects of a design; thereafter, a number of devices are presented. Finally a brief outlook is given.

5.2. Design

5.2.1. Choice of the Electro-optic Coefficient

The electro-optic effect in a polymer is used to change the refractive index of that material (and with this index, also the properties of the entire device). If the electro-optic polymer has uniaxial anisotropy, the linear electro-optic effect can be described, neglecting higher order nonlinearities [30], by the simplified relationship.

$$\Delta n_i = -\frac{1}{2} n_i^3 r_{ik} E_k \tag{5}$$

where n_i is the refractive index, Δn_i is the change in the refractive index experienced by an optical wave polarized in direction i ($i = 1, 2, 3$), E_k is an externally applied electric field in direction k ($k = 1, 2, 3$), and r_{ik} is the linear electro-optic coefficient that corresponds to the electric field in direction k. If the material was poled in the z-direction, the largest electro-optic coefficient is r_{33}. Magnitudes of more than 60 pm/V at 1318 nm have been reported [1], which are about twice as much as for $LiNbO_3$. The only other nonzero coefficients are $r_{13} = r_{23} \approx r_{33}/3$

In waveguide devices, the propagation of supported modes of order m is described by their effective indices N_m. These indices are a function of the refractive indices of the materials in the system (including the electro-optic material) and the dimensions of the waveguide. In most devices, only single-mode waveguides are allowed when use is made of the electro-optic effect. The change of the effective index ΔN is a little smaller but of the same order of magnitude as Δn. For most applications, the intent is to obtain a phase shift that, by a certain device geometry (e.g., a Mach–Zehnder interferometer), can be translated into an intensity change. The phase shift $\Delta\varphi$ is related to the length l of the electro-optic section and the wavelength λ_0 in vacuum by

$$\Delta\varphi = 2\pi\, \Delta N l/\lambda_0 \tag{6}$$

The voltage needed to obtain a phase shift of π is called V_π.

5.2.2. Polarization Dependence

The preceding description makes it clear that the large differences between the r_{ik} coefficients may lead to polarization-dependent behavior. Moreover, the mere fact of poling the polymer introduces a strong birefringence. Theoretically, polarization independence can be obtained by poling the material in the propagation direction. Then both polarizations experience equal coefficients $r_{13} = r_{23}$. However, in practice, it is not clear how sufficiently large fields for poling and driving can be induced in the direction parallel to the waveguide channel. Another approach is to split the polarizations at the input of the integrated-optic circuit, then process both polarizations separately, and, finally, combine them again. Although conceptually simple, this method leads to complicated devices that may need separate electronic drivers for both polarizations. It is also not trivial to keep both sections matched with respect to wavelength-dependent behavior and loss. A different method, which was successfully demonstrated with non-electro-optic phased-array wavelength demultiplexers (as mentioned in Section 3.3.4), involves rotating the polarization of the guided lightwaves by 90°, half-way the propagation through a symmetric device. In this way, formerly TE-polarized light becomes TM and vice versa, so that, in the end, all incoming modes received the same treatment. Yet another approach, which complements the previous one, is to rotate the structure by 90° at the half-way point in the device. An example of this technique [126, 127] is given in Section 5.3.2.

5.2.3. Poling Considerations

Electro-optic polymeric materials are based on second-order optical nonlinear molecules or groups that mostly have a large dipole moment. When used in a waveguide, the electro-optic effect is not present unless an additional poling treatment is given that aligns the previously randomly oriented dipoles in a preferential direction. This can be done by applying an electric field over the electro-optic material at an elevated temperature just above the glass transition temperature of the polymer, where the dipoles can rotate freely. Upon cooling down, the field-induced order is frozen [128].

Different poling options are discussed thoroughly in the literature (e.g., [129, 4]). Two relatively new poling methods, triode poling and all-optical poling, deserve mention. Triode poling is a form of corona poling where a grid is placed between the corona needle or wires and the electro-optic material. This grid can be set to a certain potential to achieve more uniform poling [130]. The second method, optical poling [131], has the distinct advantage of not needing an externally applied electric field. The origin of the all-optical poling process is coherent resonant excitation by a strong fundamental beam that interferes with its weaker second harmonic.

5.2.4. Choice of Cladding Materials

The cladding layers perform important functions in an electro-optic device. First, they serve as optical buffer layers, preventing attenuation of the evanescent field of the propagating mode by the metal electrodes. This implies the need for a material with a well defined refractive index, as well as low optical losses. Second, the cladding layers influence the electric field applied to the core layer during the poling procedure at high temperature and later during use as an electro-optic device. The objective is to create a large electric field at the channel region in the active layer. Therefore, the electrical conductivity of the cladding materials should be high relative to that of the electro-optic layer at all temperatures. Figure 22 [132] illustrates that these demands on the electrical conductivity cannot be easily met because the value of the conductivity may change 2–3 orders of magnitude in the temperature region of interest. With a coplanar electrode configuration, the cladding layers should prevent dielectric breakdown via air. Third, the stability of the device under normal operating conditions is strongly dependent on the resistivity and permittivity of the polymers used and their thermal drift [133, 134]. The last function of cladding polymers is that of a mechanical buffer: it should allow the deposition and patterning processes for electrodes and possibly etching of the material itself, for example, for a gutter to form an inverted ridge waveguide with the core layer. Therefore, in the selection of cladding polymers, the optical properties as well as electrical, chemical, and mechanical properties should be taken into account. In the literature, a wide range of materials is used, for example, polyimides, polyurethane, and acrylates. Adhesion problems are a major source of optical degradation, and careful design and processing are required to prevent dielectric breakdown, especially with a coplanar electrode configuration.

5.2.5. Electrode Configuration

Electrodes are needed for induce the electro-optic effect. Two basic configurations are used: the microstripline and the coplanar waveguide structure (see Fig. 23). Criteria for selecting a particular configuration include:

- Resistivity of the materials (the microstripline is unfavorable if the cladding materials have much larger resistivity than the core see Sec. 5.2.4).

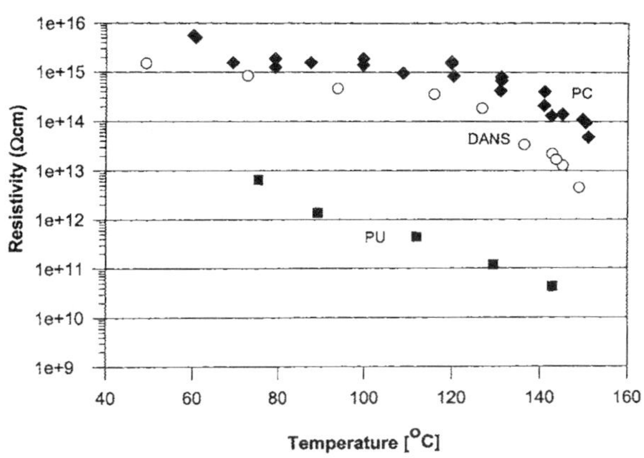

Figure 22. Resistivity of several polymer thin films as a function of temperature: DANS, an electro-optic polymer, PC, a passive cladding (both supplied by Akzo Nobel Research), and PU, polyurethane [132].

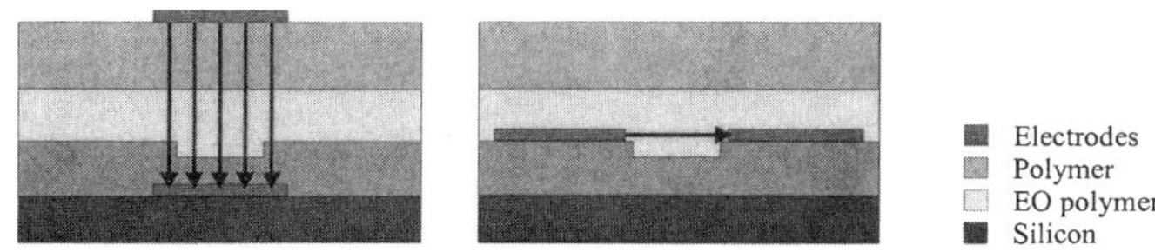

Figure 23. Cross section of the electrode structure for an electro-optic planar waveguide device with (a) microstripline structure and (b) coplanar waveguide structure. For the optical field, an inverted ridge optical waveguide is used between the electrodes in the electro-optic polymer layer.

- Stability (temporal and thermal drift due to slowly accumulating and changing charge at the interfaces between high-resistivity layers that possibly have different temperature dependences of resistivity; see Section 5.2.7).
- Breakdown voltage (see Sect. 5.2.7).
- Poling method relative to the choice of a particular electro-optic coefficient (see Sects. 5.2.1 and 5.2.3).
- The microwave waveguide design in the case of high-speed operation (see Sect. 5.2.8).
- Electrode-induced optical loss (see Sect. 5.2.6).

5.2.6. Optical Losses Caused by Electrodes

When designing an electro-optic device, a certain refractive-index change in the electro-optic material has to be realized. For example, in a Mach–Zehnder interferometer-based modulator, the refractive index should be changed by at least the amount that causes a π phase shift of the optical field so as to shift from the on to the off state. The two main design parameters are the length of the electro-optic section driven by the electrodes and the spacing between the electrodes. For a given voltage, the accumulated electro-optic phase shift is proportional to the length of this section and inversely proportional to the distance between the electrodes. Obviously, for a given voltage and a certain electro-optic phase shift, the electrodes can be shorter as their spacing decreases. On the other hand, the optical loss caused by the electrodes increases as their spacing decreases because the evanescent field of the optical mode interacts more strongly with the electrode metal. In both extreme cases of (a) a very wide and (b) a very narrow electrode spacing, the optical loss tends to infinity because in case (a) the structure needs to be infinitely long to obtain π phase shift and in case (b) the optical field is completely obstructed by the electrodes. This suggests that an optimum electrode spacing should exist. The calculated optical loss of a Mach–Zehnder modulator versus electrode spacing, given in Figure 24, clearly shows such an optimum.

5.2.7. Maximum Allowed Voltage

When a sufficiently strong electric field is applied to a material, the material electrically breaks down. Therefore, the dielectric strength of the materials used in electro-optic structures is an important parameter. For most polymers this parameter is on the order of 10^7 V/m [135]. Depending on the type of electrode structure, the weakest link in the breakdown process may be air, the cladding polymers, or the

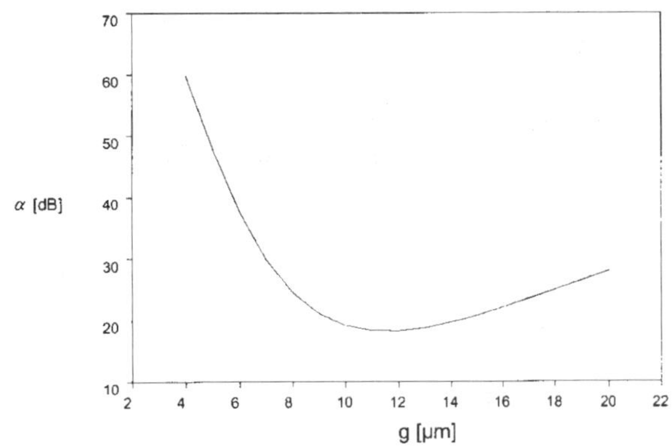

Figure 24. Total device loss α vs. electrode distance g, for a fixed electrode voltage in a polymer-based Mach-Zehnder interferometer.

electro-optic core polymer. As a practical rule, air claddings should be avoided because the breakdown voltage in air is much lower than in a polymer. For a given polymer layer the breakdown voltage may be much lower than would be expected from the bulk dielectric strength, due to nonuniformity of the layer. Several kinds of defects (voids, inclusions) may initiate early breakdown. For more details, see [136].

5.2.8. Frequency Range

The main advantage of the use of polymers as high-speed electro-optic material is the extremely high intrinsic frequency response, because the origin of the nonlinearity is purely electronic and has subpicosecond response time. Another technically relevant property is low dispersion of the polymeric materials between essentially zero and optical frequencies. One of the problems is that at high frequencies the wavelength of the modulating electric field becomes shorter than the modulator length. To achieve modulation in this case, traveling microwaves have to be used. To obtain a larger bandwidth, the optical and electrical wave must travel at the same speed along the waveguide. This means that the effective refractive indices at optical and microwave frequencies must match. In addition, the characteristic impedance (Z_0) of the microwave structure and the modulating source also must match. Tools for this kind of design task can be found in textbooks on microwave engineering (e.g., [137]). A simplified approach that often can be justified is to make a quasistatic approximation for the (quasi-)TEM fundamental mode [4].

Another problem in realizing high-speed modulators is the connection between the electrical source and the electrode structure. Above several tens of Gigahertz, coaxial cables have too much loss and the source has to be coupled to the electrode structure using hollow electrical waveguides.

5.3. Devices

The same basic structures that are employed in integrated optics with the thermo-optic effect (see Sect. 3.3.2) can be used to exploit the refractive-index change induced by an electric field in an electro-optic material (see, e.g., [30]). The main differences are due to (1) the relatively small magnitude of the electro-optic coefficient, which limits the attainable refractive-index change, (2) the possibility of push–pull operation (see, e.g., [138]), because the electro-optic effect allows both positive and negative index changes, depending on the polarity of the applied field, (3) the much higher speed of the electro-optic effect, and (4) the birefringence generally introduced by the electro-optic effect. These basic structures can be applied to realize devices in several classes, like modulators, switches, and wavelength filters. Several examples of applications of some of the basic structures follow. First we give a brief overview of the most important basic structures, and discuss their relative merits. Then, some typical examples of functional devices are given.

5.3.1. Basic Structures

Figure 25 shows schematic drawings of several basic planar waveguide structures that perform electro-optic modulating or switching functions. Also, their typical idealized

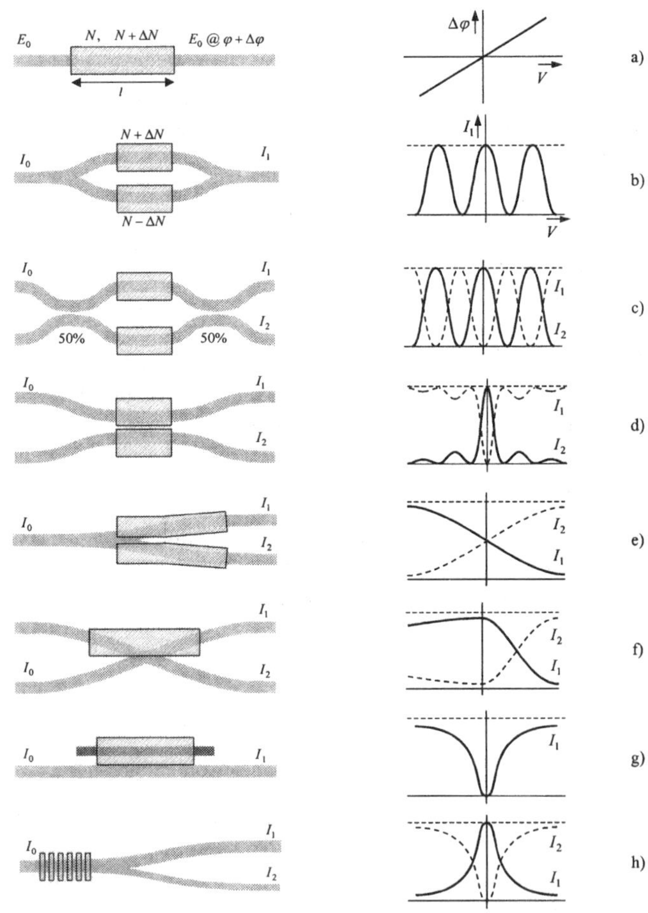

Figure 25. Basic electro-optic modulating structures. The left column shows diagrams of the optical paths (the shaded areas symbolize the regions of electro-optic interaction); the right column gives the corresponding idealized responses versus the driving voltage. (a) Phase modulator; (b) Mach-Zehnder interferometer (MZI) modulator; (c) MZI switch; (d) switched directional coupler (SDC); (e) digital optical switch; (f) total internal reflection switch; (g) SDC with lossy waveguide; (h) electro-optically induced mode-converting grating with a mode-sorting Y junction.

response curves are given. The simplest and most fundamental structure is the phase modulator (Fig. 25a), which induces a phase change according to Eq. (6). Since the phase shift depends on the wavelength, any device based on phase modulation will be wavelength-dependent, in principle. In most applications, a phase change of π radians is required. The Mach–Zehnder interferometer, which can be configured as an intensity modulator (Fig. 25b) or a switch (Fig. 25c), is the most efficient structure, but its periodical response can be a drawback because it requires an accurately controlled driving voltage. An interesting variant is an MZI switch that uses asymmetrical Y junctions instead of critical 3 dB couplers [139]. The operation of a switched directional coupler (Fig. 25d) is based on the interference of system modes in the coupling region, so its response is wavelength-dependent. However, its transfer function is not periodical because complete power transfer between the waveguides can occur only when the propagation constants of the local modes are equal (phase-matched condition). The digital optical switch (Fig. 25e) [132] is based on the mode-sorting effect in an asymmetric Y junction, where the fundamental system mode evolves into a local mode of the output branch that has the highest effective index. The electro-optic effect can be used to interchange the roles of the output branches. The theoretical switching curves are completely monotonous. This is an important property, because it could enable polarization-independent operation by "overdriving" the effect for the most sensitive orientation [140]. In the undriven state, half of the input power goes to each of the output ports. This may be undesirable for some applications. By employing an asymmetric Y junction, a preferred zero state can be imposed [141]. Another attractive feature of this kind of switch is its wavelength-independent operation (it is not based on interference). However, the magnitude of the electro-optic effect in available materials is still insufficient to obtain completely satisfactory devices.

The principle of a total internal reflection (TIR) switch is shown in Figure 25f. It consists of a waveguide crossing, where the refractive index can be lowered in one half. At the interface between the unmodified and the lowered index regions, total reflection can occur. This type of switch shows some very interesting theoretical properties like monotonous "digital" switching behavior that has a well defined zero state, which is applicable to multimode waveguides and possible polarization-independent operation. For the presently available electro-optic materials, however, the waveguide crossing angle should be very small (well below 1°), which can give rise to large cross talk between the output waveguides. Considering this small angle, the TIR switch is probably best modeled as a special case of the digital optical switch. A variant of the switched directional coupler principle that is applicable only to modulation is shown in Figure 25g. Here, one of the coupled waveguides is lossy and acts as a power sink. In the phase-matched condition, maximum power is transferred from the low-loss input waveguide to the high-loss waveguide and the output power is minimum. By electro-optically detuning the coupler, the power loss is reduced. This type of modulator can be designed to have a monotonous switching behavior toward the on state, although it is quite difficult to fabricate devices that can attain the phase-matched off state, because the two waveguides are essentially different [142, 143]. Finally, yet another switched coupler structure is shown in Figure 25h. Here, an electro-optically induced grating acts as a mode converter from the fundamental to the first-order mode. A mode splitter (in this example, an asymmetrical Y junction) guides the fundamental mode to its widest output branch, while the first-order mode evolves to the fundamental mode in the narrower branch. Without an applied voltage, there is no index contrast in the periodic electrode structure, so there is no grating, and all power remains in the fundamental mode and appears in the wider waveguide channel. By applying a voltage to the electrodes, a grating is induced that converts (part of) the input power to the first-order mode, which appears at the narrower channel output.

5.3.2. Some Recent Implementations

Although electro-optic polymers have been around for some decades, relatively few devices have been developed from proof-of-principle devices to real world applications. One exception is relatively short point-to-point analog rf link used, for example, to feed radar antennas. In this kind of application, the input polarization can be well controlled, and both imperfect extinction and moderate optical loss can be tolerated.

The most common structure in demonstrator devices is the MZI modulator, which has been implemented using various materials. The emphasis in most devices often has been on assessing material properties. For example, PMMA doped with an optimized chromophore and having an electro-optic coefficient of 58 pm/V has been employed to realize a modulator operating at 1318 nm wavelength with a half-wave voltage V_π of 0.8 V at an interaction length of 2.2 cm [1, 144]. Other modulators have been reported that have a higher V_π (2.4–3.7 V at 1300 and 1550 nm), but also a higher bandwidth (up to 100 GHz) [145]; see also [146]. For applications in short-distance communications links driven by superconducting electronics, driving voltages of only a few millivolts are available. As a possible solution, Lagali et al. [147] showed that by using resonant superconducting electrodes in combination with electro-optic polymers, ultra-low voltage modulation can be achieved.

Switches are very similar to modulators. They require at least two output ports and are optimized to have a high extinction ratio. Although extinction ratios of more than 20 dB have been measured [145, 148] without cascading devices, the realized values are still too low for most telecommunication system applications.

An interesting approach toward obtaining polarization independence is to cascade two active sections, one optimized for TE input light and the other for TM (Fig. 26).

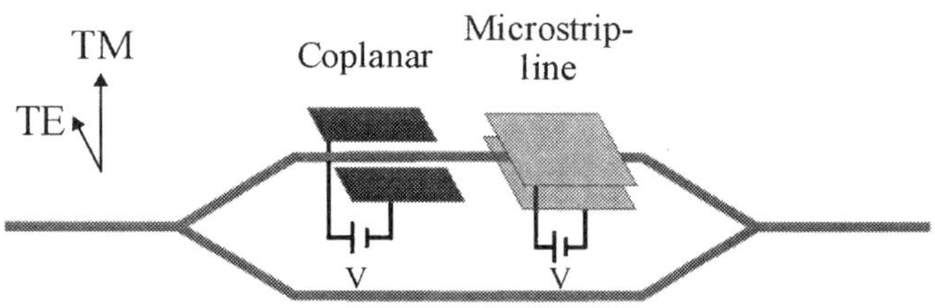

Figure 26. Coplanar and microstripline electrode configuration integrated on a MZI.

This can be done by using two different electrode configurations—a coplanar and a microstripline waveguide—and poling the electro-optic polymer locally using these electrodes [126]. Another approach is to incorporate a vertically tapered cladding to modulate both TE and TM modes [149].

In addition to the standard application of polymeric electro-optic materials in modulators, Dalton and co-workers [150] also considered some new devices: a photonically controlled rf phase shifter [151], a time stretcher for 102 GHz millimeter waves [152], and an integrated-optical spectrum analyzer [153].

5.4. Outlook

Many years of research on polymeric electro-optic devices have shown impressive results, but have not led to real breakthrough. The reason for this may be that today's communication systems work at a speed of only 10 Gbit/s per channel and only very recently have approached 40 Gbit/s. At these relatively low speeds, lithium niobate and semiconductor modulators perform relatively well. Once the speed approachs 100 Gbit/s, polymeric devices will be needed. Much work still has to be done on stability and reliability issues, but small companies like Pacific Wave Industries [154] have demonstrated that devices with a bandwidth above 40 GHz can be produced commercially.

Another application field will be the access network where low-cost solutions for the generation of the upstream data stream with several wavelengths are needed. The modulator arrays should work at a speed restricted by the limits of standard complementary metal–oxide–semiconductor electronics, presently a few gigabits per second. Electro-optic polymers offer the possibility of hybrid integration in integrated optics modules made in a reliable and low-cost technology.

In addition to the stability issue, the major problem is that it is very difficult to achieve both very low loss and very high electro-optic coefficients in the same material [28]. A high electro-optic coefficient results in more sensitive and faster devices with higher loss. There is a trade-off between loss, speed, and sensitivity of the device. In spite of the trade-off, there are some advantages of polymers over other electro-optic materials (for instance, $LiNbO_3$). The most important advantages are the following:

- Polymers have low dispersion in the index of refraction between infrared and millimeter-wave frequencies. This is because the dielectric properties of the polymers are all electronic in nature.
- Polymers can be deposited and will adhere to many substrates including semiconductors.
- Active polymers can be integrated into an optical circuit, which includes other optical polymers.

ACKNOWLEDGMENTS

We express our gratitude to all members of the Lightwave Device Group at the MESA$^+$ Institute of the University of Twente who, in the last 10 years, have contributed to the activities on polymer-based integrated optics devices: Toon Andringa, Gunter Beelen, Melis Jan Gilde, Marcel Hoekman, Hugo J. W. M. Hoekstra, (Marcel) H. M. M. Klein Koerkamp, Gijs J. M. Krijnen, Neil S. Lagali, Paul V. Lambeck, Peter Linders, Theo J. A. Popma, Erwin Rikkers, Jan B. P. van Schoot, Gabriel Sengo, Robert Wijn, and Kerstin Wörhoff. The longstanding collaboration with Akzo Nobel and later JDS Uniphase Netherlands (Mart B. Diemeer and others) is gratefully acknowledged. Part of this work was financially supported by the Dutch Technology Foundation STW, the IOP electro-optics financed by the Dutch Ministry of Economic Affairs, and the European Commission.

REFERENCES

1. Y. Shi, C. Zhang, H. Zhang, J. H. Bechtel, L. R. Dalton, B. H. Robinson, and W. Steier, *Science* 288, 119 (2000).

2. M. B. J. Diemeer, P. De Dobbelaere, and R. Flipse, in "Integrated Optical Circuits and Component, Design and Application," Chap. 8, pp. 239–302. Dekker, New York, 1999.

3. L. Eldada, and L. W. Shacklette, *IEEE J. Select. Topics Quantum. Electron.* 16, 54 (2000).

4. Ch. Bosshard, M. Bösch, I. Liakatas, M. Jäger, and P. Günter, in "Nonlinear Optical Effects and Materials" (P. Günter, Ed.), pp. 163–299. Springer, Verlag, Berlin, 2000.

5. Sulur, Polymer Waveguides for GIPOF Applications, Master Thesis, University of Twente, The Netherlands, 2001.

6. M. Kagami, H. Ito, T. Ichikawa, S. Kato, M. Matsuda, and N. Takahashi, *Appl. Opt.* 34, 1041 (1995).

7. M. Kagami, K. Hasegawa, and H. Ito, *Appl. Opt.* 36, 7700 (1998).

8. T. S. Barry, D. L. Rode, M. H. Cordaro, R. R. Krchnavek, and K. Nakagawa, *IEEE Trans. Comput. Packaging Manufact. Technol. Part B* 18, 685 (1995).

9. L. Wu, F. Li, S. Tang, B. Bihari, and R. T. Chen, *IEEE Photon. Technol. Lett.* 9, 1601 (1997).

10. D. H. Hartman, G. R. Lalk, J. W. Howse, and R. R. Krchnavek, *Appl. Opt.* 28, 40 (1989).

11. L. Robitaille, C. L. Callender, and J. P. Noad, *IEEE Phot. Technol. Lett.* 8, 1647 (1996).

12. M. Hikita, R. Yoshimura, M. Usui, S. Tomaru, and S. Imamura, *Thin Solid State* 331, 303 (1998).

13. T. Watanabe, N. Ooba, S. Hayashida, T. Kurihara, and S. Imamura, *J. Lightwave Technol.* 16, 1049 (1998).

14. D. Fischer, and E. Voges, *Electron. Lett.* 33, 1626 (1997).

15. R. Yoshimura, M. Hikita, S. Tomaru, and S. Imamura, *J. Lightwave Technol.* 16, 1030 (1998).

16. B. L. Booth, *J. Lightwave Technol.* 7, 1445 (1989).

17. R. R. Shannon and J. C. Wyant, "Applied Optics and Optical Engineering," Vol. VII, Academic Press, Orlando, 1979.

18. Y. Koike, T. Ishigure, and E. Nihei, *J. Lightwave Technol.* 13, 1475 (1995).
19. S. Imamura, R. Yoshimura, and T. Izawa, *Electron. Lett.* 27, 1342 (1991).
20. T. Kaino, "Polymers for Lightwave and Integrated Optics," Chap. 1, p. 1. Dekker, New York, 1992.
21. T. C. Kowalczyk, T. Kosc, K. D. Singer, P. A. Cahill, C. H. Seager, M. B. Meinhardt, A. J. Beuhler, and D. A. Wargowskil, *J. Appl. Phys.* 76, 1505 (1994).
22. K. Wörhoff, A. Driessen, P. V. Lambeck, L. T. H. Hilderink, P. W. C. Linders, and T. J. A. Popma, *Sens. Actuat.* 74, 9 (1999).
23. T. Matsuura, J. Kobayashi, S. Ando, T. Maruno, S. Sasaki, and F. Yamamoto, *Appl. Opt.* 38, 966 (1999).
24. C. W. Phelps, T. S. Barry, D. L. Rode, and R. R. Krchnavek, *J. Lightwave Technol.* 15, 1900 (1997).
25. Master Bond Inc., http://www.masterbond.com.
26. L. Eldada, S. Yin, R. A. Norwood, and T. J. Yardley, *Proc. SPIE* 3234, 161 (1997).
27. J. Kobayashi, T. Matsuura, S. Sasaki, and T. Maruno, *Appl. Opt.* 37, 1032 (1998).
28. W. H. Steier, A. Chen, S. S. Lee, S. Garner, H. Zhang, V. Chuyanov, L. R. Dalton, F. Wang, A. S. Ren, C. Zhang, G. Todorova, A. Harper, H. R. Fetterman, D. Chen, A. Udupa, D. Bhattacharya, and B. Tsap, *Chem. Phys.* 245, 487 (1999).
29. C. C. Teng, in "Nonlinear Optics of Organic Molecules and Polymers" (H. S. Nalwa and S. Miyata, Eds.), p. 441. CRC Press, Boca Raton, 1997.
30. R. M. de Ridder, A. Driessen, E. Rikkers, P. V. Lambeck, and M. B. J. Diemeer, *Opt. Mater.* 12, 205 (1999).
31. R. Spreiter, Ch. Bosshard, F. Pan, and P. Günter, *Opt. Lett.* 22, 564 (1997).
32. M. S. Wong, F. Pan, M. Bösch, R. Spreiter, C. Bosschard, and P. Günter, *J. Opt. Soc. Am. B* 15, 426 (1998).
33. T. Watanabe, N. Ooba, Y. Hida, and M. Hikita, *Appl. Phys. Lett.* 72, 1533 (1998).
34. J. D. Lytle, "Polymeric Optics in Handbook of Optics," Vol. II. McGraw–Hill, New York, 1995.
35. Y. Kokubun, N. Funato, and M. Takizawa, *IEEE Photon. Technol. Lett.* 5, 1297 (1993).
36. M. Usui, M. Hikita, T. Watanabe, M. Amano, S. Sugawara, S. Hayashida, and S. Imamura, *J. Lightwave Technol.* 14, 2338 (1996).
37. L. Lüer, H. J. Engelhaaf and D. Oelkrug, *Opt. Mat.* 9, 454 (1998).
38. D. Elliott, "Micro-lithography Process: Technology for IC Fabrication." McGraw–Hill, New York, 1986.
39. F. D. Egitto, L. J. Matienzo, and H. B. Schreyer, *J. Vac. Sci. Technol. A* 10, 3060 (1992).
40. H. D. Buer, W. Ehrfeld, M. Harder, T. Paatzsch, M. Popp, and I. Smaglinski, *Synth. Met.* 15, 13 (2000).
41. D. Ficher, D. Ullmann, L. Muller, and E. Voges, "Proceedings of the ECIO Conference on Simple Polymer Technologies for Multimode Integrated Optics," 1995.
42. R. R. Krchnavek, G. R. Lalk, and D. H. Hartman, *J. Appl. Phys.* 66, 5156 (1989).
43. A. Mukherjee, B. J. Eapen, and S. K. Baral, *Appl. Phys. Lett.* 65, 3179 (1994).
44. J. C. Seferis, "Polymer Handbook," Wiley, New York, 1999.
45. L. Eldada, C. Xu, K. M. T. Stengel, W. Shacklette, and J. T. Yardley, *J. Lightwave Technol.* 14, 1704 (1996).
46. R. A. Soref, J. Schmidtchen, and K. Petermann, *IEEE J. Quantum Electron.* 27, 1971 (1991).
47. R. Moosburger, G. Fischbeck, C. Kostrzewa, and K. Petermann, *Electron. Lett.* 32, 544 (1996).
48. G. Beelen, X. J. M. Leijtens, P. Linders, G. J. M. Krijnen, A. Driessen, M. B. J. Diemeer, A. F. Bakker, A. J. T. de Krijger, and R. M. de Ridder, "Proceedings of the IEEE/LEOS Annual meeting," San Francisco, 1999, pp. 545–546.
49. K. Worhoff, B. J. Offrein, P. V. Lambeck, G. L. Bona, and A. Driessen, *IEEE Photon. Technol. Lett.* 11, 206 (1999).
50. M. B. J. Diemeer, L. H. Spiekman, R. Ramasoedj, and M. K. Smit, *Electron. Lett.* 32, 1132 (1996).
51. G. Beelen, G. J. M. Krijnen, E. Rikkers, P. W. Linders, H. F. Bulthuis, A. F. Bakker, X. Leijtens, M. Diemeer, A. Driessen, J. van den Heuvel, R. M. de Ridder, "Proceedings of the Annual Symposium of the IEEE/LEOS Benelux Chapter," Gent, Belgium, 1998, p. 105.
52. Y. Ando and H. Hanafusa, *IEEE Photon. Technol. Lett.* 4, 1041 (1992).
53. R. S. Fan and B. Hooker, *J. Lightwave Technol.* 17, 466 (1999).
54. U. Streppel, P. Dannberg, C. Wächter, A. Bräuer, and P. Nicole, "Proceedings of ECIO," 2001, pp. 113–116.
55. B. L. Booth, in "Polymers for Electronic and Photonic Applications," (C. P. Wong, Ed.), pp. 549–599. Academic Press, New York, 1993.
56. H. J. W. M. Hoekstra, *Opt. Quantum. Electron.* 29, 157 (1997).
57. C2V Software, P.O. Box 318, 7500 AH, Enschede, The Netherlands; http://www.c2v.nl/software.
58. H. W. Weber, R. Ulrich, E. A. Chandross, and W. J. Tomlinson, *Appl. Phys. Lett.* 20, 143 (1972).
59. N. Keil, J. Krauser, B. Strebel, and H. H. Yao, in "Laser in Engineering" (W. Waidelich, Ed.), p. 629. Springer-Verlag, Berlin, 1992.
60. M. B. J. Diemeer, *Opt. Mater.* 9, 192 (1998).
61. M. B. J. Diemeer, F. M. M. Suyten, E. S. Trommel, A. McDonach, J. M. Copeland, L. W. Jenneskens, and W. H. G. Horsthuis, *Electron. Lett.* 26, 379 (1990).
62. J. H. Kim, E. J. Kim, H. C. Choi, C. W. Kim, J. H. Cho, Y. W. Lee, B. G. You, S. Y. Yi, H. J. Lee, K. Han, W. H. Jang, T. H. Rhee, J. W. Lee, and S. J. Pearton, *Thin Solid Films* 341, 192 (1999).
63. N. Agarwal, S. Ponoth, J. Plawsky, and P. D. Persans, *Appl. Phys. Lett.* 78, 2294 (2001).
64. Y. Y. Maruo, S. Sasaki, and T. Tamamura, *Appl. Opt.* 34, 1047 (1995).
65. X. M. Zhao, S. P. Smith, S. J. Waldman, G. M. Whitesides, and M. Prentiss, *Appl. Phys. Lett.* 71, 1017 (1997).
66. R. Moosburger, R. Hauffe, U. Siebel, D. Arndt, J. Kropp, and K. Petermann, *IEEE Photon. Technol. Lett.* 11, 848 (1999).
67. M. B. J. Diemeer, J. J. Brons, and E. S. Trommel, *J. Lightwave Technol.* 7, 449 (1989).
68. M. H. M. Klein Koerkamp, M. C. Donckers, B. H. M. Hams, and W. H. G. Horsthuis, "Proceedings of the Integrated Photonics Research Conference," 1994, Vol. 3, pp. 274–276.
69. K. Propstra, T. Hoekstra, A. Borreman, M. Diemeer, T. Hoekstra, and P. Lambeck, "Proceedings of ECIO," 1997, Post deadline paper PD3, p. 618.
70. U. Siebel, R. Hauffe, and K. Petermann, *IEEE Photon. Technol. Lett.* 12, 40 (2000).
71. N. Keil, C. Weinert, W. Wirges, H.H. Yao, S. Yilmaz, C. Zawadzki, J. Schneider, J. Bauer, M. Bauer, K. Lösch, K. Satzke, W. Wischmann, and J. V. Wirth, *Electron. Lett.* 36, 430 (2000).
72. A. Borreman, T. Hoekstra, M. Diemeer, H. Hoekstra, and P. Lambeck, "Proceedings of ECOC," 1996, Oslo, Chap. 5, p. 59.
73. R. Hauffe, U. Siebel, and K. Petermann, *IEEE Photon. Technol. Lett.* 13, 200 (2001).
74. T. Goh, M. Yasu, K. Hattori, A. Himeno, M. Okuno, and Y. Ohmori, *J. Lightwave Technol.* 19, 371 (2001).
75. L. Eldada, J. Amin, and R. W. Sharps, *IEEE Photon. Technol. Lett.* 12, 819 (2000).
76. L. H. Spiekman, M. B. J. Diemeer, T. H. Hoekstra, and M. K. Smit, *Tech. Dig. Integrated Photonic Res.* 6, 36 (1996).
77. M. A. Cowin, M. Owen, I. H. White, and R. V. Penty, *Electron. Lett.* 35, 1464 (1999).
78. S. Toyoda, A. Kaneko, N. Ooba, M. Hikita, H. Yamada, T. Kurihara, K. Okamoto, and S. Imamura, *IEEE Photon. Technol. Lett.* 11, 1141 (1999).
79. S. Toyoda, N. Ooba, A. Kaneko, M. Hikita, T. Kurihara, and T. Maruno, *Electron. Lett.* 36, 658 (2000).
80. J.-V. Viens, C. L. Callender, J. P. Noad, and L. Eldada, *IEEE Photon. Technol. Lett.* 12, 1010 (2000).

81. N. Keil, H. H. Yao, C. Zawadzki, J. Bauer, M. Bauer, C. Dreyer, and J. Schneider, *Electron. Lett.* 37, 579 (2001).
82. M. B. J. Diemeer, E. Rikkers, D. T. van Veen, G. J. M. Krijnen, M. J. Gilde, J. L. Joppe, A. F. Bakker, M. K. Smit, R. M. de Ridder, A. Driessen, and A. M. J. Koonen, "Proceedings of the 2nd Symposium of the IEEE/LEOS Benelux Chapter," Eindhoven, 1997, p. 33.
83. C. Kostrzewa, R. Moosburger, G. Fischbeck, B. Schüppert, and K. Petermann, *IEEE Photon. Technol. Lett.* 9, 1487 (1997).
84. L. Eldada, S. Shing Yin, C. Poga, C. Glass, R. Blomquist, and R. A. Norwood, *IEEE Photon. Technol. Lett.* 10, 1416 (1998).
85. L. Eldada, R. Blomquist, M. Maxfield, D. Pant, G. Boudoughian, C. Poga, and R. A. Norwood, *IEEE Photon. Technol. Lett.* 11, 448 (1999).
86. G. Beelen, P. Linders, C. G. H. Roeloffzen, A. Driessen, M. B. J. Diemeer, X. J. M. Leijtens, A. F. Bakker, and A. J. T. de Krijger, "Proceedings of ECOC," Nice, 1999, Post deadline paper PD 58.
87. D. T. van Veen, F. N. Krommendijk, B. H. Sikken, and A. M. J. Koonen, "Proceedings of ECOC," Nice, 1999, Vol. 2, p. 48.
88. H. Takehashi, Y. Hibino, and I. Nishi, *Opt. Lett.* 17, 499 (1992).
89. K. Wörhoff, P. V. Lambeck, and A. Driessen, *J. Lightwave Technol.* 17, 1401 (1999).
90. C. G. H. Roeloffzen, F. Horst, B. J. Offrein, R. Germann, G. L. Bona, H. W. M. Salemink, and R. M. de Ridder, *IEEE Photon. Technol. Lett.* 12, 1201 (2000).
91. R. M. de Ridder, K. Wörhoff, A. Driessen, P. V. Lambeck, and H. Albers, *IEEE J. Select. Top. Quantum Electron.* 4, 930 (1998).
92. C. Liguda, G. Böttger, A. Kuligk, R. Blum, M. Eich, H. Roth, J. Kunert, W. Morgenroth, H. Elsner, and H. G. Meyer, *Appl. Phys. Lett.* 78, 2434 (2001).
93. D. R. Beltrami, J. D. Love and F. Ladouceur, *Opt. Quantum Electron,* 31, 307 (1998).
94. O. Parriaux, P. Roth, and G. Voirin, pp. 295–310 in "Critical Reviews of Optical Science and Technology" (S. I. Najafi, Ed.), Vol. 53, p. 295. Baker, London, 1994.
95. S. Musa, N. S. Lagali, Sulur, G. Sengo, and A. Driessen, "Proceedings of the 5th Symposium of the IEEE/LEOS, Benelux Chapter," Delft, The Netherlands, 2000, p. 95.
96. R. T. Chen, S. Tang, T. Jannson, and J. Jannson, *Appl. Phys. Lett.* 63 1032 (1993).
97. D. Israel, R. Baets, M. J. Goodwin, N. Shaw, M. D. Salik, and C. J. Groves-Kirkby, *J. Lightwave Technol.* 13, 1057 (1995).
98. I. Ilic, R. Scarmozzino, R. M. Osgood, Jr., J. T. Yardley, K. W. Beeson, M. J. MacFarland, and K. M. T. Stengel, *IEICE Trans. Commun.* **E80-B**, 135 (1997).
99. S. Musa, N. S. Lagali, G. J. M. Krijnen, and A. Driessen, "Proceedings of the 13th IEEE/LEOS Conference," Puerto Rico, 2000, p. 468.
100. L. Aronson, B. E. Lemoff, L. A. Buckman, and D. W. Dolfi, *IEEE Photon. Technol. Lett.* 10, 1489 (1998).
101. S. Musa, N. S. Lagali, B. Docter, G. Sengo, G. J. M. Krijnen, and A. Driessen, "Proceedings of the 10th European Conference on Integrated Optics", Paderborn, Germany, 2001, p. 248.
102. H. Noda, T. Namioka, and M. Seya, *J. Opt. Soc. Am.* 64, 1031 (1974).
103. C. Winkler, J. D. Love, and A. K. Ghatak, *Opt. Quantum Electron,* 11, 173 (1979).
104. M. Hikita, S. Tomaru, K. Enbutsu, N. Ooba, R. Yoshimura, M. Usui, T. Yoshida, and S. Imamura, *IEEE J. Select. Top. Quantum Electron.* 5 1237 (1999).
105. J. Moisel, "Proceedings of the 13th IEEE/LEOS Conference," Puerto Rico, 2000, p. 567.
106. D. Israel, R. Baets, N. Shaw, and M. J. Goodwin, Technical Digest, MOC GRIN. Kawasak, 1993.
107. D. B. Keck, A. J. Morrow, D. A. Nolan, and D. A. Thompson, *J. Lightwave Technol.* 7, 1623 (1989).
108. J. P. G. Bristow, *SPIE Proc.* 1178, 103 (1989).
109. C. M. Olsen, *J. Lightwave Techno.*, 9, 1742 (1991).
110. F. E. Kiamilev, P. Marchand, A. V. Krishnamoorthy, S. C. Esener, and S. H. Lee, *J. Lightwave Techno.* 9, 1674 (1991).
111. D. Israel, R. Baets, M. J. Goodwin, N. Shaw, M. D. Salik, and C. J. Groves-Kirkby, *Appl. Opt.* 36, 5091 (1997).
112. A. Simon and R. Ulrich, *Appl. Phys. Lett.* 31, 77 (1977).
113. L. B. Soldano and E. C. M. Pennings, *J. Lightwave Technol.* 13, 615 (1995).
114. G. Giaretta, W. White, M. Wegmuller, and T. Onishi, *IEEE Photon. Technol. Lett.* 12, 199 (2000).
115. M. Duser and P. Bayvel, *Electron. Lett.* 36, 57 (2000).
116. H. W. Yen, H. R. Friedrich, R. J. Morrison, and G. L. Tangonan, *Opt. Lett.* 6, 639 (1981).
117. T. Suhara, J. Viljanen, and M. Leppihalme, *Appl. Opt.* 21, 2195 (1982).
118. J. Liu and R. T. Chen, *J. Lightwave Technol.* 17, 354 (1999).
119. R. Watanabe and K. Nosu, *Appl. Opt.* 19, 3588 (1980).
120. B. E. Lemoff, L. B. Aronson, and L. A. Buckman, *Electron. Lett.* 34, 1014 (1998).
121. A. Driessen, H. J. W. M. Hoekstra, F. C. Blom, F. Horst, G. J. M. Krijnen, J. B. P. van Schoot, P. V. Lambeck, Th. J. A. Popma, and M. B. Diemeer, *Opt. Mater.* 9, 329 (1998).
122. G. I. Stegeman and A. Miller, in "Photonics in Switching" (J. E. Midwinter, Ed.), Vol. 1, p. 81. Academic Press, New York, 1993.
123. G. I. Stegeman and W. E. Toruellas, *Philos. Trans. Roy. Soc. London Sec A* 354, 707 (1996).
124. Ch. Bosshard, in "Nonlinear Optical Effects and Materials" (P. Günter Ed.), Chap. 7. Springer-Verlag, Berlin, 2000.
125. A. Driessen, H. J. W. M. Hoekstra, F. Horst, G. J. M. Krijnen, B. J. Offrein, J. B. P. van Schoot, P. V. Lambeck, and Th. J. A. Popma, *IEE Proc. Optoelectron.* 145, 227 (1998).
126. A. Donval, E. Toussaere, R. Hierle, and J. Zyss, *J. Appl. Phys.* 87, 3258 (2000).
127. A. Donval, E. Toussaere, R. Hierle, and J. Zyss, *Synthetic Metals* 115, 21 (2000).
128. M. A. Firestone, M. A. Ratner, and T. J. M. Marks, *Macromolecules* 28, 6296 (1995).
129. H. S. Nalwa, and S. Miyata, "Nonlinear Optics of Organic Molecules and Polymers," CRC Press, Boca Raton, FL, 1997.
130. P. A. Ribeiro, D. T. Balogh, J. A. Giacometti, *IEEE Trans. Dielect Electron. Insul.* 7, 572 (2000).
131. R. Piron, E. Toussaere, D. Josse, S. Brasselet, and J. Zyss, *Synthetic Metals* 115, 109 (2000).
132. E. Rikkers, R. M. de Ridder, and A. Driessen, "Proceedings of the Symposium of the IEEE/LEOS Benelux Chapter, Enschede, 1996, p. 214.
133. H. C. Ling, W. R. Holland, and H. M. Gordon, *J. Appl. Phys.* 70, 6669 (1991).
134. Y. Shi, W. Wang, W. Lin, D. J. Olson, and J. H. Bechtel, *Appl. Phys. Lett.* 71, 2236 (1997).
135. J. Brandrup and E. H. Immergut, "Polymer Handbook", Wiley, New York, 1975.
136. H. R. Zeller, *IEEE Trans. Electron. Insul.* **EI-22**, 115 (1987).
137. K. C. Gupta, R. Garg, I. Bahl, and P. Bhartia, "Microstriplines and Slotlines," Artech House, Boston, 1996.
138. W. Wang, Y. Shi, D. J. Olson, W. Lin, and J. H. Bechtel, *IEEE Photon. Technol. Lett.* 11, 51 (1999).
139. W. Y. Hwang, M.-C. Oh, H.-M. Lee, H. Park, and J. J. Kim, *IEEE Photon. Technol. Lett.* 9, 761 (1997).
140. S.-S. Lee and S.-Y. Shin, *Electron. Lett.* 33, 314 (1997).
141. C. Duchet, C. Brot, L. Sarrabay, M. Di Maggio, and D. Haux, "OFC/IOOC Technical Digest," 1993, p. 114.
142. A. Driessen, H. M. M. Klein Koerkamp, and Th. J. A. Popma, *Fiber Integrated Opt.* 13, 445 (1994).
143. H. M. M. Klein Koerkamp, An Integrated-Optics Intensity Modulator Based on Mode Coupling, Ph.D. Thesis, University of Twente, the Netherlands, 1992.
144. Y. Shi, W. Lin, D. J. Olson, and J. H. Bechtel, *Appl. Phys. Lett.* 77, 1 (2000).

145. M.-C. Oh, H. Zhang, A. Szep, V. Chuyanov, W. H. Steier, C. Zhang, L. R. Dalton, H. Erlig, B. Tsap, and H. R. Fetterman, *Appl. Phys. Lett.* 76, 3525 (2000).
146. D. Chen, H. R. Fetterman, A. Chen, W. H. Steier, L. R. Dalton, W. Wang, and Y. Shi, *Appl. Phys. Lett.* 70, 3335 (1997).
147. N. S. Lagali, D. J. W. Klunder, G. J. Gerritsma, and A. Driessen, *Nonlinear Opt.* 25, 253 (2000).
148. D. An, Z. Shi, L. Sun, J. M. Taboada, Q. Zhou, X. Lu, S. Tang, H. Zhang, W. H. Steier, A. Ren, and L. R. Dalton, *Appl. Phys. Lett.* 76, 972 (2000).
149. M.-C. Oh, W.-Y. Hwang, H.-M. Lee, S.-G. Han and Y.-H. Won, *IEEE Photon. Technol. Lett.* 9, 1232 (1997).
150. See the Dalton website at: http://depts.washington.edu/eooptic/
151. S.-S. Lee, A. H. Udupa, H. Erlog, H. Zhang, Y. Chang, C. Zhang, D. H. Chang, D. Bhattacharya, B. Tsap, W. H. Steire, L. R. Dalton, and H. R. Fetterman, *IEEE Microwave Guided Wave Lett.* 9, 357 (1999).
152. D. H. Chang, H. Erlig, M. C. Oh, W. H. Steier, L. R. Dalton, and H. R. Fetterman, *IEEE Photon. Technol. Lett.* 12, 537 (2000).
153. A. Yacoubian, V. Chuyanov, S. M. Garner, W. H. Steier, A. S. Ren, and L. R. Dalton, *IEEE J. Select. Topics Quantum Electron* 6, 810 (2000).
154. http://www.pacificwaveind.com/

CHAPTER 6

High Capacity Transmission Systems

Giok-Djan Khoe, Henrie van den Boom, and Idelfonso Tafur Monroy
COBRA Institute, Eindhoven University of Technology, P.O. Box 513, Eindhoven, The Netherlands

CONTENTS

ISBN: 1-58883-012-8
Copyright © 2004 by American Scientific Publishers
All rights of reproduction in any form reserved.

1. INTRODUCTION

The demand for high data-rate communication in the consumer's premises and office areas is rapidly increasing. This demand is driven by broad-band applications, such as fast internet access and compressed digital video-based services among others. Transmission techniques based on polymer optical fibers (POF) have emerged as promising candidates for a cost-effective and future-proof solution. Data transmission over POF is an attractive technology for short distance scenarios such as local area networks (LAN), in-house networking, and fiber-to-the-home. This chapter covers high capacity transmission over POF, including its enabling technologies and experimental details.

Conventional technologies for wiring in local area networks are based mainly on copper cables and glass fiber, both single and multimode. These transmission media are not suited for the high capacity demand of future telecom and data services. For instance, copper-based technologies suffer from strong susceptibility to electromagnetic interferences and a limited capacity for digital transmission. Conventional silica-based fibers are costly due to precise connecting, and dedicated installation and handling. An alternative technology is the use of optical polymer fiber for data transmission. This fiber could be used for data transmission in many applications areas, ranging from in-house networking to the avionic environment. Compared to copper (coaxial cable and twisted pair), POF guarantees (EMC), no cross talk, small volume, and light weight. For data transmission, POF offers higher bandwidth at longer distances. In comparison to multimode glass optical fiber (MM GOF), POF offers ease of use and improved safety. Due to a large core, POF has a large tolerance to misalignments. Termination is simplified and low-cost connectors can be used, so installation is cheaper. In general, POF fibers have the potential to provide high bandwidth for data transmission at short distances (up to 100 m). In addition, POF is not brittle but ductile and will stretch under increased tension rather than break. Cutting POF is much easier as compared to thick glass fiber, and even a thick bundle of POF is more flexible than a bundle of glass fiber. Recent developments have demonstrated the feasibility of transmission at data rates up to 11 Gbit/s over 100 m, as well as the possibility of capacity enhancement by using wavelength division multiplexing (WDM) of the data channels.

This chapter is organized as follows. Section 1 presents an introductory section on the transmission properties of POF and enabling technologies for POF-based transmission systems. Section 2 presents a list of record transmission experiments that demonstrate the feasibility of poly(methyl methacrylate) (PPMA) POF technology for high capacity systems. A set of experimental trials, including WDM transmission, is described in detail. Section 3 is dedicated to transmission experiments, including WDM transmission, and application using perfluorinated graded-index POF. Section 4 presents a comparison of the characteristics of POF and large core silica-based multimode fibers for data transmission. Section 5 presents a list of applications that can benefit from the use of POF in the physical layer. Finally, summarizing conclusions are drawn in Section 6.

1.1. Short Historical Note

Optical polymer fibers started to be developed in the 1960s. Although they were not intended for data transmission, but for illumination purposes, they pre date glass optical fiber. The first type of commercial optical polymer fibers had a step-index configuration of the refraction index throughout the core and cladding. Polysterene was used as the core material for those fibers whose transmission losses were on the order of thousands of dB/km. Further development took place in the mid-1960s when polymer optical fibers; that used PMMA for the core material were produced. This type of fiber showed improved transmission loss characteristics. Despite this improvement, the attenuation was still considerably high, around 500 dB/km into the 650 nm wavelength region. In 1977, PMMA optical fibers were reported that had transmission loss below 300 dB/km in the visible and near infrared spectral region [1].

Further progress in the technology and manufacturing process of step-index POFs were achieved by Nippon Telegraph and Telephone (NTT) and Mitsubishi Rayon in the 1970s. Specifically, improvements to purification issues and manufacturing technology were achieved. For instance, in 1982, NTT reported the production of PMMA core fiber that had a low attenuation (55 dB/km) at 560 nm. In the 1980s, polymer optical fibers started to be specifically designed for data transmission. These fibers had improved attenuation characteristics. For instance, in 1983, reported *Kaino et al.* [2] a POF fiber that was loss on the order of 20 dB/km. This fiber [2] was produced by using deuterated PMMA (PMMA-d8) as the core material. All these fibers had a step-index profile and suffered from bandwidth limitations. Bandwidth improvement can be achieved by manufacturing fibers that have a graded-index instead of a step-index profile. The next important achievement in the development of polymer fibers was the introduction of graded-index POF. The first graded-index POF was reported in 1982 by Koike et al. [3] from Keio University. Graded-index POF showed larger bandwidth than the step-index counterpart, but, the graded-index fiber at that time showed higher transmission loss than step-index fiber. The reason for this loss was excess scattering due to the phase separation of the copolymer used in the manufacturing process. In 1994, a graded-index POF that had enhanced bandwidth and attenuation characteristics was reported [4]. The fiber was produced by a novel method called interfacial-gel polymerization. The attenuation of that fiber was measured to be 90 dB/km at a 572 nm wavelength. The bandwidth was 1 GHz/km, which was on the order of 400 times larger than that previously reported for step-index fibers. This result opened the way for broad-band data transmission over POF. Although grading the index profile of PMMA-POF enhanced characteristics, its bandwidth and attenuation still limited the transmission distances and capacity of systems used that POF. Further improvement in POF fibers has made it possible to reduce attenuation and to expand the bandwidth. Improved attenuation characteristics have been achieved by using amorphous perfluorinated polymers for the core material. This type of fiber has been named perfluorinated (PF) POF. In 1995, Keio University and Asahi Glass reported the production of PF-based GI POF that operates at 1300 nm wavelength with low attenuation and high bandwidth. Perfluorinated POF shows improved characteristics for data transmission: low attenuation, high bandwidth, favorable mechanical properties, and long time stability. Improved perfluorinated graded-index (GI) POFs have reportedly show low attenuation of 15 dB/km at 1.0–1.3 μm wavelengths [5]. Since June 2000, a cable of graded-index PF POF is been commercially available from Asahi Glass Company [6]. This cable consists of two fibers of that 120 μm core diameter and 500 μm cladding diameter. The fiber attenuation is on the order of 50 dB/km in the broad spectral window between 800 and 1300 nm. The bandwidth product is in the order of 2 GHz km [6]. A summary of the history of POF development is given in Table 1.

1.2. Classification of Plastic Optical Fibers

In principle, polymer optical fibers can be produced to have small core diameters that satisfy the condition for single-mode operation. However, as explained a later, POFs that have large cores (multimode operation) have several advantageous features for application in cost-sensitive broad-band telecommunications applications. The subsequent discussion refers to large core diameter multimode POFs. We can classify POF into two groups depending on the refractive index profile: step-index and graded-index POFs. Furthermore, we distinguish between POFs that use different material for the core; see Figure 1. Three major materials have been used for the core of step-index POF: polystyrene, PMMA, and polycarbonate. Because of improved attenuation and mechanical characteristics, PMMA is the preferred material in commercially available step-index POF. Graded-index POFs have been produced using PMMA and perfluorinated polymers for the core material. Perfluroinated POFs show improved attenuation and bandwidth characteristics compared to the PMMA counterpart. For data transmission, graded-index POFs are preferred due to their larger bandwidth. For comparison reasons, we present a general overview of the characteristics of both PMMA and PF graded-index POFs.

Typical characteristics of PMMA GI POF are [7]:

- Bandwidth–distance product at 650 nm: 2 GHz × km
- Attenuation at 650 nm: 150 dB/km
- Attenuation at 850 nm and 1300 nm: larger than 1000 dB/km
- Higher cost due to graded-index (GI) profile compared to step-index fibers
- Best reported transmission capacity and distance: 200 m at 2.5 Gbit/s.

Figure 2 presents a plot of the dependence of the attenuation of POF fiber for different wavelengths. It is relevant to note that the used perfluorinated polymers promises low attenuation over a wide range of wavelength. Let us consider now the typical characteristics of perfluorinated GI POF [6, 8]:

- Potential bandwidth at 1300 nm: 10 GHz × km
- Attenuation at 1300 nm: 50 dB/km
- Potential attenuation at 1300 nm: 0.3 dB/km
- Open spectral window for transmission: 500–1300 nm
- Best reported transmission data rate and distances: 550 m at 2.5 Gbit/s and 100 m at 11 Gbit/s.

Table 1. Evolution of POF.

Organization	Core material	Refractive index profile	Minimum attenuation (dB/km)	Wavelength (nm)	Year
DuPont	PMMA	SI	500	650	1968
Toray	PS	SI	1100	670	1972
DuPont	PMMA-d8	SI	180	790	1977
Mitsubishi Rayon	PMMA	SI	300	650	1978
NTT	PMMA	SI	55	568	1982
NTT	PS	SI	114	670	1982
Keio University	MMA-co-VPAc	GI	1070	670	1982
NTT	PMMA-d8	SI	20	650	1983
Mitsubishi Rayon	PMMA	SI	110	570	1983
Asahi Chemical	PMMA	SI	80	570	1986
Fujitsu	PC	SI	450	770	1986
NTT	P5F3DSt	SI	178	850	1986
Hitachi	Thermoset resin	SI	600	650	1987
Keio University	PMMA	GI	130	650	1990
Hoechst Celanese	PMMA	SI	130	650	1991
Keio University	PMMA-d8	GI	56	688	1992
Bridgestone	Silicone	SI	800	650	1993
Keio University, Asagi Glass Co.	Perfluorinaled polymer	GI	50	1300	1995

Source: Y. Koike et al. *IEEE J. Lightwave Technol.* 13, 1475. Copyright © 1995, IEEE.
PMMA, poly(methyl methacrylate); PS, polystyrene; PMMA-d8, deuterated PMMA; PC, polycarbonate; P5F3DSt, pentafluorotrideuterostyrene; MMA-co-VPAC, copolymer of MMA and vinyl Pentyl acetate.

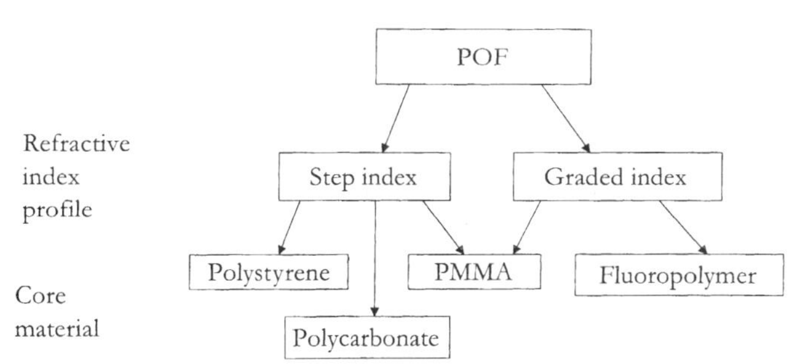

Figure 1. Classification of POF depending on the profile of the index of refraction through the core and the cladding, and on the material of the fiber core.

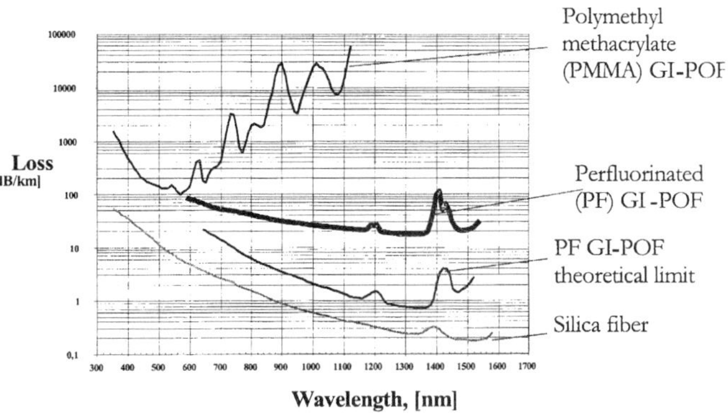

Figure 2. Attenuation of GI POF as a function of wavelength.

As can be seen from a comparison of the characteristics perfluorinated GI POF have the prospect to offer low attenuation and large bandwidth in data transmission applications over a broad range of wavelengths. The capability to support wavelength division multiplexing in the broad spectral region of 550–1300 nm makes POF especially attractive for use in high capacity transmission systems.

1.3. Applications

Application areas that may benefit from the use of POF for data transmission include those that cover short transmission distances, typically up to 100 m. Further, applications require low and medium data rates, up to 0.1 Gbit/s. POF is an alternative in areas where cost is a driving factor, in addition to ease of handling and installation. For instance, POF may be considered for

- In-home networking
- LAN
- Fiber to the flat
- Wireless LAN backbones
- High speed office LANs, horizontal wiring
- Airline entertainment

In areas that require enhanced capacity, parallel fibers or multiplexing several wavelength channels can be used to increase capacity. Application areas where EMC is required, such as the automotive and aerospace industries, may benefit form the use of POF fibers. The advantage of introducing POF for in-home networking is strongly linked to the prospect of its installer-friendiness and low cost. POF will serve as the physical backbone for transmission of broadband services, such as digital audio and video is POF on-demand also ductile allowing bending during installations, and thermal stable, offers an untapeable transmission environment. Standardization bodies are considering POF for the physical layer in several applications. The ATM forum has developed a standard for asynchronous transmission over POF fibers [9].

1.4. Bandwidth of POF

Apart from attenuation, an important characteristic of POF as a transmission medium is the bandwidth. The bandwidth is a measure of the transmission capacity of a fiber data link. Several factors determine the bandwidth of polymer optical fiber, including chromatic dispersion, modal dispersion, launching condition, and intramodal effects. The bandwidth characteristic is usually specified by the 3 dB bandwidth-distance product. The 3 dB bandwidth is determined by analyzing the POF transfer function. The transfer function relates the output signal to the input version in an optical fiber transmission line. The transfer function is mainly determined by the fiber dispersion. Dispersion in POF can be divided into two main factors; the chromatic dispersion and modal dispersion. Chromatic dispersion is related to the dependence of the index of refraction on the wavelength. Therefore, the various spectral components of each mode propagating in the fiber travel at different speeds, depending on the wavelength. This results in pulse broadening, that is dispersion. Modal dispersion is related to spreading of the pulse as a result of differences in propagation delay among the modes as well as dispersion from intermodal effects such as power mixing between modes, mode dependent loss, excitation of modes, launching conditions of light into the fiber, and the effects of microbendings, among others. It is interesting to note that the material dispersion of PF fiber is smaller than that of silica fiber [10], so on optimized PF GI POF can outperform a graded-index silica fiber at the optimum operating wavelength.

It is of relevance to improve the bandwidth characteristics of POF. It is worth mentioning that, given a certain operating wavelength, bandwidth can be enhanced by optimizing the graded-index profile for that chosen wavelength. Moreover, it is relevant for the bandwidth characteristics of POF to control parameters like index profile and fiber geometry during production, but this is challenging and increase production costs. Several techniques to reduce modal dispersion have been investigated, including restricted launch conditions and mode filtering [11]. It is important to note that effects such as microbendings and external pressure on the fiber, are difficult to control. However, parameters like the geometry of the fiber and the launching conditions can be used to obtain improved bandwidth characteristics. An accurate analysis of dispersion in POF is required to establish bandwidth limitations and find ways to improve its characteristics. A comprehensive theory of dispersion in graded-index optical fibers was presented in [11], it includes most of the effects that influence the bandwidth of GI fibers, both chromatic and modal contributions. Modal delay, distributed loss, and mode coupling also were taken into account.

1.5. Potentials

Polymer optical fibers offer the potential to serve as the transmission medium for broad-band telecommunications service, for instance, in in-home networks and for LAN networking environments. There are several alternative technologies such as copper-based, silica single-mode fiber, and multimode glass fiber technologies. However, these technologies suffer from several drawbacks. Let us examine the advantages and disadvantages of POF with respect to these technologies. Compared to copper-based technologies like coaxial cables and twisted pair, POF guarantees EMC and the absence of cross talk, has smaller volume, is less bulky, is more flexible, and has light weight. For data transmission, POF offers higher bandwidth at longer transmission distances. Compared to multimode glass optical fiber, POF is easier to handle termination can be realized faster and cheaper. The typical large core of polymer fiber allows for greater tolerance on misalignments, which results in the possibility of making cheaper connectors. For comparison, let us examine the power loss due to lateral (axial) misalignment of connecting two graded-index (parabolic case) multimode fibers that have different core diameters. Assuming uniform modal power distributions, calculations for a misalignment of 25 μm yields a loss of 1.76 dB for a 62.5 μm core diameter multimode fiber (MMF). For the case of PF GI POF with a core diameter of 200 μm the 25 μm displacement results only in 0.48 dB loss. Another advantage of large core PF GI POF observed with respect to the bandwidth degradation due to modal noise at misalignments in fiber-to-fiber connections, namely, PF GI POF shows very short time delay at the wide area of the core. This is in contrast to the case of multimode silica fiber that has a 50–62.5 μm core, where small displacements cause severe bandwidth degradation [12].

With respect to data transmission, POF has the potential of high bandwidth and less problems with modal dispersion. Although, the attenuation and bandwidth characteristics of POF are inferior compared to standard single-mode fiber, POF offers the mentioned advantages of ease of coupling, termination, and flexibility. In the case of standard single-mode fibers, specialized precision mechanisms have to be used for coupling and handling. We can see that POF data transmission characteristics at the current state of the art do not pair those of standard single-mode fiber, but they are superior to those of copper-based technologies. Furthermore, installation and termination of POF are easier and promise low costs compared to single-mode fiber and multimode glass fiber. It is worth noting that connector devices for POF are not only easier to assemble compared to those for single-mode glass fiber, but also easier than those for high frequency coaxial cabling. Termination of coaxial cabling requires more skilled handling, for instance improper termination will causes large loss and considerably increased ingress noise that results in considerable degradation of system performance. On the other hand, POF connectors can be assembled easily, for instance, using a low cost polymer ferrule [12]. Coupling loss using this method has been measured to be on the order of 0.8 dB at 850 nm wavelength [12]. Large core glass fiber shows lower attenuation than POF, but the core size is restricted to 200 μm due to the inherent inflexibility of glass. In this situation, POF again offers its advantages of ease of handling and termination, and tolerance to misalignments.

Ethernet is a widely deployed networking technology. Gigabit Ethernet standard, operating at 1.25 Gbit/s, supports a range of transmission lengths: 100 m over copper wire 550 m over multimode glass fiber with a 50 or 62.5 μm core, and 5 km over single-mode fiber. Experiments performed by our group have demonstrated 1.25 Gbit/s Ethernet transmission successfully over PF GI POF, reaching a distance of 990 m with good bit-error rate performance. These experiments show that PF GI POF can be used for Gigabit Ethernet applications in short to medium link distances. The transmission results presented in this chapter are record experiments. They show the feasibility of POF for high capacity transmission. The reported experiments use a receiver with avalanche photodiode (APD) photodetector, which is not exactly a cost competitive option in relation to conventional PIN photodiode receivers. A receiver with a conventional, less costly, PIN photodiode also can be used at the expense of lower receiver sensitivity. Assuming a degradation of 11 dB in receiver sensitivity (16.9 dBm receiver sensitivity at 2.5 Gbit/s using a pin detector was reported in [13]) with respect to the APD receiver, we estimated the following reachable transmission distances at 2.5 Gbit/s: at the 840 nm operating system over PF GI POF with attenuation of 43.6 dB/km, a distance of 280 m could be feasible; for a system operating at 1310 nm with a fiber attenuation of 31 dB/km, a transmission distance of 430 m could be reached. However, for POF technology to be competitive in the customer's premises, it will have to operate with low-cost components, including a light sources, receiver modules, and WDM devices.

We would like to remark that the transmission end shows less technical problem than the reception end. This also applies for the case of WDM transmission systems. WDM multiplexing can be performed simply by butt joining the pigtailed fibers from the laser sources, for instance, as reported in our experiments. It remains, therefore, of high importance to further develop POF-based transmission systems and low-cost, reliable transceiver modules, connectors, and WDM (de)multiplexing devices. Light sources based on vertical-cavity surface emitting lasers (VCSELs) are promising solutions. Perhaps the use of plastic–polymer lens systems for fiber to photodetector coupling could realize compact and reliable receivers. These challenging issues, among others, are open for further research and development.

1.6. POF Data Link and System Devices

In this section, we review the components being used in experimental POF data links, such as light source, detectors, and couplers, etc. A generic block diagram for a POF data transmission link is presented in Figure 3a. The main systems building blocks are the sources of light, photodetectors, and light source to the fiber and the fiber to photodetector

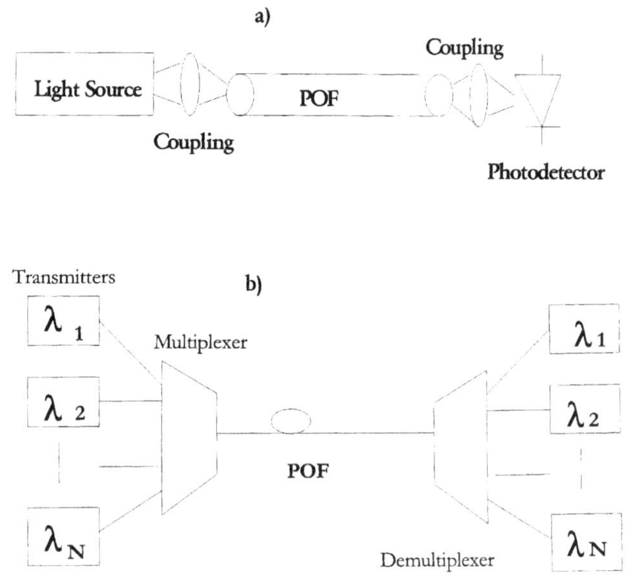

Figure 3. Diagram of general POF data transmission link: (a) single channel; (b) several WDM channels over POF.

coupling devices. In advanced systems like in LAN applications and taping devices, wavelength (de)multiplexer are required; see Figure 3b.

1.6.1. Light Sources

Most of the existing multimode silica fiber links use broadband light sources such as light emitting diodes (LEDs) and Fabry–Perot lasers. These sources also have been used in many reported experimental transmission links over polymer fibers. Vertical-cavity surface emitting lasers have been used for multimode fiber links and for experimental POF data trials. VCSELs are reported to become a low-cost light source for LAN applications [14]. They have the potential to achieve high modulation rate and bandwidth, operate at low threshold current, allow the possibility to be manufactured in arrays, operate at the low-loss transmission wavelengths of POF at 850, 650, and 1300 nm. Research is ongoing to develop integrated sources with POF. Work on integrating a monolithic array of VCSELs with POF was reported in [15]. It is believed that VCSELs could provide a source for fiber optical distribution of radio frequency signals. Analog modulation of VCSELs operating at 850 and 650 nm have been reported [16, 17]. Effort is focused on the development of light source devices for operation with POF, the most promising candidates are VCSEL devices.

1.6.2. Coupling and Connectors

An important element in a POF link is the coupling, that is, analog to the fiber and coupling to the detector. In laboratory trials, lenses are commonly used for this task. However, for deployment of practical links, effective coupling devices must become available. Large core polymer fiber enables the use of injection-molded plastic connectors, which would replace the complicate lens alignment systems and decrease the cost of the system [10]. Perhaps a solution for efficient fiber-to-photodetector coupling can be found in the integration of a polymer lens in a receiver module. Simple and low cost fiber-to-fiber connection and fiber termination devices also are required for POF-based networks. At present, POF connectors are available that use a low-cost plastic ferrule [12].

1.6.3. Photodetectors

As just mentioned, coupling from the large core POF to the detector is an important issue. Therefore, photodetectors with from a large sensitive detection area and high speed are preferred for POF data links. This is a challenging issue because detectors that bared a large sensitive area are usually slower than small area detectors. Experimental trials that use, large area APDs and metal-semiconductor-metal receivers (MSM) have been reported. Both Si and GaAs technology has been considered for POF photodetectors. Realization of a AlGaInP multi-quantum-well laser diode operating at 650 nm wavelength has been reported for POF transmission applications Research also has been dedicated to the integration of a detector receiver for POF fibers with a core of 250 μm. High speed operation up to 600 Mbit/s has been demonstrated experimentally [19].

1.6.4. Wavelength (De)Multiplexers

Further capacity enhancement can be reached by the use of wavelength division multiplexing. To employ this technique, POF compatible multiplexers and demultiplexers, must become available. In laboratory trials, a set of dielectric thin-film filters has been used to demultiplex of optical channels in a POF data link [20]. Multiplexing of channels was achieved by butt joining a bundle of small core fibers (pig-tailed to the light sources) to a single, but large, core POF fiber. A promising technology under consideration for WDM POF (de)multiplexers is based on the use of polyimide waveguides [21]. For the case of multimode glass fiber link as a compact WDM transceiver based on the use of VCSELs, and polymer waveguide demultiplexers and multiplexers has been proposed [14]. Perhaps similar transceiver modules will become available for POF data links.

1.7. Review of Data Transmission Experiments

In this section, we present a review of transmission experiments using POF as the transmission medium. The experimental trials reported in Table 2 have demonstrated the feasibility of POF links for high capacity transmission. These trials represent record achievements in data transmission over POF and are in chronological order; see Table 2. We can see how POF based transmission has evolved in the recent past years to show the feasibility of high data rate transmission. The summary in Table 2 is by far not complete, but it clearly shows the improvements in reached transmission capacity and link distances.

Table 2. List of experimental transmission records using POF.

Date	Bit rate (Gbit/s)	Distance (m)	Wavelength (nm)	Fiber material
1993	0.531	100		PMMA (SI)
1994	2.5	100	650	PMMA (GI)
1994	1.0	30	670	PMMA (GI)
2-1998	2.5	200	645	PMMA
8-1998	5	200	1310	PF
10-1998	2.5	300	645	PF
11-1998	2.5	550	1310	PF
1-1999	2.5	550	840	PF
9-1999	3 × 2.5	200	645, 840, 1300	PF
3-2000	2 × 2.5	328	840, 1300	PF
7-2000	2 × 2.5	456	840, 1300	PF

Source: G. D. Khoe et al. "Proceedings of the International Optoelectronics Symposium," Kyoto University, June 7–9, 2000, Vol. 55, No. 2, pp. 9–16.

1.8. Concluding Remarks

Polymer optical fiber have reached a development phase were they are a promising technology for delivery of broadband telecommunication services at the customer's premises. The introduction of improved POF such as perfluorinated graded index has opened the road for high capacity transmission over POF at link distances up to 1 km. Transmission rates up to 11 Gbit/s over 100 m of POF have been experimentally demonstrated [22]. Furthermore, using WDM transmission techniques record bit-rate distance product of 2.28 Gbit.km/s has been achieved operating at 840 and 1300 nm wavelengths. Moreover, transmission of 1.25 Gbit/s Gigabit Ethernet traffic has been demonstrated over nearly 1 km of POF. New techniques for capacity enhancement of POF are under investigation [23]. For instance, dispersive multiplexing has been explored for multimode fibers [24]. This technique has been demonstrated for low bit rates but it is scalable to high rate by using of high speed electronic processing [24].

POF-based systems, apart from high capacity, have important advantages over copper and single-mode fiber technology such as the prospect of low cost, and ease of connection and termination.

2. PMMA GI POF SYSTEMS

This section presents transmission systems that use based graded-index polymer optical fiber (PMMA GI POF). Single channel as well as wavelength division multiplexing transmission systems are described. The systems operate at bit rates up to 2.5 Gbit/s.

2.1. Single Channels Systems

PMMA POF-based fibers that use graded-index profile were introduced they showed enhanced bandwidth compared to their step-index counterpart [4]. The introduction of PMMA GI POF opened the path to high capacity transmission over POF-based systems and networks. Experimental work has explored the limits of PMMA GI POF-based systems. A transmission experiment using one wavelength channel and its enabling technology are described in detail

2.1.1. Experimental Set-up

A 2.5 Gbit/s system experiment over 200 m that use a PMMA GI POF and a visible light laser at 650 nm wavelength will be described. Figure 4 is a block diagram of the experiment. As it can be seen, the setup is very basic. The transmitter side consists of a laser driver and a laser diode, which is coupled to the GI POF by means of a lens system. At the receiverside, light from the GI POF is coupled into a photodiode by means of a second lens system. In the receiver, the detected signal of the photodiode is amplified and filtered, and the data and its clock are recovered. To test this digital system, the bit-error rate (BER) was measured with a bit-error rate test set, which consists of a pseudo random binary sequence pattern generator at the transmitter side and a BER detector at the receiver side.

Key elements used in the experiment are a silicon APD receiver with a record sensitivity of −29 dBm at 2.5 Gbit/s and 10^{-9} bit-error rate, a PMMA GI POF that has low attenuation at 0.164 dB/m and a large core diameter of 0.5 mm (Mitsubishi Rayon), and a laser (NEC) that has a modulated spectral width of 0.4 nm at an average output power of 6.8 dBm. The experiment was carried out using a nonreturn to zero (NRZ) PRBS with a pattern length of $2^{23} - 1$.

One important feature in the APD receiver was the combination of the APD with two limiting amplifiers (HP) in a low-impedance front-end configuration. In principle, the APD can be set to a higher gain combination with just one amplifier, but that will cause degradation due to excess noise. Another significant improvement is the use of a data and clock recovery (IC; Lucent) instead of a direct clock connection and decision circuit to the BER equipment. To avoid electrical reflections, the APD, limiting amplifiers, and clock recovery circuit were mounted as compact as possible.

Another key to the result is a new version of the GI POF as compared to those reported earlier in the literature. The loss of the GI POF decreased from 0.2 to 0.164 dB/m by improving the homogeneity along the fiber by minimizing the fiber outer and core diameter variation, and asymmetry

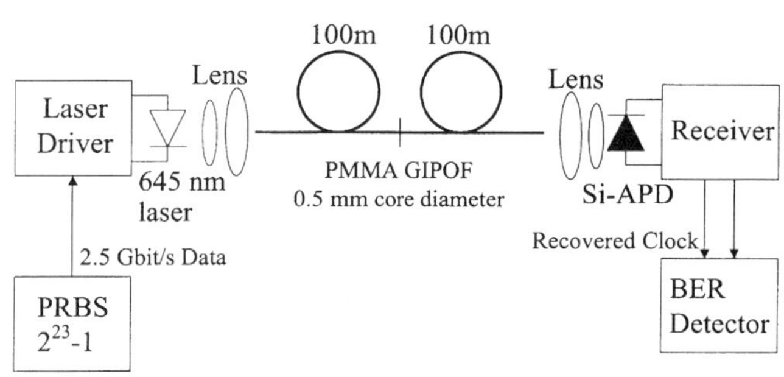

Figure 4. Block diagram of 2.5 Gbit/s transmission over 200 m PMMA GI POF.

of the index profile and its change along the fiber. This is achieved by more precise control of the preform preparation, the drawing process, and the cabling process.

Also contributing to the achievement are measures taken to avoid reflections and to improve optical coupling efficiency between laser, fiber, and APD. Laser light is launched by means of a lens doublet that offers a numerical aperture (NA) of 0.55 at the laser side and 0.16 at the fiber side. At the receiver side, a lens doublet is used that has an NA of 0.25 at the fiber side and 0.55 at the APD side. The diameter of the active area of the APD was 230 μm. Coupling losses for both the laser and APD sides were less than 0.3 dB. The lenses are anti reflection (AR) coated to less than 1% reflection. The fiber end at the laser side was set at an angle of 4° to avoid back reflections into the laser. The coupling optics was manually adjusted to achieve maximum coupling efficiency.

2.1.2. System Performance

The eye diagram before and after transmission over 200 m of GI POF are nearly identical, indicating a sufficient bandwidth in this situation (see Fig. 5). The BER against received average power at the input of the APD, with and without 200 m GI POF, was measured (see Fig. 6). The received power was changed by altering the distance between the laser and the GI POF. The sensitivity of the receiver was −29 dBm at 2.5 Gbit/s for a BER of 10^{-9}. The laser output power was 6.8 dBm, so the available power budget was 35.8 dB.

The attenuation of the 2 × 100 m GI POF was 32.8 dB. A 2 dB receiver sensitivity degradation was observed, due to modal dispersion of the fiber (see Fig. 6). The total coupling losses were 0.6 dB, so a power budget of 35.4 dB was needed (see Table 3). Moreover, the optical output spectrum of the modulated laser at an average output power of 6.8 dBm was measured (see Fig. 7). The width of the spectrum 3 dB below the maximum value is 0.4 nm, which limits pulse broadening due to dispersion of the fiber.

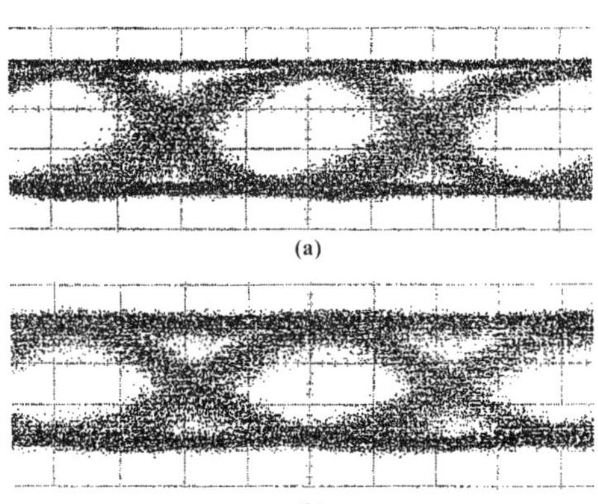

Figure 5. Eye diagram of (a) back-to-back and (b) after 200 m GI POF transmission.

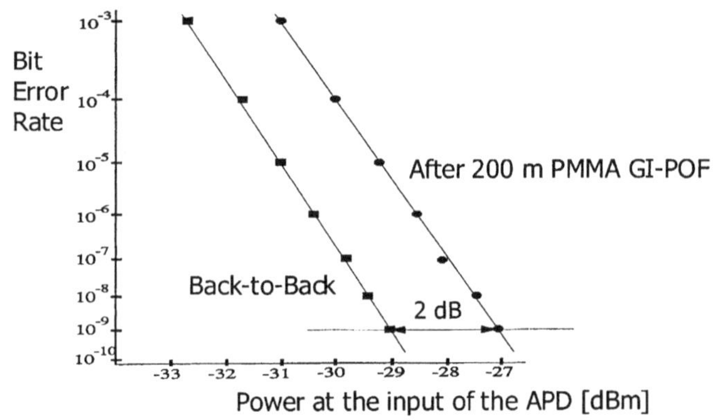

Figure 6. Bit-error rate measurement results for a 2.5 Gbit/s transmission system over 200 m of PMMA GI POF.

2.1.3. Discussion of Results

It is of interest to analyze why our results outperform those reported so far. In addition to the receiver sensitivity, we believe that both the method of excitation as well as the spectral characteristics of the exciting source play a significant role. Due to the tilting of the launching beam, back reflections from the fiber's input surface into the laser source are avoided. Semiconductor lasers subject to back-reflected waves in optical communication systems may undergo different and complicated states of behavior. Unless the reflected light is well monitored, optical feedback is often detrimental, because it enhances noise and introduces multiple nonlinearity into the emission characteristics that degrade the signal-to-noise ratio at the receiver. Obviously, the fact that such an impairment was avoided may have contributed to the achievement of a good system bit-error rate.

Another explanation for the present performance can be found in connection with the spectral characteristics of the exciting source. The measured spectral width was 0.4 nm as indicated earlier (see Fig. 8). This small value certainly balances the effect of the material dispersion of the fiber. The influence of the wavelength was checked by referencing to the existing literature. It is seen that the bandwidth is significantly enhanced for the 0.4 nm compared to the 2 nm case reported in the literature. The dispersion behavior of the fiber may be even better under the conditions of our experiment. Indeed, the exciting beam was nearly parallel, meaning that the numerical aperture was not overfilled as in the case of the curves in the literature. However the fiber should exhibit less modal dispersion because the number of propagated modes is

Table 3. Power budget.

Attenuation 200 m PMMA GI POF (164 dB/km)	32.8 dB
Laser to GI POF loss	0.3 dB
GI POF to APD loss	0.3 dB
Dispersion fiber	2.0 dB
Needed power budget	35.4 dB
Laser output power	6.8 dBm
Sensitivity receiver (BER = 10^{-9})	−29.0 dBm
Available power budget	35.8 dB

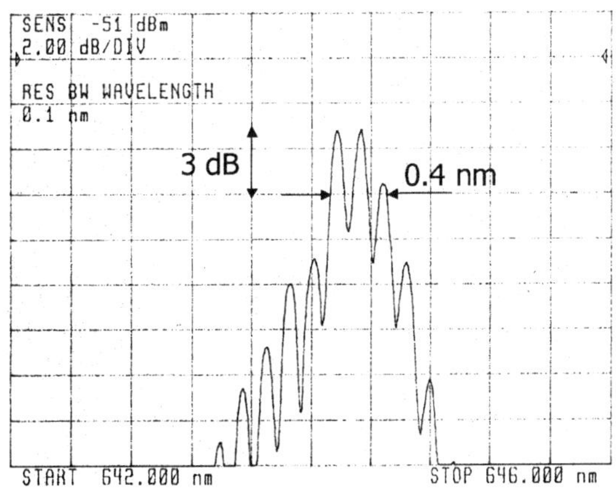

Figure 7. Measured optical spectrum of a modulated laser used in the 200 m PPMA GI POF transmission experiment.

less than can be excited under full launch condition, and this may lead to a further bandwidth improvement. The curves suggest that transmission experiments above 5 and 10 Gbit/s over 200 m may be possible if the fiber is graded among the optimum profile region. Because the intrinsic bandwidth of the laser was limited to 3 GHz, these experiments could not be carried out.

2.2. WDM Systems

A powerful method to increase the transmission capacity of an optical link is to send a number of different wavelengths along a single fiber simultaneously, applying wavelength division multiplexing. This principle is shown in Figure 8. To combine signals with different transmitter wavelengths, a wavelength multiplexer has to be used at the sending side. To separate the wavelength channels at the receiver side, a wavelength demultiplexer has to be used.

2.2.1. Multiplexing

Investigation in the possibilities of a wavelength division multiplexing system over PMMA graded-index polymer optical fiber as begun. A great advantage of the large core diameter of the GI POF is that joint coupling of two or more glass fiber laser diode pigtails can be used to multiplex signals but. The coupling efficiency versus lateral displacement of a single-mode fiber, which couples light into a graded-index fiber, has been calculated and measured. The result can be seen in Figure 9. A displacement of 70% of the core radius of the graded-index fiber gives hardly any degradation.

A realized multiplex configuration using but coupling is shown in Figure 10, in which two multimode glass input fibers, both 140 μm in diameter, are connected with one, 750 μm in diameter GI POF. This wavelength multiplexer works quite well with less than 1 dB loss. The multiplexer simplicity is a good example of the advantages of the large core of the polymer optical fiber.

2.2.2. Demultiplexing

The wavelength demultiplexer is a more complex subsystem than the multiplexing section. Several configurations were considered, resulting in the user of the so-called Littrow configuration shown in Figure 11. This configuration uses a reflective grating to separate the two wavelengths. The lens, on the one hand, collimates the incoming beam that originates from the GI POF and, on the other hand, focuses the separated beams or their respective output. Here, short pieces of step-index (SI) polymer optical fiber (POF) have been used. Because lasers with wavelengths of 645 and 675 nm were available, these two wavelengths were chosen. The difficulty in this set up lies in the large diameters of the fibers used and, thus, the large distances between fibers and lens, however, measurements using white light as the light source showed a very promising result. The transmission characteristics from input GI POF fiber to output SI POF fibers were measured. Figure 12 shows the wavelength regions covered by the two output fibers. As it can be seen in Figure 12, at the two center wavelengths of 645 and 675 nm the cross-talk level is much lower than −25 dB. Because the spectra of the lasers are very narrow (see, e.g., the optical spectrum of the 645 nm channel in Figure 13), the two wavelength channels are not influenced by cross talk.

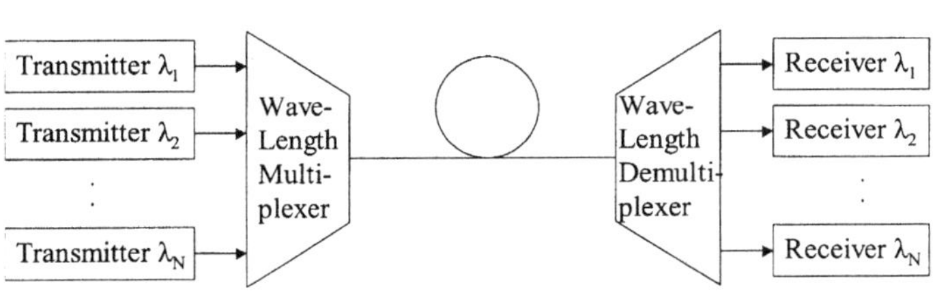

Figure 8. Principle of WDM. Several channels that operate at different wavelengths propagate along a single POF.

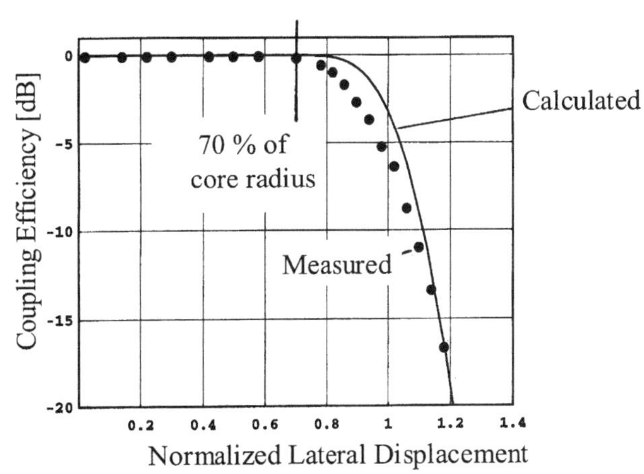

Figure 9. Coupling efficiency of standard single-mode fiber to graded-index fiber versus lateral displacement.

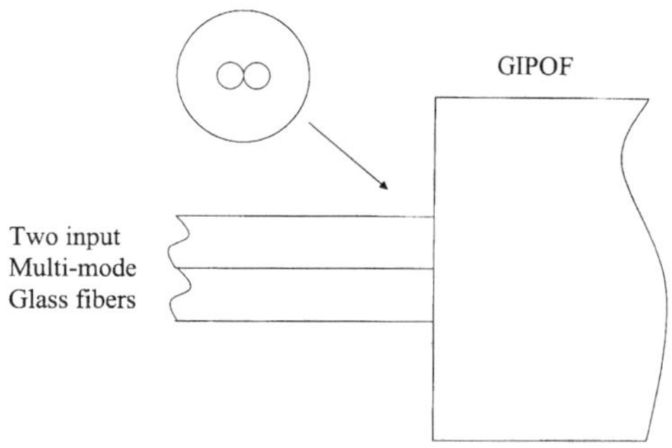

Figure 10. Multiplexer configuration for a WDM system using GI POF.

The loss introduced in this configuration is less than 5 dB and can be reduced further by some adaptations of the lens and grating. The 2.5 Gbit/s 645 nm transmitter described in the previous paragraph and a 675 nm transmitter that use a standard visible light laser have been used in the WDM transmission experiment. Because the 675 nm standard visible light laser was not developed for high data rate modulation, only 622 Mbit/s could be achieved at this wavelength. Simultaneous 2.5 Gbit/s and 622 Mbit/s WDM transmission over a single PMMA GI POF has been achieved.

2.3. Concluding Remarks

Polymethyl methacrylate-based GI POF has a low-loss wavelength region around 650 nm. Under single channel operation, a transmission distance of 200 m was reached at a bit rate of 2.5 Gbit/s. Applying WDM can further enhance transmission capacity via POF. Using butt coupling of glass optical fiber laser pigtails, wavelength channels easily can be multiplexed and coupled into the GI POF. A realized grating demultiplexer split up wavelength channels at 645 and 675 nm was described.

3. PERFLUORINATED GI POFS

In this section, record transmission systems using perfluorinated graded-index polymer optical fiber are reviewed. Single channel as well as wavelength division multiplexing transmission systems are described. Systems that operate at bit rates up to 5 Gbit/s are demonstrated using wavelengths of 645, 840, and 1310 nm. Transport of 1.25 Gbit/s Giga-

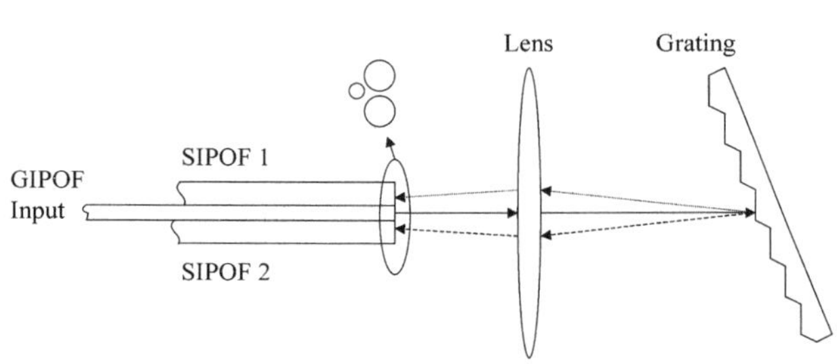

Figure 11. Demultiplexer configuration for a WDM system using GI POF.

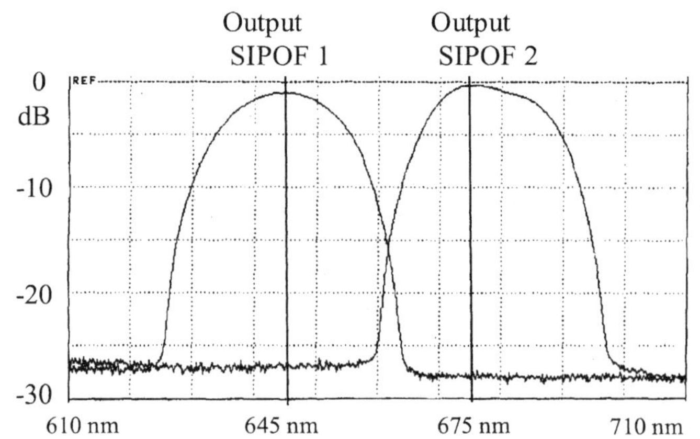

Figure 12. Measured results of the spectral characteristics of the realized demultiplexer.

bit Ethernet data is also demonstrated experimentally over nearly 1 km of PF GI POF.

3.1. Single Channel Transmission

The introduction of PF GI POF has brought POF–based technology to a stage of development where it is a strong candidate for the physical layer in several applications like in-home LAN and short distance transmission scenarios. PF GI POF offers large bandwidth and low attenuation over a broad spectral range between 500 and 1300 nm [5]. The experimental trials reported in this section explored the limits and feasibility of high capacity transmission over POF-based systems. Experimental details are presented in the following subsections.

3.1.1. 2.5 Gbit/s Transmission Over PF GI POF

At 840 and 1310 nm wavelength, distances of 550 m have been achieved at a bit rate of 2.5 Gbit/s. These results were obtained through the use of improved low-loss fiber, very sensitive large active area avalanche photodiode receivers, laser sources with a small spectral width, launching into the fiber with only a few modes, and improved optical interconnection between the components.

Perfluorinated polymer-based GI POF has a low-loss wavelength region from 500 to 1300 nm. The 550 m GI

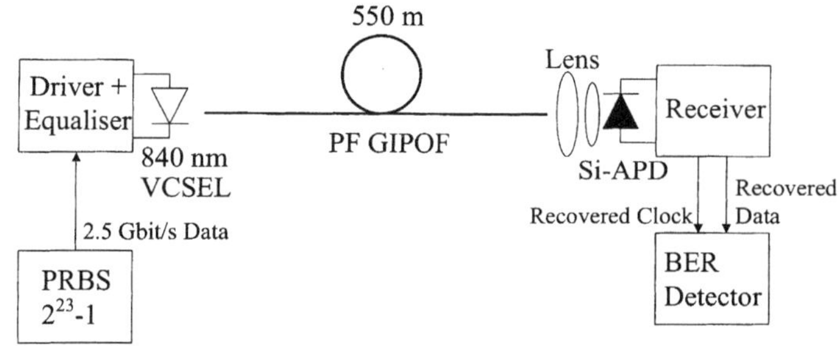

Figure 13. Experimental setup for a transmission system at 840 nm over 550 m of GI POF.

POF we used has an attenuation of 43.6 dB/km at 840 nm and 31 dB/km at 1310 nm. With the 2.5 Gbit/s experiment at 1310 nm, a transmission distance of 550 m was reached by using a very sensitive large active area (APD) receiver and low-loss interconnection between the GI POF and the APD receiver. In the 2.5 Gbit/s experiment at 840 nm, a VCSEL was used in combination with a silicon APD receiver. Figure 13 shows a block diagram of the 550 m experiment at 840 nm.

Key elements used in the 840 nm experiment are a VCSEL with a high bandwidth of 2 GHz, a silicon APD with a large active area of 230 μm in diameter, and GI POF with a low attenuation of 43.6 dB/km at 840 nm as mentioned before. To compensate for the insufficient bandwidth of the VCSEL, an electrical equalizing circuit was used. Key elements in the 1310 nm experiment are a distributed feedback (DFB) laser with a high bandwidth of 5 GHz, an InGaAs APD with an active area of 80 μm in diameter, and GI POF with a low attenuation of 31 dB/km (Asahi Glass Company) at 1310 nm. (see Fig. 14). Both experiments were carried out using a NRZ PRBS with a pattern length of $2^{23} - 1$. Also contributing to the achievement are measures taken to avoid reflections and to improve optical coupling efficiency between laser and fiber, and fiber and APD. At the transmitting side in the 840 nm experiment, light is launched directly from the VCSEL component into the GI POF. Because the VCSEL component already was provided with a lens, which was not optimized for coupling with the GI POF, the coupling loss was 1 dB. In the 1310 nm experiment, the light from a single-mode fiber (SMF) pigtail of the DFB laser was launched into the large core of the GI POF by means of a butt coupling with losses less than 0.1 dB. At the receiver side in both experiments, a lens doublet was used that had a NA of 0.25 at the fiber side and 0.55 at the APD side. GI POF-to-APD coupling loss was less than 0.3 dB. The lenses are AR coated to less than 1% reflection. The coupling optics was manually adjusted to achieve a maximum coupling efficiency.

3.1.2. System Performance

The rate BER as a function of received average power at the input of the APD was measured in a back-to-back configuration and after 550 m GI POF transmission (see Fig. 15) The back-to-back measurements were carried out with a short piece of a few meters of GI POF between transmitter and receiver. In case of the 840 nm system, the received power was changed by altering the distance between the laser and the GI POF.

In the 840 nm experiment, the average output power of the VCSEL was 1.3 dBm and the sensitivity of the receiver was -28.6 dBm at 2.5 Gbit/s for a BER of 10^{-9}, so the available power budget was 29.9 dB. The attenuation of the 550 m GI POF was 24.0 dB. The power penalty due to multimode dispersion of the fiber and model noise was 4.5 dB (see Fig. 15). The total coupling losses were 1.3 dB, so a power budget of 29.8 dB was needed (see Table 4). The measured BER as a function of received average power at the input of the APD with and without 550 m GI POF of the 1310 nm system is shown in Figure 16. With these measurements, a variable attenuator was used after the laser diode to vary the optical power.

In the 1310 nm experiment, the average output power of the DFB laser was 0.4 dBm and the sensitivity of the receiver was -28.4 dBm at 2.5 Gbit/s for a BER of 10^{-9}, so the available power budget was 28.8 dB. The attenuation of the 550 m GI POF was 16.3 dB at 1310 nm. The power penalty due to multi-mode dispersion of the fiber and model noise was 4.4 dB (see Fig. 16). The total coupling losses were again 0.4 dB, so a power budget of 2.11 dB was needed (see Table 5). Because of the difference of 7.7 dB between available and needed power, a distance of 750 m probably can be reached.

The modulated spectral width was smaller than 1 nm for the 840 nm laser source (see Fig. 17) and smaller than 0.1 nm for the 1310 nm laser source. These small values certainly balances the effect of the material dispersion of the fiber. Dispersion is further avoided by the launching condition of the experiments. In the case of the 1300 nm experiments, the SMF pigtail of the laser source was butt jointed to the GI POF exciting only a few modes. In the case of the 840 nm experiment, also only a few modes were excited, because the exciting beam was nearly parallel, meaning that

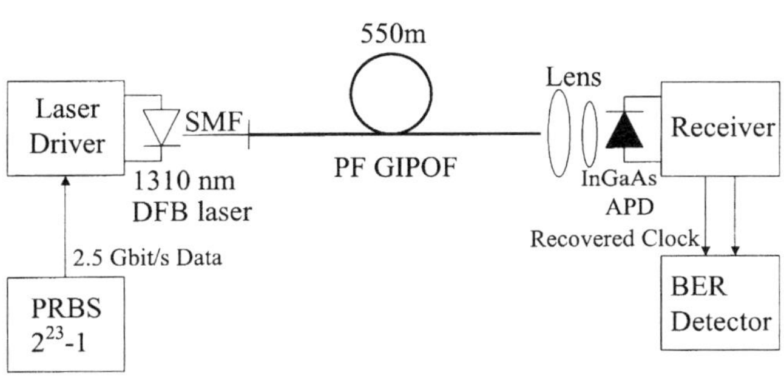

Figure 14. Experimental setup for data transmission over GI POF at 1310 nm wavelength.

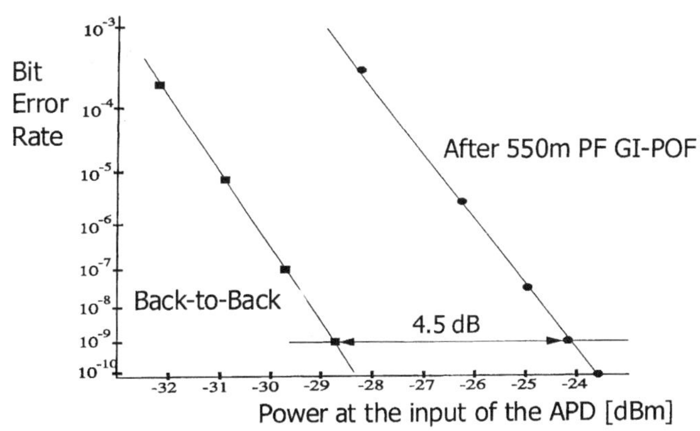

Figure 15. Bit-error rate for a system operating at 2.5 Gbit/s over 550 m PF GI POF at 840 nm.

Table 4. Power budget of the 840 nm transmission system.

Attenuation 550 m PF GI POF (43.6 dB/km)	24.0 dB
Laser to GI POF loss	1.0 dB
GI POF to APD loss	0.3 dB
Dispersion fiber	4.5 dB
Needed power budget	29.8 dB
Laser output power	1.3 dBm
Sensitivity receiver (BER = 10^{-9})	−28.6 dBm
Available power budget	29.9 dB

Table 5. Power budget of the 1310 nm transmission system.

Attenuation 550 m PF GI POF (31 dB/km)	16.3 dB
Laser to GI POF loss	0.1 dB
GI POF to APD loss	0.3 dB
Dispersion fiber	4.4 dB
Needed power budget	21.1 dB
Laser output power	0.4 dBm
Sensitivity receiver (BER = 10^{-9})	−28.4 dBm
Available power budget	28.8 dB

the numerical aperture was not overfilled. So the fiber should exhibit less modal dispersion because the number of propagated modes is less than in the case of full launch condition, and this may lead to further bandwidth improvement.

A PF GI POF 2.5 Gbit/s transmission system at 645 nm light with a distance of 300 m will be discussed (see Fig. 18). For this experiment, the transmitter and receiver modules of the 2.5 Gbit/s over 200 m PMMA GI POF were used. Figure 19 shows the measured BER versus received average power at the input of the APD with and without 300 m GI POF. The sensitivity of the receiver was −29 dBm at 10^{-9} BER. The laser output power was 6.2 dBm, so the available power budget was 35.2 dB. The attenuation of the 3 × 100 m GI POF was 32.3 dB. The power penalty due to modal dispersion of the fiber was 1.0 dB. The total coupling losses were 0.6 dB, so a power budget of 34.2 dB was needed.

3.1.3. Transport of 1.25 Gbit/s Gigabit Ethernet Data

A perfluorinated polymer-based GI POF transmission system that achieved a record distance of nearly 1 km will be described. The experiment was carried out at a wavelength of 840 nm and a bit rate of 1.25 Gbit/s, which is used for Gigabit Ethernet applications. The Ethernet protocol is used on 80–85 of the world's LAN-connected PCs and workstations because it has been adapted to meet the changing needs of its customers while remaining simple, interoperable, robust and low cost. The first generation Ethernet was defined in 1985 and worked at a data rate of 10 Mbit/s. Fast Ethernet, which became a standard in 1995, uses the same Ethernet frame format but with 10 times the transmission speed, namely 100 Mbit/s. Gigabit Ethernet is the most important addition to the Ethernet protocol; it is capable of transmission of 1,000 Mbit/s, so it is again 10 times faster. The Gigabit Ethernet physical layer uses 8B10B block coding to add extra data clock information, to reduce the low frequency components, and for error detection. In this coding technique, 8 bits of data are encoded to 10 bits, which increases the link transmission rate to 1250 Mbit/s. The Gigabit Ethernet standard supports a maximum length of 100 m over copper write, 550 m over multimode glass optical fiber with a core diameter of 62.5 or 50, 50 km and 5 km over single-mode fiber. This 1 km experiment shows that GI POF, which is easy and cheap to install, is a very strong candidate for short to medium distance Gigabit Ethernet applications.

Key elements used in the transmission experiment are an improved low-loss GI POF launched with only a few modes, a VCSEL with a small spectral width and a high bandwidth of 2 GHz, and a silicon APD with a large active area of 230 μm in diameter (see Fig. 20). The BER was measured as a function of received average power at the input of the APD with the 990 m GI POF and with only a short piece (a few meters) of GI POF between he transmitter and the receiver. The total link was built up from three separate samples of 330 m each. The measurements were carried out using a (NRZ) (PRBS) with a pattern length of $2^{23} - 1$. The average output power of the VCSEL was 1.1 dBm and the sensitivity of the receiver was −31.3 dBm for a BER of 10^{-9} and a bit rate of 1.25 Gbit/s, so the available power budget was 32.4 dB (see Table 6). The total attenuation of

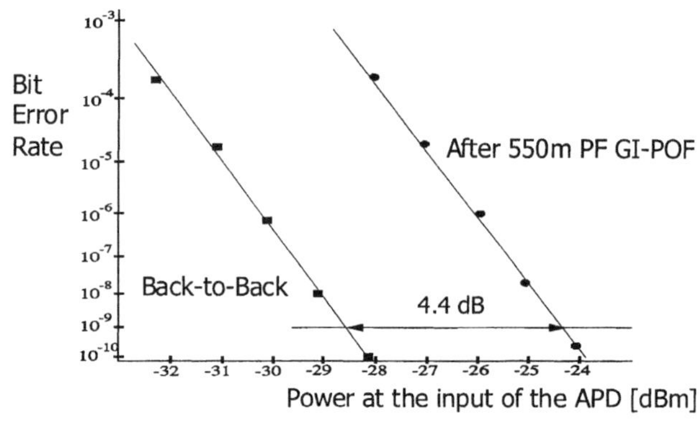

Figure 16. Experimental setup for data transmission over GI POF at 1310 nm wavelength.

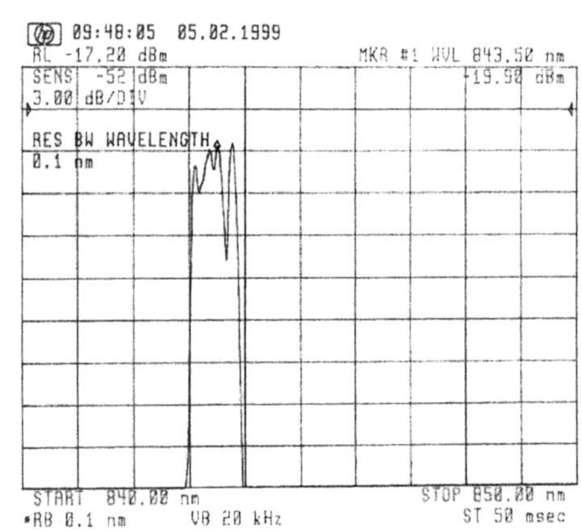

Figure 17. Optical spectrum of modulated 840 nm VCSEL.

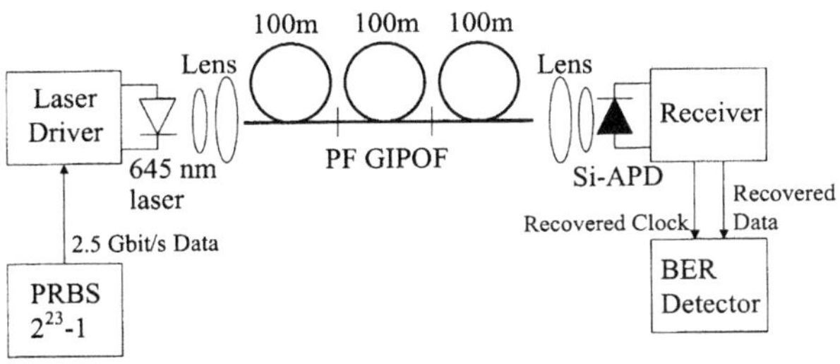

Figure 18. Setup of 2.5 Gbit/s transmission over 300 m PF GI POF at 645 nm wavelength.

the 990 m GIPOF link at 840 nm wavelength was 26.9 dB. The power penalty due to multimode dispersion of the fiber and to modal noise was 1.1 dB. The total coupling loss from VCSEL to GIPOF and from GIPOF to APD was 2.6 dB, so a total power budget of 30.6 dB was required.

3.1.4. 5 Gbit/s Transmission

Using the setup presented in Figure 21, an error-free transmission experiment at a bit rate of 5 Gbit/s was carried out over 200 m PF GIPOF at 1300 nm. Because a large area APD with sufficient bandwidth for 5 Gbit/s was not available, a pin photodiode which gives lower receiver sensitivity had to be used.

3.1.5. 2.5 Gbit/s Transmission Employing a MSM Detector

Using a MSM detector developed by the Electronic Devices Group at Eindhoven University of Technology a 2.5 Gbit/s over m GIPOF was carried out at 840 nm. A block diagram of this setup is shown in Figure 22. The MSM detector has a large active area of 100 × 100 mm and is, therefore, easy to couple with a large core fiber. Electrically the MSM detector was coupled with an amplifier that had an input impedance of 50 W. The receiver sensitivity was −6 dBm for a BER of 10^{-9}. This sensitivity can be improved by using a transimpedance amplifier.

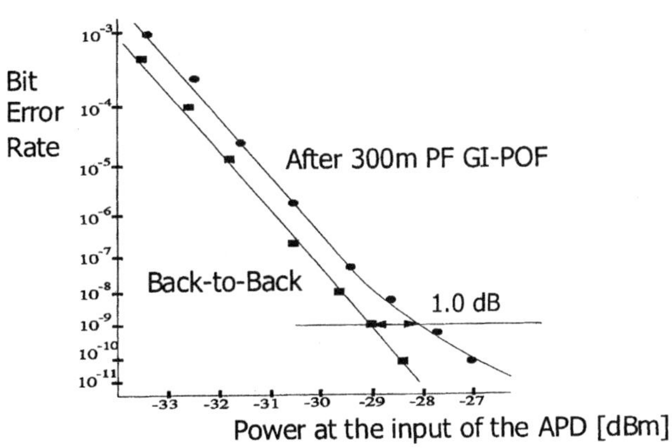

Figure 19. Bit-error rate of 2.5 Gbit/s transmission over 330 m PF GI POF at 645 nm.

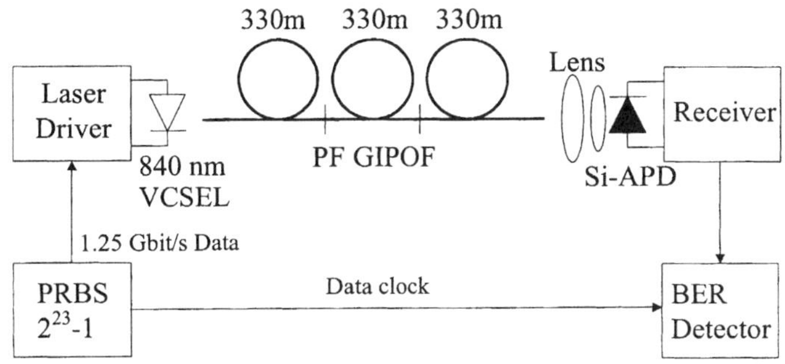

Figure 20. 1.25 Gbit/s transmission over nearly 1 km GI POF using an 840 nm VCSEL and a Si APD receiver.

3.2. WDM Experiments

Applying WDM can further enhance transmission capacity via POF; see Section 2.2. WDM also increases the flexibility in POF networks. Perfluorinated polymer-based GIPOF has a low-loss wavelength region from 500 to 1300 nm, so a broad wavelength range is available for the implementation of coarse WDM, as opposed to dense WDM. Contrary to dense WDM, coarse WDM can use low-cost technology to separate the wavelength channels.

3.2.1. Demultiplexing

As a start development of POF-based WDM systems, a demultiplexer for splitting up the 645, 840 and 1310 nm, wavelengths has been realized with planar interference filters. In Figure 23, the principle of operation of the demultiplexer is shown. First the light from the input GI POF is transformed to a parallel beam by means of lens 1. Interference filter 1 deflects the light in the 645 nm wavelength region. The other wavelengths are passed through. To decrease cross talk, an extra filter 2 has been used that is transparent only the 645 nm wavelength region. Lens 2 focuses the light at the photodiode of the 645 nm receiver. The light in the 840 and 1310 nm wavelength regions, which passed through filter 1, is split up by filter 3. Light in the 840 nm wavelength region is deflected by filter 3, filtered by filter 4, and focussed on the detector of the 840 nm receiver by lens 3. The remaining 1310 nm light is focused on the 1310 nm detector by lens 4. The measured insertion losses for all three wavelengths from GI POF input to photodetectors are smaller than 1.6 dB. Measured cross-talk levels are

Table 6. Power budget of the 1.25 Gbit/s over nearly 1 km GI POF.

Attenuation 990 m GI POF	26.9 dB
Laser to GI POF loss	1.0 dB
GI POF to APD loss	1.6 dB
Fiber dispersion penalty	1.1 dB
Needed power budget	30.6 dB
Laser output power	1.1 dBm
Sensitivity receiver (BER = 10^{-9})	−31.3 dBm
Available power budget	32.4 dB

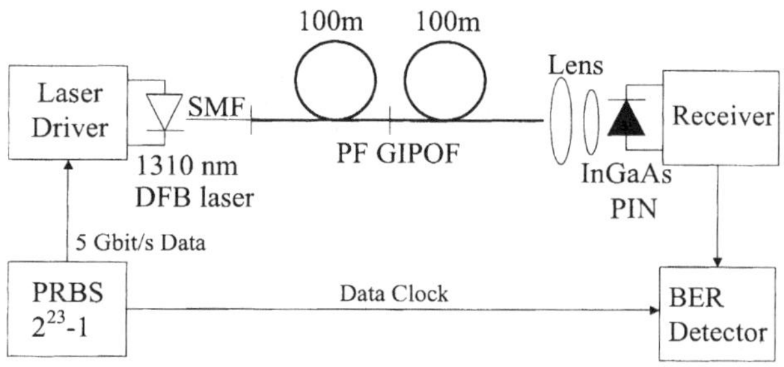

Figure 21. Experimental setup for 5 Gbit/s transmission over 200 m PF GI POF setup.

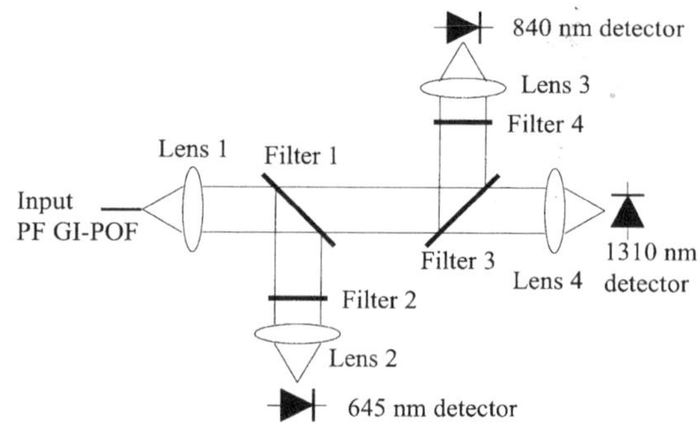

Figure 23. Principle of operation of the 645–840–1310 nm demultiplexer.

lower than −30 dB. Figure 24 is a photograph of the demultiplexer, including the three receiver modules. On the left, the GI POF can be seen.

3.2.2. WDM Transmission Experiments

The realized demultiplexer has been used for a three channel 2.5 Gbit/s over 200 m GI POF WDM experiment with a record bit rate times distance product. A block diagram of the setup is shown in Figure 25. For this experiment, the transmitters and receivers described in the previous section were used. The measured BER was smaller than 10^{-10} for all three wavelengths. Moreover, a two channel, 2.5 Gbit/s, WDM transmission experiment over 456 m GI POF was carried out using 840 and 1310 nm wavelengths. A block diagram is shown in Figure 26. Because the fiber samples have an attenuation of more than 100 dB/m at 640 nm, this wavelength could not be used. To multiplex both wavelength channels, a multimode glass fused fiber 850/1300 nm WDM was used. The step-index pigtails of the 850/1300 nm multiplexer has a core diameter of 50 μm and the losses of the multiplexer are 1.5 and 0.7 dB, respectively, for the 840 and 1310 nm channels. The 456 m PF GI POF consists of 328 and 128 m pieces. The total losses of the GI POF are 24.2 and 25.9 dB, respectively, for the 840 and 1310 nm wavelengths. To split up the wavelength channels, a demultiplexer developed specifically for the wavelengths 645, 840, and 1310 nm was used, so in this experiment the 645 nm output was not been used. The measured BER was smaller than 10^{-11} for all three wavelengths.

3.3. Concluding Remarks

Perfluorinated polymer-based GI POF has a low-loss wavelength region from 500 to 1300 nm, so a broad wavelength range is available for the implementation of single channel and coarse WDM systems. Single channel systems at bit rates of 2.5 Gbit/s over transmission distances of 300, 550, and 550 m have been reached at 645, 840, and 1310 nm wavelength, respectively. At a bit rate of 1.25 Gbit/s, used for Gigabit Ethernet, a system span of nearly 1 km was reached. Applying WDM can further enhance transmission capacity via POF. WDM also increases the flexibility in POF networks. A three channel system operating at 2.5 Gbit/s over 200 m GI POF and a two channel system at 2.5 Gbit/s, over 456 m of GI POF using WDM transmission have been described.

4. POF VERSUS LARGE CORE GLASS OPTICAL FIBER

Standard graded-index multimode fibers with a core diameter of 50 or 62.5 μm are widely used in local area networks. In the past few years, graded-index polymer optical fiber (GI POF) has gained interest for use within customer premises because of the large core and high bandwidth, and future possibilities to reduce the high attenuation. This created interest to investigate the transmission capacity of large core (100–200 μm) silica graded index multimode (GIMM) optical fibers. After a review of the properties of GI POF,

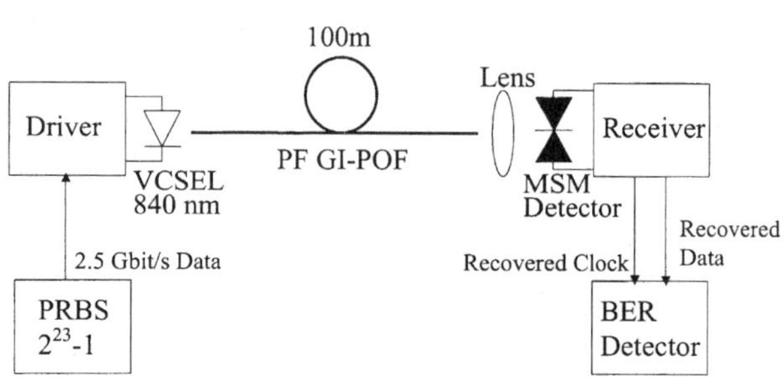

Figure 22. Experimental setup for 2.5 Gbit/s over 100 m PF GI POF using an MSM detector.

Figure 24. Photograph of the realized demultiplexer including the three receiver modules.

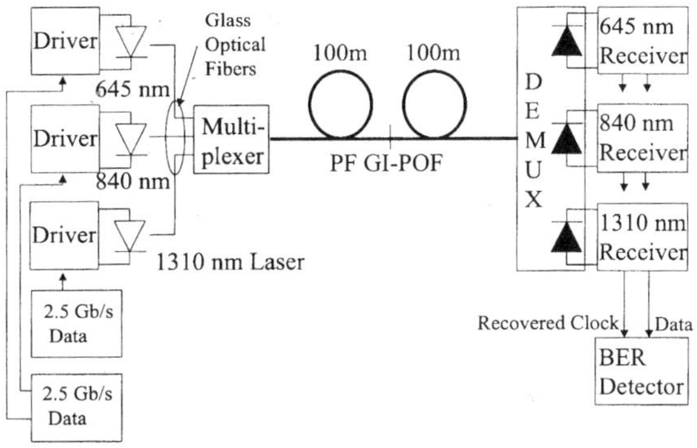

Figure 25. Block diagram of the three channel 2.5 Gbit/s WDM transmission experiment over 200 m GI POF.

measurement results on large core GIMM fibers are discussed and compared with GI POF.

4.1. Review of GI POF

There are two types of GI POF that are practical for communications; (PMMA) GI POF and PF-based GI POF.

4.1.1. Characteristics of PMMA GI POF:

- Bandwidth at 650 nm: 2 Ghz. km
- Attenuation at 650 nm: 150 dB/km
- Attenuation at 850 and 1300 nm: >1000 dB/km
- Higher cost due to GI profile compared with SI profile
- Reliability GI profile (bandwidth) at high temperature under investigation
- Realized transmission experiment at 645 nm: 2.5 Gbit/s over 200 m [25]
- Core diameter of GI POF in this experiment 500 μm

4.1.2. Characteristics of PF GI POF

- Potential bandwidth at 1300 nm: 10 GHz. km
- Realized attenuation at 1300 nm: 50 dB/km
- Potential attenuation at 1300 nm: 0.3 dB/km
- Open spectral window: 500–1300 nm
- Higher material cost compared with PMMA

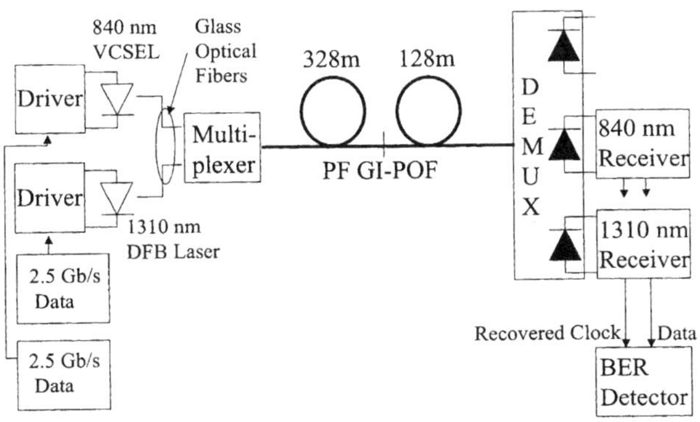

Figure 26. Block diagram of the two channels system at 2.5 Gbit/s over 456 m PF GI POF WDM transmission experiment.

- Reliability GI profile (bandwidth) at high temperatures under investigation
- Realized transmission experiment at 1300 nm: 2.5 Gbit/s over 550 m
- Realized transmission experiment at 840 nm: 2.5 Gbit/s over 550 m
- Realized transmission experiment at 650 nm: 2.5 Gbit/s over 300 m
- Core diameter of GI POF in these experiments: 170 μm

There are very optimistic reports on the potential market share of GI POF, but this still has to be developed further.

4.2. Characteristics of Large Core Silica GIMM Fibers

Presently silica GIMM fibers are produced mainly in two types: 50/125 and 62.5/125 core–cladding diameter. The attenuation is about 2.5 dB/km at 850 nm and 0.5 dB/km at 1300 nm wavelength. The fibers show a high modal bandwidth, certainly when produced for that purpose by excellently equipped plasma chemical vapor deposition (PCVD). The quality of the profile shape is very important in relation to bandwidth. Although the profile shapes of GI POF are not perfect, some reports still refer to high bandwidth values. From this point of view, it is interesting to know the bandwidth behavior of large core silica GIMM fibers, because that technology is available and present attenuation values are by far better than those of current PMMA GI POFs. A series of investigations have been made using core sizes of 93, 148, and 185. These fibers were made from a standard 50 μm GIMM PCVD preform, drawn to different cladding diameters (125, 200, 250 μm). With these large core fibers, it is impossible to excite all transmission modes at the transmitter side with a laser to obtain an overfilled-launch condition. This condition gives a worst-case bandwidth. The bandwidth can be increased by excitation of a small number of low order modes, using a launch from a single-mode fiber [26]. This technique was used for the bandwidth measurements, transmission experiments, and theoretical analysis.

4.3. Bandwidth and Data Transmission

Using single-mode fiber launching, the bandwidth of the three fibers with core–cladding diameters of 93/125, 148/200, and 185/250, each with a length of 2 km was measured with a 3 GHz optical network analyzer at 1300 nm wavelength (see Fig. 27). At the receiver side, a lens coupling was used to detect all the power out of the fiber. The 3 dB bandwidths are 3.0, 2.4, and 2.3 GHz, and are comparable with the bandwidth of graded-index fiber of standard diameters. An example of the bandwidth measurements can be seen in Figure 28.

Additionally, transmission experiment were carried out at 1300 nm wavelength using the setup depicted in Figure 29. Again, a lens coupling was used at the receiving end to detect all the power out of the fiber to avoid modal noise because

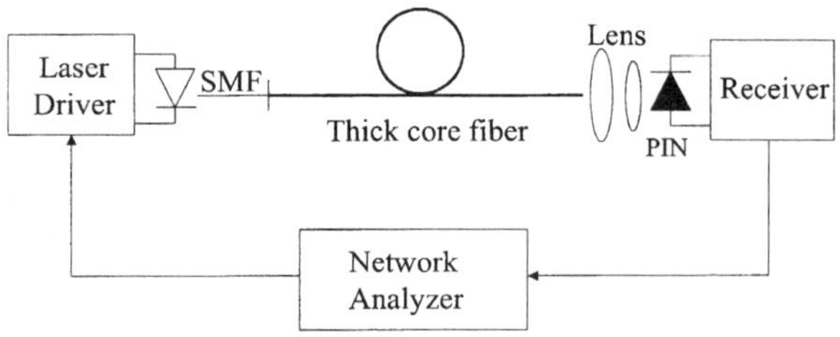

Figure 27. Bandwidth measurement setup with a network analyzer.

of masking effects. The attenuation of the 2 km large core GIMM fiber is less than 1 dB at 1300 nm, so the fiber introduces almost no power loss. According to the bandwidth measurements and calculations, the bandwidth is sufficient, so for all three fibers at 2.5 Gbit/s transmission, a bit-error rate lower than 10^{-12} was measured. Even with two combined 2 km fibers, which is equal to 4 km transmission distance, the BER was lower than 10^{-12}. In this case, the fiber with the largest diameter is situated at the end of the link to avoid masking effects. A simulation model of the bandwidth of these fibers, including limited fiber launching, mode coupling, chromatic dispersion, and differential mode attenuation, has been developed [27]. The measurement results are in accordance with the simulation results.

4.4. Discussion and Concluding Remarks

It has been shown that 2.5 Gbit/s transmission over 4 km of large core (148 and 185 μm) graded-index silica fiber can be realized easily. Because of the high attenuation of graded-index polymer optical fiber, only distances of some hundred meters have been reached at 2.5 Gbit/s.

The present status for large core silica and polymer graded-index fiber is as follows:

- Maximum transmission distances of the large core graded-index silica fibers are much larger compared with the graded index polymer fibers.
- There is a large difference in attenuation between silica and polymer fibers.

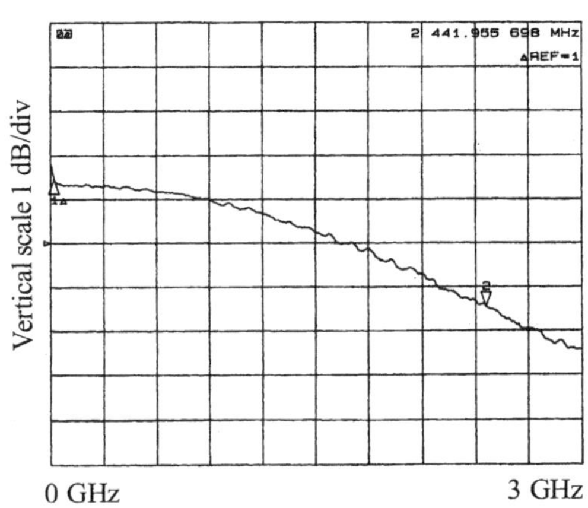

Figure 28. Measured bandwidth 2 km 148/200 large core GIMM fiber Vertical scale: 1 dB/div.

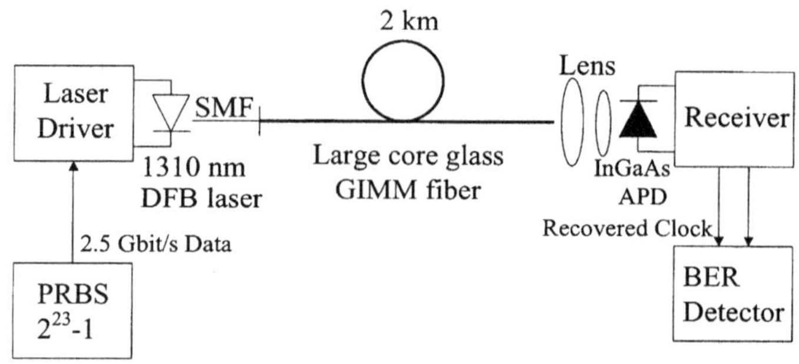

Figure 29. Transmission setup for 2.5 Gbit/s over 2 km large core glass GIMM fiber.

- The diameter of silica fibers is limited because of the inherent inflexibility of glass materials.
- Because of the difference in mechanical properties of silica and polymer, the handling techniques are different.
- Other material combinations for silica optical fibers and different handling are options for further study.

5. APPLICATIONS FOR POF BASED SYSTEMS

Polymer optical fiber technology has reached a state of development where we can identify several areas of possible application. Applications that can benefit from POF as a transmission medium span short distances and support data traffic up to Gigabits per second. Due to the prospect of ease of handling and termination, applications that require such features are good candidates for POF based systems. Applications where EMC is required can benefit from using POF. A promising aspect of POF is perhaps the prospect of lowest price systems. Taking into consideration these factor and the performance results of POF-based systems, we outline some areas of application. In Figure 30, we depict the possible application areas for POF-based systems based on the achievable transmission data rate over several link distances.

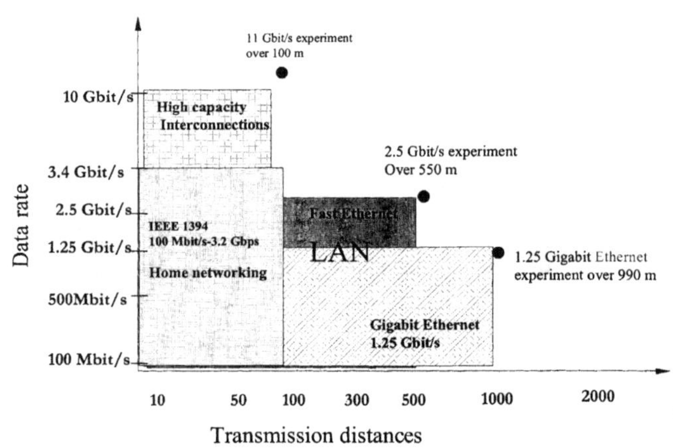

Figure 30. Application of POF-based transmission systems.

5.1. Home and Office LAN

We group in this category applications that have bit rates ranging from 100 Mbit/s up to gigabits per second but over short distances, up to 100 m. One possible scenario is in-home networks, where connectivity of electronic appliances to each other is foreseen, for instance, connecting several domestic appliances to a residential gateway. Another possible scenario is intraoffice connections for delivery of telecommunication services from an exchange to several office users, for instance, applications covered in the IEEE 1934 standard may use POF for thc physical layer. The advantages of using POF in this area, apart from high capacity, are linked to its ease of handling and termination: a bundle of POF fiber is much easier to install in a home or an office through duct work than a coaxial cable. Multimedia applications, video, and voice over IP and interactive internet services can use the POF infrastructure in the above-mentioned scenarios.

5.2. Wireless-POF LAN Backbone

The application area concerns a home or office backbone to provide connectivity from wireless appliances to multiple access points. POF could be used as the transmission medium for such applications, carrying data at bit rate up to several hundred Megabits per second over distances up to 100m. In Figure 31, we schematically represent this application in a home environment. IEEE standard 802.11 could benefit from this application (infrared and radio frequency) as could Hiperlan/IEEE 1394/USB2.0.

5.3. Gigabit Ethernet

Transport of 1.25 Gbit/s Ethernet traffic over 990 m has been demonstrated experimentally. We may consider POF for the links between LAN subnetworks using Gigabit Ethernet. Distances envisioned may vary from 100 to 550 up to 1000 m at a bit rate, for instance, of 1.25 up to 10 Gbit/s using a single channel system. Capacity could be enhanced by using multichannel wavelength division multiplexing transmission. A conceivable scenario is the connectivity of network terminals, and hubs and switches for office LAN or to connecting terminals to a main switch or server in the classrooms of university campuses and educational institutions.

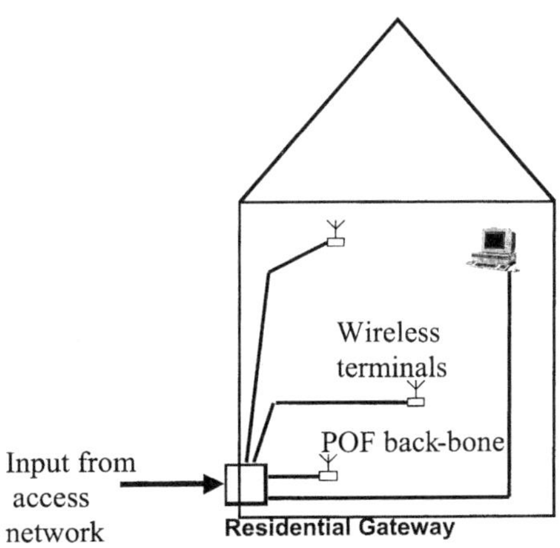

Figure 31. Wireless POF backbone for home or office networking.

6. CONCLUSIONS

The transmission distances reached by POF-based systems are increasing very fast. At bit rates of 2.5 Gbit/s, system spans of 300 m at 645 nm wavelength, and 550 m at 840 nm and 1310 nm wavelength have been reached using perfluorinated graded-index POF. Using WDM transmission, system capacities have been further enhanced. For instance, we reported a three channels $\times$ 2.5 Gbit/s GI POF WDM transmission over 200 m experiment and a two channels $\times$ 2.5 Gbit/s over 328 m experiment with record bit-rate distance products. These experiments show the feasibility of high capacity transmission over POF. Transmission of 1.25 Gbit/s Gigabit Ethernet data has been demonstrated over nearly 1 km of perfluorinated GI POF.

It has also been shown that 2.5 Gbit/s transmission over 4 km of large core (148 and 185 μm) graded-index silica fiber can be realized easily. Maximum transmission distances of the large core graded-index silica fibers are much greater compared with the graded-index polymer fibers. There is a large difference in attenuation between silica and polymer fibers. The diameter of silica fibers is limited because of the inherent inflexibility of glass materials. Because of the difference in mechanical properties of silica and polymer, the handling techniques are different.

The experimental results reported in this chapter clearly show the applicability of polymer optical fiber for several applications at the customer premises, such as local area networks, wireless-POF backbones, and in-home networking. We believe that the record results reported here are important milestones that may encourage the development of polymer fiber systems and networks. However, for POF-based systems to be introduced successfully for customers' premises networks, they will have to be cost competitive, future proof, and user-friendly. Therefore, further development of POF-based transmission systems and the development of low-cost, reliable transceiver modules, connectors, and WDM (de)multiplexing devices is very important.

REFERENCES

1. M. H. Schleinitz, *Proceedings of the International Wire Cable Symposium*, 1977, Vol. 26, pp. 352–355.
2. T. Kaino, K. Jinguji, and S. Nara, *Appl. Phys. Lett.* 42, 567 (1983).
3. Y. Koike, Y. Kimoto, and Y. Ohtsuka, *Appl. Opt.* 21, 1057 (1982).
4. T. Ishigure, E. Nihei, and Y. Koike, *Appl. Opt.* 33, 4261 (1994).
5. Y. Koike, "Proceedings of the International POF Technical Conference," Cambridge, MA, September 2000, pp. 1–5.
6. Asahi Glass Company, Lucina: Graded Index-Cytop Optical Fiber, Technical Note T009E, July 2000.

7. Y. Koike, T. Ishigure, and E. Nihei, *IEEE J. Lightwave Technol.* 13, 1475 (1995).
8. Y. Koike and T. Ishigure, *IEEE J. Lightwave Technol.* 18, 178 (2000).
9. The ATM Forum, version 1.1. AF-PHY-0079.001, January 1999.
10. Y. Koike, "Proceedings of the 22nd European Conference on Optical Communication," Oslo, 1996, Paper MoB3.1.
11. G. Yabre, *IEEE J. Lightwave Technol.* 18, 166 (2000).
12. K. Koganezawa and T. Onishi, "Proceedings of the International POF Technical Conference," Cambridge, MA, September 2000, pp. 19–21.
13. T. Ishigure, E. Nihei, S. Yamazaki, K. Kobayashi, and Y. Koike, *Electron. Lett.* 31, 467 (1995).
14. L. B. Aronson, B. E. Lemoff, L. A. Buckman, and D. W. Dolfi, *IEEE Photon. Technol. Lett.* 10, 1489 (1998).
15. J. D. Lambkin et al., *Mater. Sci. Semicond. Process.* 3, 467 (2000).
16. C. Carlsson, M. Ghisoni, and A. Larsson, "Proceedings of the International Topical Meeting on Microwave Photonics," 1999, pp. 137–140.
17. C. Carlsson, H. Martinsson, J. Vuksic, J. Halonen, and A. Larsson, "Proceedings of the 17th IEEE International Semiconductor Laser Conference," 2000, pp. 17–18.
18. F. Miyasaka et al., "OFC'95, Technical Digest," 1995, paper ThG6, pp. 257–258.
19. M. Hayashi, M. Tsuji, K. Makita, T. Nyu, S. Sugita, and K. Taguchi, *IEEE Photon. Technol. Lett.* 8, 833 (1996).
20. G. D. Khoe, Y. Yoike, T. Ishigure, P. K. v. Bennekom, H. P. A. van den Boom, W. Li, and G. Yabre, "Proceedings of the 25th European Conference on Optical Communication," 1999, pp. II/274–II/277.
21. J.-I. Mizusawa, Proceedings of the International POF Technical Conference, Cambridge, MA, September 2000. pp. 88–92.
22. G. Giaretta, W. White, M. Wegmueller, R. V. Yelamarty, and T. Onishi, "Proceedings of OFC/IOOC'99," February 1999, pp. PD14-1 PD14-3.
23. A. M. J. Koonen, A. Ng'oma, H. van den Boom, I. Tafur Monroy, P. Smulders, and G. D. Khoe, "Proceedings of the International POF Technical Conference," Amsterdam, September 27–30, 2001.
24. H. R. Stuart, "Proceedings of OFC 2000," Baltimore, Mar. 7–10, 2000.
25. W. Li, G. Khoe, H. van den Boom, G. Yabre, H. de Waardt, Y. Koike, S. Yamazaki, K. Nakamura, and Y. Kawaharada, *Microwave Opt. Technol. Lett.* 20, 163 (1999).
26. L. Raddatz, I. H. White, D. G. Cunningham, and M. C. Nowell, *J. Lightwave Technol.* 16, 324 (1998).
27. G. Yabre, *IEEE/OSA J. Lightwave Technol.* 18, 668 (2000).
28. G. D. Khoe, W. Li, G. Yabre, H. P. A. van den Boom, and P. K. van Bennekom, "Proceedings of the International Optoelectronics Symposium," Kyoto University, June 7–9, 2000, Vol. 55, No. 2, pp. 9–10.

CHAPTER 7

Plastic Optical Fiber Passive Devices

Liliana R. Kawase
Fluminense Federal University, Av. Passos da Pátria 156, CP 1126025, CEP 24241-970, Niterói, RJ, Brazil

CONTENTS

1. INTRODUCTION

This chapter presents some plastic optical fiber (POF) passive devices, their properties, and the existing technologies to obtain them. For each type of device, the most suitable technology obtains the following characteristics:

- Low insertion loss
- Mass production reliability
- Low-cost effective production
- Thermal and mechanical stability

Sometimes all not of these characteristics can be obtain simultaneously, but, depending on the final device application, some of them can be relaxed. Telecommunications, industrial applications, decoration, and signaling systems present different requirements that must be considered in the final performance-cost trade-off.

Most of the devices presented here are not yet available on the market, because POF technology is quite recent. However, due to the efforts of many research groups, this picture should modify in a short time.

2. OPTICAL COUPLERS

Earlier optical communication systems were based on simple point-to-point interconnections. In this way, just a simple interconnection with an optical fiber between a transmitter and a receiver was enough. However, increasingly fast growth in optical distributed systems and also wavelength division multiplexing (WDM) systems can be observed [1].

Another interesting area of POF application is signalization, lighting, and decoration systems, where it is also necessary to distribute or redirect incoming light to other outgoing

ISBN: 1-58883-012-8
Copyright © 2004 by American Scientific Publishers
All rights of reproduction in any form reserved.

fibers. Automotive, entertainment, and sensor industries represent other areas of potential applications for POF technology [2]. In all cases, it is necessary to split and/or combine optical signals in two or more fibers as shown schematically in Figure 1. To do so, it is necessary to use optical passive devices, such as bidirectional couplers, splitters, and star couplers. In the silica fiber industry these passive devices are very well known and are fabricated by many major companies. The same is not true in the POF area, where many such devices are still under development by different research groups [3–5].

POF couplers basically can be divided into two major categories. In one category, a coupler uses POF to divide or combine optical signal; in the other, the light division-combination is made by another type of technology. Samples of both categories are detailed presently.

2.1. POF Techniques

This section presents an overview of some techniques used to assemble POF directional couplers using POF itself. These methods also are known as micro-optical (MO) approaches [6–8]. None of these techniques can be labeled as the best one: each presents advantages and disadvantages that must be analyzed for low insertion loss, mass producibility, and low-cost assembly.

2.1.1. Twisting and Fusion

This technique is very well known in fabricating silica fiber couplers and can also be used in POF couplers to obtain even $N \times N$ star couplers. In this case, two (or more) stripped POF segments are twisted together and securely fixed. In the next process step, heat is applied to the fiber bundle while it is twisted and pulled so that all the fibers are in contact at the moment of fusion as shown in Figure 2. In the POF case, the heat source can be low temperature, because the thermal deformation temperature is about 110 °C [for polylmethyl methacrylate (PMMA)] [9]. However, this technique does present a great disadvantage for POF applications: light transmission quenching can occur due to heat yellowing caused by material impurities in the commonly used PMMA POF. To avoid such problem, high purity PMMA-cored POF must be used. The behavior of perfluorinated (PF) polymer-based POF (Lucina^{TM}, a trademark of Asahi Glass Company) [10] exposed to heat is still under investigation.

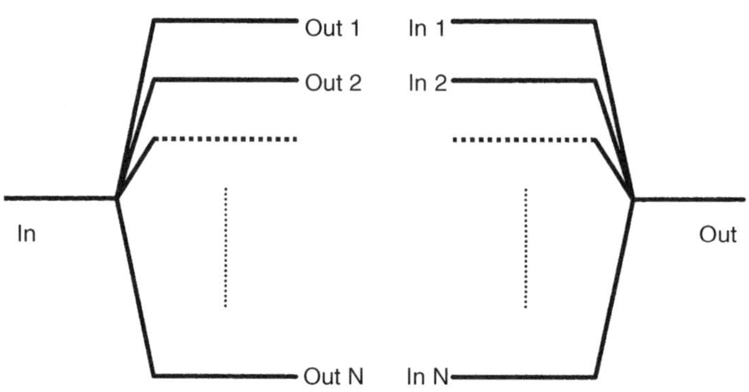

Figure 1. Split-combine optical coupler.

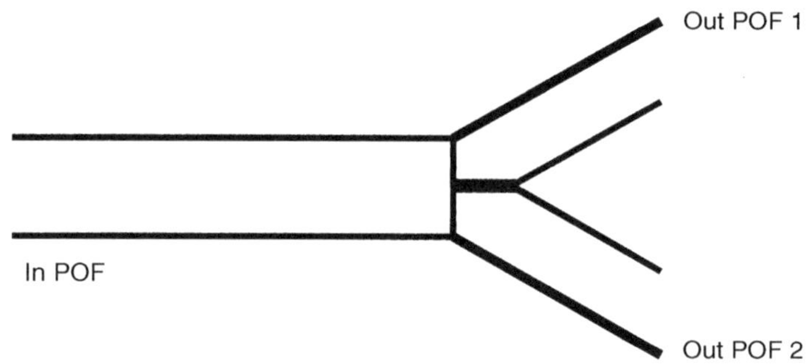

Figure 2. Coupler based on the twisting and fusion technique.

2.1.2. Side Polishing

The first silica couplers were made by side polishing techniques. Although the twisting and fusion technique yields better results for silica, side polishing is still a good option for POF couplers. In this case, two POF segments are bonded in a bent surface and polished until the core-cladding interface appears. The polished depth can be controlled to achieve a desired split ratio. After this step, the polished surfaces are placed in contact and fixed.

This technique presents many advantages because jacketed POF can be used and high precision coupling adjustments are possible. This method permits the use of simplex, duplex, and even ribbon cables for coupler assembly.

2.1.3. Chemical Etching

The chemical etching technique is quite similar to the previous method in that two POF segments are bonded into a bent surface. The difference is in the technique applied to pare down the desired surface. In the present case, chemical solvents are used to etch the bent POF to an immersed volume. Acetone, chloroform ($CHCl_3$), and methyl isobutyl ketone are examples of chemical agents that can be used in water solution to etch the fiber. After this step, the exposed cores are fixed butt to butt.

This technique has the disadvantage that the jacket must be removed in the middle of the POF segment prior to the etch process, leading to scratches on the fiber surface. Other problems to be taken into account are the fiber depth below the solution surface, solution concentration, and control of the etching time.

2.1.4. Cutting and Gluing

This technique for fabricating POF couplers appears to be the simplest. It is based on simply cutting the POF at a certain angle and gluing it to the following POF segment as shown in Figure 3. Depending on the angle and the number of following fibers, it is possible to obtain the desired

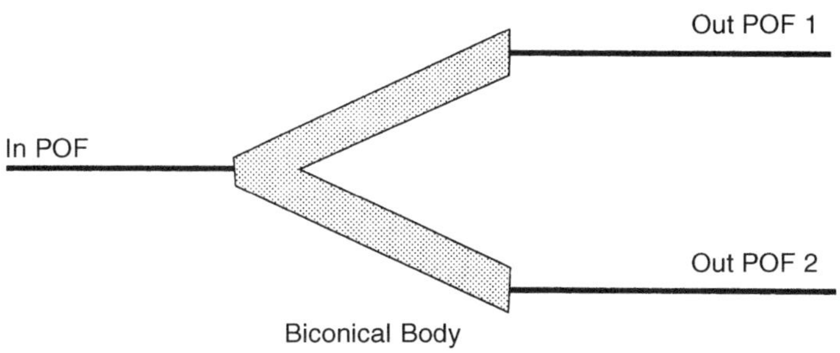

Figure 3. Coupler based on the cutting and gluing technique.

coupler. The process itself is not all that simple, because the fiber must have a high degree of flatness and any type of misalignment must be avoided.

2.1.5. Thermal Deformation

In the thermal deformation technique, two POF ends are shaped into a semicircular form by thermoplastic deformation via a hot-plate flattening technique. These two semicircles are inserted into a longitudinal cone-shaped ferrule and the resulting circular end face is polished to guarantee flatness. In a second ferrule that has an eccentric hole, a nondeformed POF is positioned. Finally, both ferrules are inserted into a third external ferrule. By rotating the eccentric ferrule inside the external ferrule, it is possible to fine adjust the splitting ratio. An excess loss of about 1.5 dB for 1×2 splitters has been reported [4].

2.2. Other Techniques

This section presents an overview of some other techniques used to assemble POF directional couplers, where POF is not involved in the coupling process itself.

2.2.1. Molding

In the case of molding, the coupler is assembled by a polymer waveguide, which can be fabricated by different techniques. Molding is known as an integrated optical (IO) approach [3, 5, 11–13]. Usually a lithography method is used to fabricate the mold, thereby permitting replication on a mass production scale. Other methods use ultra precision milling machines in combination with fine polishing. After mold fabrication, the POFs are positioned into grooves (usually V grooves, which yield better alignment), and a core material is injected into the mold. The last part of this process is curing of the core material, which is usually done by ultraviolet light.

2.2.2. Biconical Body

In this case, light splitting is performed by a biconically shaped external component as shown in the axial section in Figure 4. The device itself is made of polymer material and is delimited by two coaxial conical surfaces. By positioning an input POF at the cone base, the resulting output is

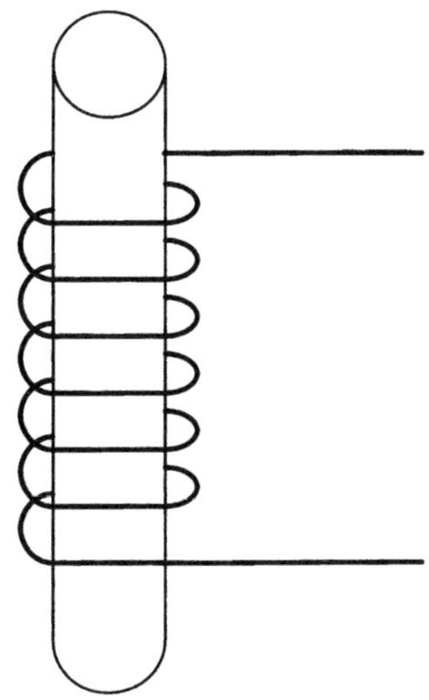

Figure 4. Axial section of a biconical coupler.

a ring light. Depending on the cone aperture, a number N of POFs can be positioned at the circular surface. For better light coupling, the circular surface thickness must be equal to the external diameter of the output POF [14].

2.2.3. Reflective Body

This technique is very similar to the previous one. In this case, the device used to split the light is a cylindrical polymer rod. Both end faces are polished to ensure flat surfaces. The rod diameter (a) corresponds to the external diameter of the ring formed by the N output POFs positioned in a ring format. The input POF is positioned in the middle of a metal circular piece that has diameter a and a mirror quality polished surface. The N output POFs are arranged in a ring format and placed around a metal disk. The metal disk is also polished to obtain mirror quality and its diameter corresponds to a minus twice the POF diameter.

By positioning both prepared metal disks at the polymer rod end faces, it is possible to obtain a device that is able to split the incoming light between N outgoing POFs. An insertion loss of 4.3 dB and uniformity of 0.68 dB were measured for such a device [14]. Due to its high insertion loss, this device is most suitable for lighting, decoration, and signalization applications.

3. FILTERS

Large core optical fiber exhibits a large number of propagating modes. The presence of such modes can cause modal noise in telecommunication systems. This problem is very significant for high-speed telecommunication applications based on large core silica fibers, where modal noise seriously affects the bit error rate. In the case of POFs, the number of these propagating modes can be a thousand times higher compared to standard silica multimode fiber. However, because of the presence of millions of propagating modes, modal noise is no longer a problem [15]. For graded-index (GI) POFs with a 600 μm core diameter, no modal noise was observed, even for 200 μm misalignment [16]. This performance shows that POFs can be used to advantage in short-distance high-speed telecommunication systems.

On the other hand, for many sensor applications, the presence of millions of propagating modes can lead to misunderstanding of the measure to be performed. To overcome this problem, it is crucial to filter these higher modes. The technique to obtain a POF modal filter is very simple; merely wrap a POF segment around a small diameter cylinder [17].

There are no references to the literature with regard to POF mode filter insertion losses. In many lab applications, once the filter is integrated into the system, it leads to reasonably stable POF-based sensors, even under vibration and manipulation.

4. SCRAMBLERS

In some applications, most of which are related to sensors, it is necessary not to filter higher orders modes, but to scramble them. This treatment is done to guarantee that all propagating modes carry the same amount of power. The device used to accomplish this is known as a scrambler. Mode scramblers have been designed and assembled that consist of a polyester plate that has induced microbends and a POF segment that is positioned to compress against the microbends. A diagram of a scramble is shown in Figure 5. The number and the curvature of the microbends are designed to follow a previous model [18]. Although there are no literature references with regard to insertion loss, once this device is integrated into a system, it leads to stable setups, as with mode filters. At the time of this writing, this device was not commercially available.

5. TAPERS

For some applications, such as light-emitting diode (LED) to POF coupling, POF to small area photodetector coupling, step-index (SI) to GI POF coupling, collimation, and evanescent wave optical sensors, it is necessary to reduce the fiber diameter. A device that performs this task is known as a taper. There are many ways to obtain tapers; each has advantages and disadvantages. The three most commonly used techniques are polishing, chemical etching, and temperature-tension control.

In the first technique, the taper is obtained by polishing the fiber faces. Because the fiber is initially a round section, a D-shaped section results. This technique is more often used in sensor applications, where the lapping film size (coarse or fine) can be adjusted relative to the sensor characteristics [17]. This technique has the disadvantage of creating a flat surface. To regain the round shape, it is necessary to rotate the fiber, which can introduce irregularities in the final device and cause difficulties in mass production.

POF tapers also can be obtained using chemical etching, which is accomplished by soaking POF segments in some chemical solutions [19]. The content and concentration of each solution and the soaking time depend on the POF material used (PMMA or CYTOP) and on the desired final taper shape. Controlling the time and rate of immersion generates tapers that have radial profiles that vary linearly with length or that have discrete steps [20].

The third technique basically consists of two simultaneous steps: controlled heating while stretching the POF [21]. This fabrication method has two subsystems: one controls the applied temperature and the other controls the applied longitudinal tension. The heating system surrounds the fiber to ensure homogenous heat application. In the tensioning system, the fiber is held between two posts, so it can be stretched longitudinally. When the fiber reaches its melting point, it is elongated to the desired configuration as shown in Figure 6. The great advantage of this method is that mass production with reproducibility is possible. Preliminary results show that this method exhibits very low losses (a mean of 0.35 dB).

All the techniques presented possess the POF advantages of polymer softness and very low melting temperature (typically 180 °C), although these feature can be disadvantageous for some applications. At the time of this writing, tapers were not commercially available.

6. LENSES

For many applications it is necessary the use a fiber-integrated lens (FIL), which is a device that fits the fiber size and is used to focus or collimate the fiber output light beam. The classic application for FILs is light coupling between POF and an optoelectronic device such as LEDs, lasers, and

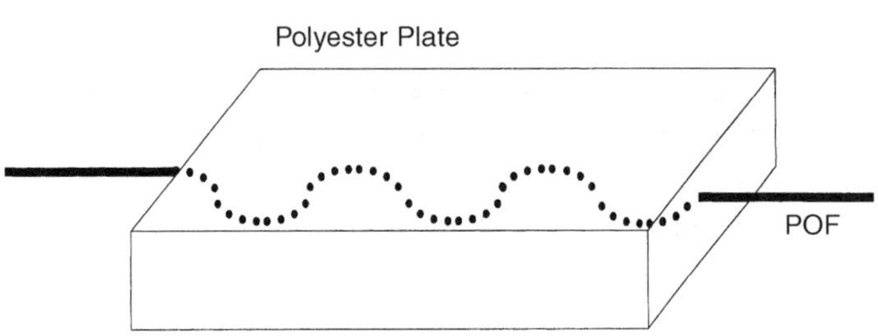

Figure 5. POF modal scrambler.

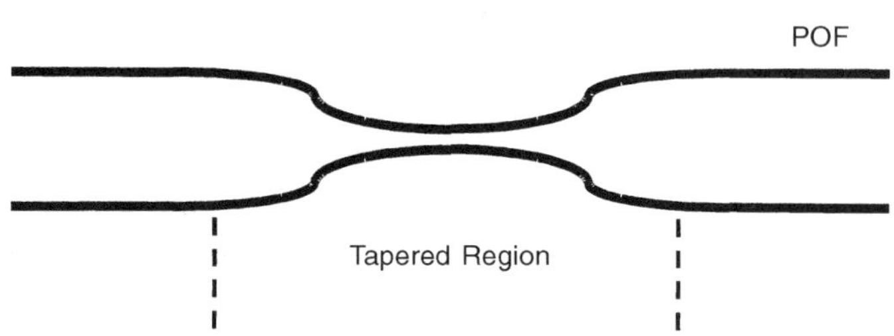

Figure 6. POF taper.

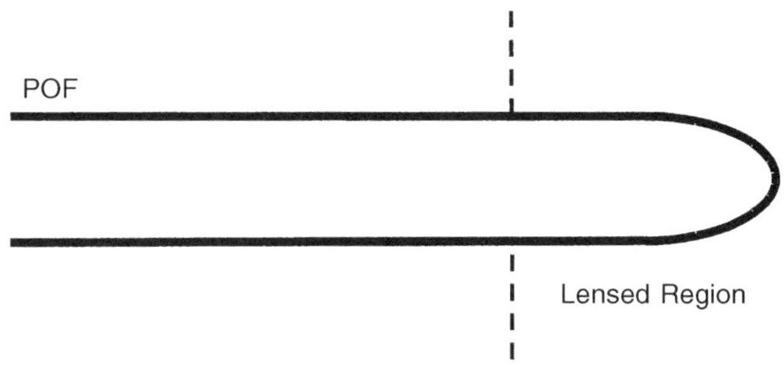

Figure 7. Lensed POF.

photodetectors. The most popular FIL is the graded index lens, where the desired output beam depends on the lens pitch. When GI fiber is used for lens applications, it is called as rod lens. The uses of POF in lenses have been studied by some groups [22, 23]. Depending on the fiber length (different pitches), many imaging characteristics can be achieved, such as inverted real or virtual images and erected real or virtual images. The rod lens can be directly coupled at the POF end face. Rod lenses also can be assembled in an array, where single row and dual row are the most popular configurations. Rod Scope® [24] is an industrial technique developed to produce plastic rod lenses and arrays. The application of such devices is related to printers, fax machines, image scanners, copy machines, and so forth.

Another way to obtain FILs for POF applications is to shape the desire lens format at the fiber end itself. This can be done using different techniques. One technique uses chemical etching, where the POF end face is sculpted by a time and depth controlled setup in a chemical solution. Another technique uses a hot mold, where the inverse of the desired POF shape is sculpted into a metal mold and the POF is pushed against the hot mold for few minutes. This technique is used more often with SI POF and yields better results for mass production. The last technique presented here is the polishing method, where the POF lens is obtained by polishing its end face with different grain sizes of lapping films. Regardless of the technique used, the expected result should resemble Figure 7, where a pointed POF is shown schematically.

7. BRAGG GRATING POF PASSIVE DEVICES

After the development of Bragg gratings in silica fiber, many fiber optical devices based on this technique became available in the telecommunications market. Some examples of such devices are add-drop WDM, dense WDM multiplexing and demultiplexing, power equalization filters, and dispersion compensation gratings. The growing interest in POF applications in the area of telecommunications has increased the efforts to develop Bragg grating technology for in POFs as well. The Optical Communications Group at the University of New South Wales (Sydney, Australia) first announced the creation of Bragg gratings in polymer fibers [25] and first announced the application of Bragg gratings for WDM devices [26].

Bragg gratings in POF are more tunable than in silica fibers and, more importantly, demonstrate no thermal hysteresis effect when the grating heats up or cools down [27]. Application of a temperature gradient of about 50 °C causes a variation of 10 nm in the grating wavelength in POF, whereas only 1 nm is measured in silica fiber for the same temperature range. Similar results can be obtained by stretching [28]. These results show the growing potential of POF Bragg grating devices for many practical engineering applications.

Studies also have shown the great potential of Bragg grating PF POF [29, 30], once such material (e.g., CYTOP™) [31] can produce lower material dispersion. This characteristic imbues this material with great potential applications to high-speed transmission links, WDMs, and sensors. Currently no POF Bragg grating devices are available commercially, but some devices based on this technology should be announced in the near future.

8. TAPERS

For some applications, such as short haul data communication, home network, and video transmission systems, WDM on POF technique is very important. Recently, low cost WDM multiplexer and demultiplexer were announced [32]. The WDM multiplexer is based on interference filters and achromatic lens, and it is also showed the possibility of built a WDM demultiplexer using injection moulded technique.

REFERENCES

1. K. Uehara and J. Mizusawa, "Proceedings of the 7th IPOFC," Berlin, 1998, pp. 19–26.
2. D. Seidl, P. Merget, J. Schwarz, J. Schneider, R. Weniger, and E. Zeeb, "Proceedings of the 7th IPOFC," Berlin, 1998, pp. 205–211.
3. S. Lehmacher, G. Pompe, S. Rudolph, S. Kalveram, M. Jöhnck, and A. Neyer, "Proceedings of the 7th IPOFC," Berlin, 1998, pp. 271–272.
4. C. Ernst, R. Hohmann, M. Loddoch, C. Marheine, and H. Kragl, "Proceedings of the International POF Technical Conference," Cambridge, MA, 2000, pp. 64–67.
5. B. G. You, S. Y. Yi, D. B. Kim, J. H. Cho, J. H. Kim, E. J. Kim, W. H. Jang, H. J. Lee, and T. H. Rhee, "Proceedings of the Asia-Pacific Polymer Optical Fibre Workshop," Sydney, 2001, pp. 95–96.
6. L. R. Kawase, J. C. dos Santos, L. P. C. da Silva, R. M. Ribeiro, J. Canedo, and M. M. Werneck, "Proceedings of the International POF Technical Conference," Cambridge, MA, 2000, pp. 68–71.
7. J. I. Vinado and J. Mateo, "Proceedings of the 5th IPOFC," Paris, 1996, pp. 130–137.
8. T. Kobayashi, K. Tsuchiya. I. Takiguchi, and H. Kayama, "Proceedings of the 8th IPOFC," Chiba, Japan, 1999, pp. 221–225.
9. B. S. Kawasaki and K. O. Hill, *Appl. Opt.* 16, 1794 (1977).
10. M. Naritomi, "Proceedings of the Asia-Pacific Polymer Optical Fibre Workshop," Sydney, 2001, pp. 38–59.
11. T. Paatzsch, H.-D. Bauer, M. Popp, I. Smaglinski, and W. Ehrfeld, "Proceedings of the 7th IPOFC," Berlin, 1998, pp. 324–331.
12. W. Ehrfeld and H. Lehr, *Radiat. Phys. Chem.* 45, 394 (1995).
13. S. Feistner, H. Lichotka, and H. Poisel, "Proceedings of the 10th IPOFC", Amsterdam, Nederland, 2001, pp. 251–256.

14. P. Faugeras, J. Marcou, S. Louis, C. Maire, and P. Dufresne, "Proceedings of the 7th IPOFC," Berlin, 1998, pp. 191–196.
15. Y. Koike, "Proceedings of the 22nd ECOC," 1996, Vol. 1, p. 41.
16. Y. Koike, "Proceedings of the 7th IPOFC," Berlin, 1998, pp. 1–7.
17. R. M. Ribeiro, J. Canedo, A. P. Ferrreira, M. M. Werneck, and L. R. Kawase, "Proceedings of the International POF Technical Conference," Cambridge, MA, 2000, pp. 157–163.
18. R. Attia and J. Marcou, *Opt. Eng.* 39, 299 (2000).
19. D. F. Merchant, P. J. Scully, and N. F. Schmitt, Eurosensors XII, Southampton (1998).
20. R. J. Bartlett, S. L. Caulder, R. P. Chandy, D. F. Merchant, R. Morgan, P. Scully, and J. Moores, "Proceedings of the 7th IPOFC," Berlin, 1998, pp. 245–246.
21. J. Mateo, M. A. Losada, and I. Garcés, "Proceedings of International POF Technical Conference," Cambridge, MA, 2000, pp. 72–76.
22. Y. Uodu and N. Hirota, "Proceedings of the 7th IPOFC," Berlin, 1998, pp. 99–105.
23. Y. Koike, *Polymer* 32, 1737 (1991).
24. Y. Uodu and T. Ishimaru, "Proceedings of MOC'95," 1995, paper PD6.
25. Z. Xiong G. D. Peng, B. Wu, and P. L. Chu, *IEEE Photon. Technol. Lett.* 11, 352 (1999).
26. P. L. Chu, G. D. Peng, and H. Y. Liu, "Proceedings of the Asia-Pacific Polymer Optical Fibre Workshop," Sydney, 2001, pp. 76–94.
27. Z. Xiong, G. D. Peng, B. Wu, and P. L. Chu, *IEEE Photon. Technol. Lett.* 3, 352 (1999).
28. P. L. Chu, G. D. Peng, and Z. Xiong, "Proceedings of the 8th IPOFC," Chiba, Japan, 1999, pp. 242–245.
29. L. R. Kawase, R. M. Ribeiro, J. C. Santos, C. M. Gibo, H. Y. Liu, G. D. Peng, and P. L. Chu, "Proceedings of the 10th IPOFC," Amsterdam, 2001, pp. 132–136.
30. H. Y. Liu, G. D. Peng, and P. L. Chu, "Proceedings of the Asia-Pacific Polymer Optical Fibre Workshop," Sydney, 2001, pp. 9–14.
31. N. Yoshihara, "Polymer Optical Fiber Conference," September 1997, paper Tub-2.
32. W. Tschekalinskiji, S. Junger, and N. Weber, "Proceedings of the 11th IPOFC", Tokyo, Japan, 2002, pp. 139–140.

CHAPTER 8

Second-Order Nonlinear Effects in Optical Fibers

I. C. S. Carvalho
Departamento de Física, Pontifícia Universidade Católica do Rio de Janeiro, Gávea, P.O.Box 38071, 22452-970, Rio de Janeiro, Brazil

W. Margulis
Acreo, Electrum 236, Stockholm 16440, Sweden

B. Lesche
Departamento de Física, UFJF, Campus Universitário, ICE, 36036-300, Juiz de Fora, MG, Brazil

CONTENTS

ISBN: 1-58883-012-8
Copyright © 2004 by American Scientific Publishers
All rights of reproduction in any form reserved.

1. GENERAL INTRODUCTION

1.1. Presentation

The field of nonlinear optics was developed in the 1960s as a consequence of the invention of the laser. Nonlinearities were viewed as a way to exploit the high-power coherent light generated by the newly available sources, and to control and alter its properties. It is presently possible to convert the wavelength of the emitted light with efficiency close to 100% by various nonlinear mechanisms and using diverse media. Through such conversion, one can see, for instance, laser pointers that emit green light while the laser source emits in the infrared or ultraviolet (UV) radiation generated from a visible laser. It is also possible to control the spatial divergence of a beam though nonlinear effects, as well as control the temporal spread of short pulses as they propagate in a waveguide. This latter effect is nowadays exploited in advanced high-speed optical communication systems, where the duration of the pulses is of crucial importance and the temporal spreading must be limited. The interaction of laser radiation with internal or externally applied electric fields also can lead to the control of properties of the propagating light, such as its polarization, and this is used in electro-optic modulators, as discussed later in this chapter.

A large variety of materials have been purified or artificially created with remarkable nonlinear optical properties. These include ferroelectric crystals, such as KTP and $LiNbO_3$, and semiconductor systems, such as GaAs, InP, and InGaAsP, which are the base of hundreds of commercially available devices, such as lasers, modulators of various types, and frequency converters. Nonlinear effects depend on the optical intensity and on the interaction length between the radiation and the material. Consequently, it is useful to confine the optical beam in a waveguide and let it propagate along the nonlinear material without spreading spatially. This is particularly necessary when the nonlinearity is weak.

We concentrate in this chapter on the use of silica-based telecommunication-like optical fibers as our nonlinear material, the best of all optical waveguides. Telecommunication fibers are cheap, easily available, and have extremely low attenuation. A light beam of wavelength 1.5 μm typically can travel more than 10 km before its intensity drops by a factor of 2. Furthermore, as guided wave devices, fibers keep the optical field confined to a very small area ($\sim$50 μm^2) along the entire propagation length, because the radius of the core of typical fibers measures only $\sim$4 μm. This is to be compared with the unguided case, where the beam keeps a small transversal cross section (confocal length) over only a few millimeters. Also, silica glass has very high damage threshold. Therefore, even though silica has very low nonlinearity, the long effective interaction lengths make it possible to observe very efficient frequency conversion, spectral broadening, pulsewidth compression, and other interesting and important effects.

We focus our attention on second-order nonlinear optical processes. We discuss purely optical nonlinear effects (such as wavelength conversion) and effects where an externally applied voltage changes the refractive index of a fiber (such as the linear electro-optical effect). In both cases, the nonlinear interaction is based on the ability to artificially record an electric field in glass. This field is nearly permanent at room temperature because glass is an excellent electrical insulator. After discussing, in Section 1, some fundamentals that are required for comprehension of the literature review that follows, we present, in Section 2, the case where the second-order nonlinearity is entirely induced by light, by focusing laser radiation into an optical fiber. There we describe that not only is a spatially periodic field engraved, but also that the period ensures efficient nonlinear optical conversion.

We then describe, in Sections 3 and 4, the discovery in the context of nonlinear optics that a strong external electric field in the presence of heat or UV radiation can move charges in silica-based glass in such a way as to result in very strong frozen fields. We aim to show throughout the chapter that not only is the range of potential applications of second-order nonlinearities in fibers very large, but also the physics involved in the processes discussed is very rich. We argue that by learning about such processes in silica, one can learn about other low-order materials such as plastics, liquids, and gases. This increases our understanding of Nature and allows us to exploit such second-order nonlinearity for the benefit of our fellow beings.

1.2. Optical Fibers: Fabrication and Properties

The field of optical communications expanded tremendously, driven by the appearance of cable TV, the Internet, e-mail, and other IT technologies. The reader might have had the chance to see and handle an optical fiber, because its use is worldwide. Commercial optical fibers are generally surrounded by a large amount of plastic or rubber for mechanical support, that is, the fibers are enclosed in optical cables. In the laboratory, however, fibers tend to look like thin fishing line. In fact, inside an acrylic-based primary coating, fibers are a long cylinder of solid glass that is so thin that it is flexible. The outer diameter of the glass is usually 125 μm. Light is guided in the core of the fiber. For most applications, and in particular in nonlinear optics, the core is so small that the light can travel only in a single transverse mode, as further discussed subsequently. Typical single-mode fibers have a core diameter $\sim$8 μm. (When wavelengths shorter than the cutoff are used, these fibers can support a few propagating modes, but are colloquially still referred to as single-mode fibers.) A glass cladding with lower refractive index surrounds the glass core, so that internal reflection occurs at the core–cladding interface. These are called step-index fibers, because the refractive index suffers a steplike discontinuity from the cladding to the core.

It is an amazing technological achievement that highly transparent fiber can be fabricated with lengths of tens of kilometers with a core diameter of only a few micrometers (precise within 0.1 μm). The fabrication process involves making a meter-loss glass rod—the fiber preform—with a diameter of 3–10 cm. The cross section of the preform is circular, and at its center a region a few millimeters in diameter is deposited of slightly higher refractive index (which will become the fiber core), surrounded by lower refractive index glass (the cladding). The preform is heated to $\sim$2000 °C and as it melts it flows down from the oven into the drawing tower. The viscous running glass is pulled as a fiber. The drawing speed is kept constant to maintain a constant diameter and the primary coating is deposited onto the fiber while it is still being pulled. The fiber is collected on revolving drums. Modern factories draw fibers at a speed approaching 100 km/h.

To define a core, it is necessary to dope the preform. Telecommunication fibers are made of pure fused silica (SiO_2) and the core is doped with a small amount of germanium in the form of GeO_2 to increase its refractive index. Several other dopants are commonly used, such as phosphorus, aluminum, and fluorine. These elements alter the viscosity of glass and its melting point, and affect the refractive index. It is also possible to find other types of fibers, including high birefringence (hi-bi) fibers, which maintain the polarization state of the light that otherwise wanders along the fiber. Fibers that contain a large percentage of germanium tend to be highly photosensitive to UV radiation. This is also true of fibers fabricated so that the core is under great stress. Photosensitivity is highly desired for the fabrication of new fiber-based telecommunication components, such as periodic gratings in fibers, as discussed later in this chapter. The great interest in optical amplifiers for long-distance optical transmission has led to the development of rare-earth-doped fibers (in particular, Er-doped fibers). These have a large potential in nonlinear optics because of the additional resonances introduced. To make Er-doped fibers by modified chemical vapor deposition, usually a porous preform is immersed in a liquid solution of Er-containing salt. As the preform is collapsed by heating, the dopant is incorporated into the core region.

Optical fibers are far from extinguished as a research topic. They evolved from excellent optical waveguides to a rich physical system that can be manipulated in various ways. By drilling the preform before drawing the fiber, it is possible to fabricate fibers with holes, into which various materials—including metallic electrodes—can be inserted. Several fiber devices that use internal electrodes have been used to exploit second-order nonlinearities, as discussed subsequently. Likewise, it is possible to saw the fiber preforms longitudinally, so that the remaining glass rod has a nearly semicircular cross section, where the core is very close to the surface. If this preform is drawn as fibers, the cross section looks like the letter D, and such D-shaped fibers also are used in the context of second-order nonlinearities. By placing two D-shaped preforms face to face and collapsing them together, it is possible to manufacture twin-core fibers. Fibers with two cores and two holes have been demonstrated, as discussed in Section 4. By drawing a collection of closely spaced capillary silica tubes, fibers have been made with a large number of tiny holes. The presence of these holes near the core of a fiber strongly affects its index profile and, consequently, the way radiation propagates. The holes reduce the effective index, and this has been exploited to create a guiding structure without a core (holey fibers). Holey fibers have been shown to confine the optical field to such small areas that even laser beams of modest power reach intensities sufficient to produce a range of nonlinear effects efficiently. Also, if the position of the tiny holes is periodic, the index perturbation introduced can prevent diffraction, and light guiding can be accomplished by what is known as a photonic bandgap fiber. A wealth of new optical effects also appear in such fibers.

1.3. Wave Equation

When light is coupled into a fiber, it travels as a guided wave. Nonlinearities alter the characteristics of the optical field along the fiber, but almost invariably can be treated as a perturbation. We searched for an expression for the electric field $\mathbf{E} = \mathbf{E}(\mathbf{r}, t)$. The guided optical field in a fiber is described by the solution of the wave equation that follows directly from Maxwell's equations [1].

$$\begin{aligned}\nabla \cdot \mathbf{D} &= \rho_f \\ \nabla \cdot \mathbf{B} &= 0 \\ \nabla \times \mathbf{E} &= -\frac{\partial \mathbf{B}}{\partial t} \\ \nabla \times \mathbf{H} &= \mathbf{J}_f + \frac{\partial \mathbf{D}}{\partial t}\end{aligned} \tag{1}$$

In the absence of free currents and charges in the glass of the fiber, $\rho_f = 0$ and $\mathbf{J}_f = 0$. The response of the medium to the electric and magnetic fields of the optical signal is expressed by

$$\begin{aligned}\mathbf{D} &= \varepsilon_0 \mathbf{E} + \mathbf{P} \\ \mathbf{B} &= \mu_0(\mathbf{H} + \mathbf{M})\end{aligned} \tag{2}$$

To a very good approximation, glass is not magnetic ($\mathbf{M} = 0$). Using the relationship $\mu_0 \varepsilon_0 = 1/c^2$, from the preceding expressions makes it easy to obtain

$$\nabla \times (\nabla \times \mathbf{E}) + \frac{1}{c^2}\frac{\partial^2 \mathbf{E}}{\partial t^2} = -\mu_0 \frac{\partial^2 \mathbf{P}}{\partial t^2} \tag{3}$$

To have a complete description of how light propagates and changes along the fiber, it is necessary to know the relationship between the polarization $\mathbf{P}$ of the material and the electric field $\mathbf{E}(\mathbf{r}, t)$. The polarization expresses how the fiber responds to the presence of the optical field. For low light intensities, the electrons oscillate harmonically with the sinusoidal excitation and nonlinear effects are absent. The (linear) polarization $\mathbf{P}_l$ is then proportional to the electric field:

$$\mathbf{P}_l = \varepsilon_0 \chi^{(1)} \mathbf{E} \tag{4}$$

By defining, in the absence of absorption, $n^2 \equiv 1 + \chi^{(1)}$, Eq. (3) can be rewritten as

$$\nabla \times (\nabla \times \mathbf{E}) + \frac{n^2}{c^2}\frac{\partial^2 \mathbf{E}}{\partial t^2} = 0 \tag{5}$$

In this case, the only alteration of the field along the fiber is caused by dispersion, that is, the dependence of the refractive index n on the wavelength, as discussed in Section 1.4.

When higher light powers are coupled into the fiber, the oscillation of the electrons is no longer simply harmonic and the polarization is not linearly related to the electric field. Far from the resonances of glass (i.e., in the wavelength region 0.4–1.7 μm), the polarization can be expressed as the sum of a linear and a nonlinear part:

$$\mathbf{P}(\mathbf{r}, t) = \mathbf{P}_l(\mathbf{r}, t) + \mathbf{P}_{nl}(\mathbf{r}, t) \tag{6}$$

The linear term of the polarization is usually much larger and describes the response of the material proportional to the electric field in the medium, whereas the nonlinear part describes the departure from linearity and is expanded as a power series of the electric field:

$$\mathbf{P}_{nl} = \varepsilon_0(\chi^{(2)}\mathbf{EE} + \chi^{(3)}\mathbf{EEE} + \cdots) \tag{7}$$

The second- and third-order nonlinear susceptibilities ($\chi^{(2)}$ and $\chi^{(3)}$, respectively) are tensors, the elements of which give the magnitude of the nonlinearity. Highly nonlinear materials have large values of $\chi^{(2)}$ and $\chi^{(3)}$. Some simplifying assumptions already have been made here, because we assumed that a polynomial expansion is valid and that the interaction of light with matter is described solely through the electric dipole moment term. Nonlinear polarization due to magnetic dipole ($\mathbf{B} \times \mathbf{E}$) or electric quadrupole ($\mathbf{E} \cdot \nabla\mathbf{E}$ and $\nabla \cdot \mathbf{EE}$) interactions is generally neglected because it is comparatively weak, but, as discussed later, it can be relevant in the context of second-harmonic generation in fibers, and liquids and gases.

Equation (3) takes the form

$$\nabla \times (\nabla \times \mathbf{E}) + \frac{1}{c^2}\frac{\partial^2 \mathbf{E}}{\partial t^2} = -\mu_0 \frac{\partial^2 \mathbf{P}_l}{\partial t^2} - \mu_0 \frac{\partial^2 \mathbf{P}_{nl}}{\partial t^2} \tag{8}$$

and, therefore,

$$\nabla \times (\nabla \times \mathbf{E}) + \frac{n^2}{c^2}\frac{\partial^2 \mathbf{E}}{\partial t^2} = -\mu_0 \frac{\partial^2 \mathbf{P}_{nl}}{\partial t^2} \tag{9}$$

Finally, because the divergence of the field is zero, the wave equation can be written as

$$\nabla^2 \mathbf{E} - \frac{n^2}{c^2}\frac{\partial^2 \mathbf{E}}{\partial t^2} = \mu_0 \frac{\partial^2 \mathbf{P}_{nl}}{\partial t^2} \tag{10}$$

Note that the wave equation also can be represented in frequency space and, in various instances, the physical picture can be simplified greatly. For example, describing the electric field in terms of its Fourier (spectral) components as

$$\mathbf{E}(\mathbf{r}, t) = \frac{1}{2\pi}\int \mathbf{E}(\mathbf{r}, \omega) e^{-i\omega t}\, d\omega \tag{11}$$

Eq. (10) assumes the form

$$\nabla^2 \mathbf{E}(\mathbf{r}, \omega) + \frac{n^2}{c^2}\omega^2 \mathbf{E}(\mathbf{r}, \omega) = -\mu_0 \omega^2 \mathbf{P}_{nl}(\mathbf{r}, \omega) \tag{12}$$

1.4. Dispersion

Dispersion results from the fact that the refractive index depends on the frequency of the propagating wave, that is, $n = n(\omega)$. This effect can be seen with a simple glass prism in sunlight, where the various colors that compose white light separate. The fact that different wavelengths travel at different speeds in the fiber is very relevant for optical communications. Short optical pulses that carry information to be conveyed to the receiving end have a finite bandwidth (i.e., a finite spectrum). Because the spectral components do not reach the end of the fiber at the same time, pulse broadening occurs, thereby limiting the maximum pulse rate of the transmission system. The 1.3 μm wavelength region was chosen in telecommunication systems (in addition to the low attenuation) primarily due to the fact that the dispersion of standard silica-based fibers goes through zero at this wavelength. At this wavelength the group velocity is maximum. The dispersion grows rapidly away from 1.3 μm. For example, an infrared (IR) pulse at 1.06 μm and a green pulse at 0.53 μm launched together into a 1 m long piece of fiber separate by nearly 80 ps at the exit end, that is, the IR pulse leads the green pulse by ~2.4 cm. We stress that dispersion is not a nonlinear effect and it does not depend on the optical powers involved. Because $n = n(\omega)$, however, the solution for the electric field from Eqs. (10) or (12) depends on the wavelength even in the case when $\mathbf{P}_{\text{nl}}$ can be neglected.

1.5. Fiber Modes

Let us concentrate for a moment on the propagation of light in the fiber, assuming that the nonlinear polarization that expresses the appearance of nonlinear effects is negligible. After all, when the second harmonic of the laser light is generated by the fiber, it still has to propagate to the exit end to be observed. The solution of the wave equation with $\mathbf{P}_{\text{nl}} = 0$ is obtained by expressing the operator ∇^2 in cylindrical coordinates to explore the cylindrical symmetry of the fiber.

By separating the transversal amplitude distribution of the field, we find that $\mathbf{E}(\mathbf{r}, \theta)$ has to satisfy an equation that leads to particular propagation modes [1]. Therefore, the propagation of light in the fiber is possible only for particular field distributions. Physically, these distributions correspond to the arrangements in which the coherent field does not suffer destructive interference as it propagates. At the output end of a multimode fiber it is easy to see the various light patterns formed, which depend strongly on the alignment at the input end and are associated with the different modes that can be coupled. The exact modes that result from the wave equation are often grouped according to the field distribution, because many lead to nearly undistinguishable light patterns. The conventional denomination is the LP_{mn} notation: LP_{01} is the fundamental propagation mode, the only one found in single-mode fibers.

1.6. Nonlinearities

Having dealt separately with the transversal part of the electric field that leads to the existence of propagation modes, we concentrate on the longitudinal variation of the electric field of the light signal along the fiber (z direction). We assume the field to be linearly polarized. The wave equation (in time) reduces to

$$\nabla^2\mathbf{E}(z,t) - \frac{n^2}{c^2}\frac{\partial^2\mathbf{E}(z,t)}{\partial t^2} = \frac{1}{c^2}\frac{\partial^2}{\partial t^2}\left[\chi^{(2)}\mathbf{E}(z,t)\mathbf{E}(z,t) + \chi^{(3)}\mathbf{E}(z,t)\mathbf{E}(z,t)\mathbf{E}(z,t) + \cdots\right] \quad (13)$$

and in the frequency space to

$$\begin{aligned}\nabla^2\mathbf{E}(z,\omega) - \frac{n^2\omega^2}{c^2}\mathbf{E}(z,\omega) &= -\frac{\omega^2}{c^2}\int_{-\infty}^{\infty}\int_{-\infty}^{\infty}\chi^{(2)}(\omega_1,\omega_2)\mathbf{E}(z,\omega_1)\mathbf{E}(z,\omega_2) \\ &\quad\times\delta(\omega-\omega_1-\omega_2)\,d\omega_1\,d\omega_2 \\ &\quad-\frac{\omega^2}{c^2}\int_{-\infty}^{\infty}\int_{-\infty}^{\infty}\int_{-\infty}^{\infty}\chi^{(3)}(\omega_1,\omega_2,\omega_3)\mathbf{E}(z,\omega_1)\mathbf{E}(z,\omega_2) \\ &\quad\times\mathbf{E}(z,\omega_3)\delta(\omega-\omega_1-\omega_2-\omega_3)\,d\omega_1\,d\omega_2\,d\omega_3 \\ &\quad-\frac{\omega^2}{c^2}\cdots\end{aligned} \quad (14)$$

The nonlinearities are expressed in terms of a power expansion. This last equation very much resembles the structure of Newton's second law for a forced harmonic oscillator if we substitute time by three-dimensional space: ∇^2 is the second derivative that corresponds to $m\mathbf{a}$, $n^2\omega^2/c^2 = k^2$ gives the spring constant, and the right-hand side term is associated with the external force. Using this analogy, we concluded that appreciable amplitudes are obtained for the electric field only for the case of resonance, that is, when the natural (spatial) frequency of the oscillator coincides with the (spatial) frequency of the driving force (i.e., of the nonlinear polarization). This condition is known as the principle of *phase matching* and is discussed in more detail in Sections 1.8 and 1.9.

Although in this chapter we concentrate our attention on the problem of second-order nonlinearities in fibers, as we will see, such processes are generally forbidden because $\chi^{(2)}$ is zero in glass. There are only a few nonlinear processes associated with a second-order nonlinearity. Two fields mix in the material and their frequencies can either add or subtract. If the fields are identical optical waves, the sum of frequencies leads to frequency doubling (also referred to as second-harmonic generation). The difference between equal optical fields leads to the generation of dc electric field, and this process is known as optical rectification. In the more general case when two optical fields mix with unequal frequencies, we obtain sum and difference frequency mixing. When one of the fields has an optical frequency while the other is in the microwave region of the spectrum, mixing leads to the creation of side bands in the optical signal. This in turn leads to modulation of the phase and, with the use of polarizers to select the direction of vibration of the optical field, we can turn this into amplitude modulation. In this case, the material is being used as an *electro-optical modulator*. When the response of the device depends linearly on the applied field, the process is denominated as a *linear electro-optical effect*.

Whereas $\chi^{(2)}$ in glasses is generally zero, nonlinear effects normally are dominated by the $\chi^{(3)}$ term that describes a large range of effects easily observed in fibers, such as self-phase modulation (SPM), cross-phase modulation (XPM), simulated Raman scattering (SRS), simulated Brillouin scattering (SBS), and four-wave mixing (FWM). In SPM, the refractive index of the fiber is altered by the high intensity of the laser light, and this causes a phase shift that affects the spectrum of the radiation. Likewise, in XPM, the index is altered, but now an independent beam causes the effect. In SBS and SRS, phonons (vibrations) interact with the beam and energy is lost to the material. These processes can be extremely efficient, and SRS is at present being used as a gain mechanism in commercial optical amplifiers for telecommunications. In FWM, two photons disappear and two new photons are created so that the total energy is conserved. This process is also referred to as optical parametric amplification, and is now being actively studied as a way to provide for gain at various wavelengths for telecommunications. All these nonlinear effects present in optical fibers are dealt with in detail in [2] and a general overview of nonlinear optics can be found in [3].

1.7. Inversion Symmetry in Low-Order (Amorphous) Materials

As mentioned, there is a strong argument to forbid all second-order nonlinear processes in glass fibers: $\chi^{(2)}$ should be zero. **E** and **P** are polar vectors and, consequently, the second-order susceptibility transforms under space inversion as,

$$\chi^{(2)}(\mathbf{r}) \rightarrow \tilde{\chi}^{(2)}(\mathbf{r}) = \chi^{(2)}(-\mathbf{r}) \tag{15}$$

On the other hand, in a material with spatial inversion symmetry, all material constants transform under space inversion as

$$C(\mathbf{r}) \rightarrow \widetilde{C}(\mathbf{r}) = C(-\mathbf{r}) \tag{16}$$

The preceding two equations imply that $\chi^{(2)}$ vanishes in substances with space inversion symmetry. In fact, being a short-range order material (nearly amorphous), glass does not have any symmetry at all. On a macroscopic scale, however, it is expected to behave like an inversion symmetric material. Intuitively we expect that the mirror image of an amorphous material interacts with the optical fields very much in the same way as the original disordered material. Could it be that some other term in expansions (13) or (14) gives rise to second-harmonic generation in glasses? To investigate this possibility, we now use the symmetry argument without referring to the second-order susceptibility.

Imagine a collimated light beam of frequency ω and linear polarization in the x-direction propagating in vacuum (along the z direction). In the center of the beam a 180° rotation of the entire experiment around the symmetry axis of the beam is equivalent to a phase shift of π:

$$\begin{aligned} A\cos(kz - \omega t) &\rightarrow -A\cos(kz - \omega t) \\ &= A\cos[kz - \omega(t + T/2)] \end{aligned} \tag{17}$$

whereas the absolute phase is never observed in optics, we may say that a rotation by 180° is not detectable optically. We may call $A\cos(\omega t)$ an *apolar oscillation*, because it is optically indistinguishable from $-A\cos(\omega t)$. The situation is different when we have two copropagating beams with frequencies ω and 2ω. In this case, a rotation

$$\begin{aligned} &A\cos(kz - \omega t) + B\cos(k_{2\omega}z - 2\omega t) \\ &\quad\rightarrow -A\cos(kz - \omega t) - B\cos(k_{2\omega}z - 2\omega t) \end{aligned} \tag{18}$$

cannot be compensated by an overall phase shift. If A and B are different from zero, a time t always exists such that

$$\begin{aligned} &A\cos(\omega t) + B\cos(2\omega t + \alpha) \\ &\quad\neq -A\cos[\omega(t + \tau)] - B\cos[2\omega(t + \tau)] \end{aligned} \tag{19}$$

The proof of this little theorem is left as an exercise. Instead of a formal proof, we give a simple convincing example. Figure 1 shows the sum of $\cos(\omega t)$ and $\cos(2\omega t)$. It is evident that the negative and positive y axis are not equivalent and that a change of sign cannot be compensated by a phase shift. For this reason, we call a *polar oscillation* an oscillation of the kind

$$A\cos(\omega t) + B\cos(2\omega t + \alpha) \tag{20}$$

Now imagine a frequency-doubling experiment. A well-collimated beam of frequency ω enters a doubling device and emerges on the other side has a double beam with frequencies ω and 2ω. Thus, the doubling device converts an apolar oscillation into a polar one. Therefore, the device cannot be invariant under a rotation of 180° about the beam

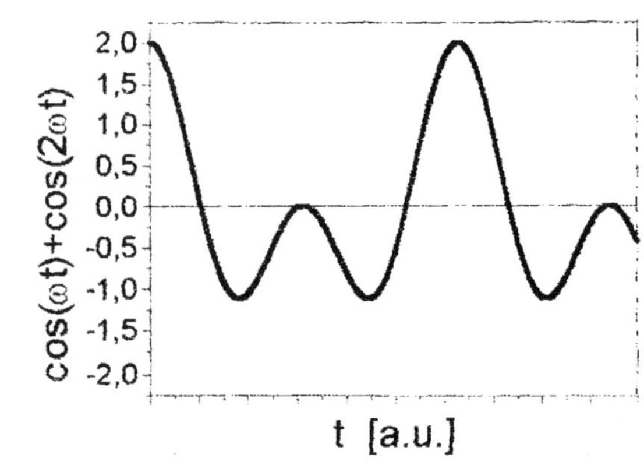

Figure 1. Example of a polar oscillation. The behavior in the lower half plane is not equivalent to that in the upper half.

axis. We concluded that frequency doubling cannot occur in liquids and gases. Note that this time the symmetry argument was based on a rotation and not a space inversion. Thus, the argument excludes frequency doubling even in chiral. However, this symmetry argument is restricted to the center of the beam. At the border of the beam, the gradient of intensity introduces a polar element, which can entail second-harmonic generation. The corresponding terms in Eq. (3) that describe this gradient-dependent second-harmonic generation correspond to electric quadrupole contributions and, as mentioned in Section 1.3, these terms are generally very weak. In fact, second-harmonic generation has been observed in gases, but it has been shown that the contributions stem from the border of the beam [4]. In optical fibers, the core–cladding interface provides a similar local breakdown of rotational symmetry and a very faint signal of second-harmonic light was known to exist even before the strange discovery of efficient second-harmonic generation in telecommunication fibers.

1.8. Third-Order Nonlinearities in Fibers

In this subsection, we look in more detail at some of the nonlinear effects expected in fibers illuminated by intense laser pulses. Let us start with the degenerate case when a single harmonic wave of frequency ω_0 is launched into a fiber. The total electric field associated with the input signal is

$$\mathbf{E}_{\text{in}}(z,t) = \frac{1}{2}\left[\mathbf{E}_0 e^{[i(k_\omega z - \omega t)]} + \mathbf{E}_0^* e^{[-i(k_\omega z - \omega t)]}\right] \quad (21)$$

Here $\mathbf{E}_0$ is the slowly varying envelope of the electric field and k_ω denotes the propagation constant at frequency ω in the material. Substitution of $\mathbf{E}_{\text{in}}$ in the wave equation (13) with $\chi^{(2)} = 0$, gives the behavior of the electric field along the fiber (z) and in time (t):

$$\nabla^2\mathbf{E}(z,t) - \frac{n^2}{c^2}\frac{\partial^2\mathbf{E}(z,t)}{\partial t^2} \cong \frac{1}{c^2}\frac{\partial^2}{\partial t^2}\left[\chi^{(3)}\mathbf{E}(z,t)\mathbf{E}(z,t)\mathbf{E}(z,t)\right] \quad (22)$$

We note that n here depends on the frequency(ω) because of dispersion, but it is independent of the input power. Let us concentrate on the right-hand side term. The cube of the electric field has several terms, including these in ωt:

$$|E_0E_0^*|\mathbf{E}_0 e^{[i(k_\omega z-\omega t)]} \quad \text{and} \quad |E_0^*E_0|\mathbf{E}_0^* e^{[-i(k_\omega z-\omega t)]} \quad (23)$$

Since $\mathbf{E}$ and $\mathbf{P}$ here both oscillate with $\exp[i(k_\omega z - \omega t)]$ and c.c., the corresponding nonlinear term on the right-hand side of Eq. (22) can be incorporated into the term on the second time derivative of the field, and Eq. (22) can be rewritten approximately as

$$\nabla^2\mathbf{E}(z,t) - \frac{n^2}{c^2}\frac{\partial^2\mathbf{E}(z,t)}{\partial t^2} - \frac{\chi^{(3)}I}{c^2}\frac{\partial^2\mathbf{E}(z,t)}{\partial t^2} = 0 \quad (24)$$

Where I is proportional to the light intensity.

By incorporating the intensity dependence into the refractive index, Eq. (24) can be solved without a polarization term, just as in Eq. (5). The refractive index then can be described as $n = n_0 + n_2 I$, where n_2 is the nonlinear coefficient associated with the intensity dependence. As mentioned in Section 1.6, such an effect alters the spectrum of the propagating beam. A time-dependent phase modulation is introduced if pulses are used, and the distorted phase corresponds to new optical frequencies that are introduced in the beam. Incorporation of the nonlinearity into the refractive index also is used in Section 3.5.

The nonlinear product $\mathbf{E}_{\text{in}}(z,t)\cdot\mathbf{E}_{\text{in}}(z,t)\cdot\mathbf{E}_{\text{in}}(z,t)$ on the right-hand side of Eq. (22) also has terms in 3ω that should lead to frequency tripling. These are

$$E_0^3 e^{[i(3k_\omega z - 3\omega t)]} \quad \text{and} \quad E_0^{*3} e^{[-i(3k_\omega z - 3\omega t)]} \quad (25)$$

As mentioned in Section 1.6, if the natural (spatial) frequency of the oscillator does not coincide with that of the driving field (the polarization wave), the nonlinear process is very inefficient because of the lack of phase matching. We remark that the 3ω term initially contained in the polarization of the material (i.e., in the driving force term) soon generates a free-propagating wave at 3ω and, therefore, at the output of the fiber, E_{out} contains a component on 3ω. Although third-harmonic generation in fibers indeed has been reported in optical fibers [5, 6], the condition of phase matching imposes a severe limitation. Note that the polarization wave at frequency 3ω in the material propagates with a velocity determined by the wave constant $3k_\omega = 3n_\omega\omega/c$. On the other hand, the free wave at the same frequency 3ω propagating in the material has a wave constant $k_{3\omega} = n_{3\omega}3\omega/c$. The refractive indices at frequencies ω and 3ω are different due to dispersion, and, consequently, the free wave generated and the generating polarization wave in the material interfere. The wave generated at one point and propagating with the velocity of the 3ω wave interferes destructively with the 3ω wave produced at another point, so that the signal grows and decreases periodically along the fiber. Energy is transferred from the fundamental to the third harmonic, and after a certain distance—the coherence length L_c—energy flows back from the third harmonic to the fundamental.

The period is given by $(k_{3\omega} - 3k_\omega)L_c = 2\pi$, that is, $L_c = \lambda_0/3(n_{3\omega} - n_\omega)$. Had the coherence length been long (e.g., $L_c \sim 100$ m), this would not be a big problem, because the harmonic signal would grow for 100 m and this would be the optimal fiber length to use. However, the coherence length for such a process for an input wavelength ~1–2 μm is only ~10 μm. Therefore, the output signal at 3ω produced by a long piece of fiber is only that generated by one of those 10 μm periods (or less). Obviously, to exploit the nonlinear interaction efficiently, it is necessary to add up constructively the contribution of every small piece of fiber along its length, that is, increase the coherence length of the process. In this case, the process is said to be phase matched. In the parallel to the forced oscillator established in Section 1.6, the driving field has to have a spatial periodicity that equals the natural spatial frequency of the system. A similar constraint exists in the case of second-harmonic generation (SHG).The period for efficient interaction between visible and infrared radiation in SHG is 20–60 μm.

So far, in this subsection we looked only at the case when a single-harmonic signal inputs the fiber. It is, however, important also to consider the case when two or three beams copropagate. After all, radiation generated should travel down the fiber and act as the input for pieces of fiber further from the input end. We consider the general case where the electric field consists of three frequencies ω_1, ω_2, and ω_3. The total field can be written as

$$\mathbf{E}_{\text{in}}(z,t) = \frac{1}{2}\{\mathbf{E}_1 \exp[i(k_1 z - \omega_1 t)] + \mathbf{E}_2 \exp[i(k_2 z - \omega_2 t)] + \mathbf{E}_3 \exp[i(k_3 z - \omega_3 t)] + \text{c.c.}_1 + \text{c.c.}_2 + \text{c.c.}_3\} \quad (26)$$

where c.c._1 corresponds to the complex conjugate of the first term, and so on. When we consider the frequencies that mix in the third-order product $\mathbf{E}_{\text{in}}(z,t) \cdot \mathbf{E}_{\text{in}}(z,t) \cdot \mathbf{E}_{\text{in}}(z,t)$, we obtain a large number of terms, related to effects such as self-phase modulation, cross-phase modulation, and four-wave mixing. Two cases are of particular interest here. In the first, one of the electric field components has frequency zero (i.e., the electric field is dc—it does not oscillate in time) and the two remaining components are the same optical field $\mathbf{E}_{\text{opt}}$ of frequency ω, for example. The nonlinear polarization then contain terms of the type

$$\frac{3}{4}\chi^{(3)}(0,\omega,\omega)\mathbf{E}_{\text{dc}}\mathbf{E}^2_{\text{opt}} \exp[i(2k_\omega z - 2\omega t)] + \text{c.c.} \quad (27)$$

Interestingly, the frequency associated with this term is 2ω, that is, the nonlinear polarization creates a free wave at double the original frequency. This process is not due to a second-order nonlinearity of the material, but in the presence of an external dc field, the inversion symmetry is broken. Here, frequency doubling can occur only while the dc field is present. The process is often denominated electric field induced second-harmonic generation (EFISH). As discussed in Section 2, a dc field in glass fibers can be recorded by purely optical means. In such a case, a dc term builds up and the second harmonic becomes possible via the third-order nonlinearity. Note that the recording of the field in the glass fiber can, in principle, precede frequency doubling (it could consist of a preparation process).

The second case of particular interest is when one of the frequencies that mix is zero (as before, corresponding to a dc field), one is a microwave frequency (for example, in the Megahertz–gigahertz range, which we denominate ω_{rf}), and one is an optical frequency (once again, represented by ω). The nonlinear polarization then contains terms of the type

$$\frac{3}{2}\chi^{(3)}(0,\omega_{\text{rf}},\omega)\mathbf{E}_{\text{dc}}\mathbf{E}_{\text{rf}}\mathbf{E}_{\text{opt}} \exp[i(k_{\omega'} z - 2\omega' t)] + \text{c.c.} \quad (28)$$

Here, a new frequency ω' is generated, that differs from ω by only $\sim 10^{-8}$–10^{-5} and side bands are introduced. If the dc field is recorded in the glass system by some preceding treatment, we obtain what is apparently a second–order nonlinearity, because only two external fields are now required for the nonlinear mixing, the third one being provided by the modified glass system. We then obtain the linear electro-optical effect, or the so-called Pockel's effect.

It is possible in these two last examples to rename as $\chi^{(2)}_{\text{eff}}$ the product of the third-order nonlinear susceptibility and the dc field:

$$3\chi^{(3)}\mathbf{E}_{\text{dc}} \equiv \chi^{(2)}_{\text{eff}} \quad (29)$$

We see then that the third-order nonlinear susceptibility can give rise to nonlinear effects normally described by the second-order nonlinearity, even if, strictly speaking, the actual value of $\chi^{(2)}$ is zero. The dc field required for this to happen in fibers and glasses can be recorded in the material by poling.

We also note that even a purely third-order nonlinear process can lead to electro-optical modulation: If the rf in the previous example is reduced to zero and becomes a dc field, we obtain a refractive index that depends on the square of the applied field—the so-called Kerr effect or quadratic electro-optical effect. The corresponding contribution to the polarization is

$$\frac{3}{2}\chi^{(3)}(0,0,\omega)\mathbf{E}^2_{\text{dc}}\mathbf{E}_{\text{opt}} \exp[i(kz - \omega t)] + \text{c.c.} \quad (30)$$

Just as in Eq. (24), it is possible to incorporate the nonlinearity induced by the field into the refractive index and solve the propagation equation in the linear regime. Here, the change in refractive index induced in the presence of an effective $\chi^{(2)}$ is

$$\delta n = \frac{1}{2} 2\chi^{(2)}\mathbf{E}_{\text{dc}} \quad (31)$$

and in the presence of only a $\chi^{(3)}$, the change in index is given by

$$\delta n = \frac{1}{2} 3\chi^{(3)}\mathbf{E}^2_{\text{dc}} \quad (32)$$

Throughout this chapter we discuss second-order phenomena in glasses; all third-order nonlinear effects that discussed play a crucial role in understanding these phenomena.

1.9. Phase Matching

There are at least six ways to search for phase matching in fibers. The simplest is when the waves involved have very similar frequencies (not the case for ω and 3ω). Even if the matching is not perfect, the refractive indices are sufficiently similar for the coherence length to be relatively large. Another possibility is to have two frequencies that are on opposite sides of the dispersion curve. As previously mentioned, the group velocity reaches a maximum at $\sim$1.3 μm in silica fibers; therefore, a shorter wavelength (say at 1.06 μm) and a longer wavelength (say at 1.6 μm) can experience equal group refractive indices. We also can exploit the fact that varying modes have different propagation velocities due to waveguide dispersion. Therefore, in some cases it is possible to compensate for material dispersion, although for harmonic generation this usually requires fibers with unusually high refractive index steps ($\Delta n \sim 0.03$). An additional phase-matching scheme in nonlinear waveguides is to collect the emission of light as a Cherenkov radiation field emitted as a cone. More common in $LiNbO_3$ waveguides, it is neither easy to guarantee that such a field will be produced efficiently in fibers nor to focus the radiation to a diffraction limited spot.

Of all phase-matching techniques, the one by far most common in bulk nonlinear crystals is that where the birefringence of the medium makes up for the material dispersion. In crystals such as KDP, KTP, and BBO, the ordinary and extraordinary indices of refraction differ so much from each other that it is possible to have the fundamental wavelength in one axis and the harmonic in the other axis, interacting along the crystal always matched in phase. This makes very efficient frequency conversion (>50%) possible in such crystals. However, birefringence in fibers (even in hi-bi fibers) is much smaller than in crystals, so that phase matching for second or third harmonic is generally not obtained by this technique.

Finally, it is possible to reach high conversion efficiencies by introducing a spatially periodic inversion in the fiber, with a period equal to the coherence length. What would have been destructive interference then becomes constructive interference. This technique, known as quasi-phase-matching requires inversion in the fiber after a few micrometers. In $LiNbO_3$, $LiTaO_3$, and others, periodic inversion of ferromagnetic domains has led to high SHG efficiencies for low power inputs. As we discuss in more detail later, in fibers this is believed to be the mechanism that makes efficient SHG possible. As described in thermal poling (Section 3.6), several experiments also have been carried out where a periodic electrode is fabricated by photolithography in D-shaped fibers. The application of a spatially periodic electric field with the correct period for phase matching (which through poling records a permanent spatially periodic electric field in the fiber) leads to efficient frequency conversion.

2. OPTICAL POLING: SECOND-HARMONIC GENERATION IN FIBERS

2.1. Introduction

In this section, an introductory review of the subject of optically induced second-harmonic generation in fibers is presented. Some of the most significant experimental results reported in the literature on the field are described, as well as a summary of the main features of models put forward to explain the process. Some review articles exist in the literature [2, 7–10]; often they concentrate on some aspects of SHG in fibers or give a brief introduction to the subject inside a wider perspective of nonlinear effects. Also, an excellent book covers the broad area of nonlinear optics in fibers including SHG [2], which we strongly recommend for those interested in the field. Finally, since the first International Workshop on Photoinduced Effects in Optical Fibers in Quebec 1991, a series of Topical Meetings of the Optical Society of America has been taking place every two years, and currently involves some 250 participants. The proceedings of every Topical Meeting gather results achieved by most of the groups that are active in the field.

Until the mid-1980s, it was generally accepted that second-order processes could not occur in fibers to any significant extent because of the intrinsic absence of $\chi^{(2)}$ in glass. Nevertheless, weak sum-frequency generation and SHG have been observed [11–13]. It was with surprise and interest that the report was received [14, 15] that some fibers, upon illumination with infrared light from a powerful Nd:YAG laser, started to produce second-harmonic light. The conversion efficiency increased by many orders of magnitude over a few hours, and what originally was an unnoticed second-harmonic (SH) signal had grown to the milliwatt average power level for a ~100 mW input. Interestingly, after blocking the IR input for a period of time and launching IR light again, the previously exposed fiber was capable of frequency doubling efficiently and immediately without having to wait again for the long preparation process. The fiber had been modified by light and could now generate second harmonic efficiently. These original fibers were tested for SHG years after, and comparable conversion efficiencies were obtained, suggesting that the preparation is nearly permanent at room temperature.

The initial belief was that some property particular to the fibers used was responsible for this effect. Those fibers had a small amount of phosphor codoping the core with germanium, and this seems to make it easier for the preparation to take place with a single strong IR input. However, it is now known that virtually any Ge-doped silica fiber normally used in telecommunications can frequency double efficiently after a preparation process where both the fundamental and the second-harmonic light from a Nd:YAG laser are launched simultaneously into the fiber. The field expanded rapidly in many different directions. Investigations were carried out to study fiber materials (role of constituents, dopants, fiber fabrication conditions) and frequency-doubling fiber properties under various treatments (heat, exposure to UV, blue–green and infrared light, electric fields, mechanical stress). Theoretical models were created and applications were demonstrated, some of which involved fiber preforms, glass waveguides, and semiconductor-doped glasses. Many interesting and creative experiments were performed in the search for clues that could lead to a general explanation for the observations.

It is probably fair to say that even at this stage most researchers who work in the area have doubts about the basic mechanisms that we behind the preparation process that leads to frequency doubling in fibers, although much is now understood. The highest conversion efficiency reported to date in the literature is 13% [16]. Although this is a relatively high value (and many orders of magnitude higher than what could be expected in fibers), the promise that this figure could soon be improved did not materialize. Even rare-earth-doped fibers that showed great potential for frequency doubling did not show conversion efficiencies over a few percent [17–23]. Some of the reasons for this are discussed throughout this section. At this stage, the conversion efficiencies that are generally achieved by optical poling of telecommunication fibers are just under the minimum required for the process to be interesting from a commercial perspective.

We subsequently describe some of the results reported in the literature. The reader should assume, unless otherwise specified, that a Nd:YAG laser was used in the experiment, usually running Q-switched and mode-locked. Although frequency doubling at other wavelengths also has been described [24–26], the great majority of studies were carried out with Nd:YAG lasers operating at 1.06 μm. The typical duration of the individual mode-locked pulses employed was 100 ps.

2.2. Is it Really Second Harmonic?

Scientists should be (and usually are) skeptical about new effects, particular those known to be forbidden by the established laws of physics. The question then arises whether this green signal that looks like the second harmonic of the laser actually came from plain frequency doubling. If this is the case, the following characteristics would be expected.

1. The output power of the visible signal depends on the square of the input power.
2. No amplification is seen when a fiber frequency doubles light.
3. The wavelength of the signal is exactly that of the second harmonic of the laser.

Experiments were carried out to test these features. After preparing a fiber for frequency doubling, the IR input was varied and the green power output was measured. The visible and IR were found to be related by the square law $P_{2\omega} \propto P_{\omega}^2$ for peak powers <5 KW [27]. For higher powers, the power law was different; a possible explanation for this is related to self-phase modulation and is discussed later. To test specifically the second feature, a Michelson interferometer was constructed so the two IR inputs with an adjustable delay could be coupled into a fiber. The superposition of the two resulting signals revealed no amplification of one due to the presence of the other, as expected in view of the quadratic power law. The most crucial experiment was the determination of the exact wavelength of the green signal generated by the fiber. To that end, IR light was coupled into a 50 cm long piece of fiber that had been prepared previously. The green and the IR outputs of the fiber were directed to the two arms of a Michelson interferometer. The green was delayed only whereas the IR signal was delayed and passed through a KTP crystal that generated second harmonic. After the interferometer, the green light from the fiber and from the crystal was directed to a screen (or a photodiode array), where interference fringes were clearly seen, indicating that the wavelengths coincided interferometrically [28]. These experiments confirmed that the fiber generates second-harmonic light after preparation and that the process is not mediated by phonons, for instance, when coherence (and energy) is lost.

2.3. The Preparation Process

As mentioned, the anomalous growth of SHG efficiency in fibers originally was observed with a single IR wavelength input of high power. Figure 2 [15] shows, on a semilogarithmic scale, the green light output as a function of time. Note that in this experiment the IR input has the double role of modifying the fiber and probing the conversion efficiency. The inset shows, on a linear scale, the visible signal measured between 6 and 9 h. The growth is seen to be exponential until it levels off, when the conversion efficiency approaches the 1% level.

The exponential growth suggests that the presence of green light in the fiber affects the preparation rate. As will be discussed later, one of the most significant results reported on the subject is the fact that the simultaneous injection of a weak *seed* of second harmonic with the powerful fundamental was sufficient to accelerate to the minutes scale the preparation process that normally takes hours [29]. Experimental setups were then devised where the two writing beams are

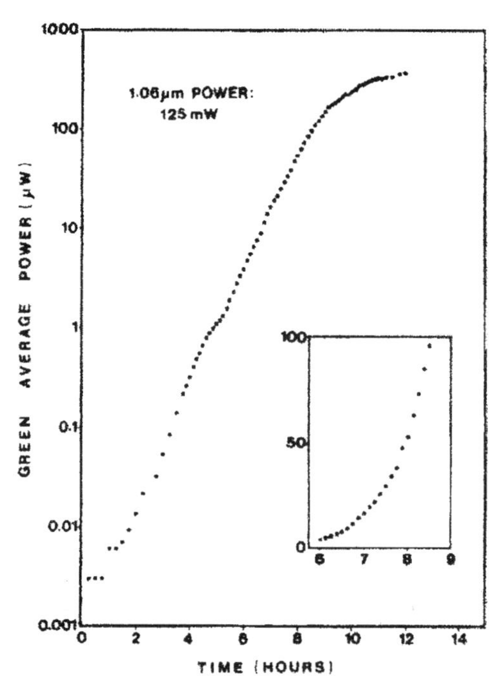

Figure 2. Second-harmonic signal generated by an optical fiber during fiber preparation.

injected from one side of the fiber for seeded preparation and reading with only IR done in real time from the other side of the fiber to distinguish the SH light injected from that generated [30].

Attempts were made to examine the dependence of the preparation rates on the input power level. For high powers, the increase in preparation rate followed the square root of the seed green power input [31]. An initial exponential [31] and quadratic growth [32] of the SH signal was claimed under two different preparation regimes. It was also generally noted that for very high green input powers, the preparation process saturates very rapidly at a low conversion efficiency level. An experiment also was carried out to measure the preparation rate for various IR beam powers. A high power law was found for the dependence of the rate coefficient on the intensity of the infrared seeding beam following I_ω^I with the exponent in the range of $x = 8$–12 [33]. There are several reports that a threshold power exists, below which fibers and preforms do not prepare for frequency doubling [34, 35].

Experiments also were carried out to examine whether the preparation process was affected by temperature [36, 37]. There is some disagreement in the literature as to the preparation rates obtained at room and liquid nitrogen temperatures (RT and LNT). An ingenious setup was used [37] where the whole fiber under test was immersed in liquid nitrogen and light from the laser was taken to it by a spliced pure silica fiber. The results showed a higher preparation rate at RT than at LNT. The thermal population of defects probably explains the observations.

2.4. Where Is the Second-Harmonic Radiation Generated?

One of the most relevant features studied was where in the fiber the SHG takes place. Experiments were performed by cutting the prepared fibers from the output end and measuring the green output. Alternatively, the weak SH light emerging from the side of the fiber was used as a means to determine the SHG efficiency [34]. The fiber length where SHG takes place was found to be limited to <1 m. It was also found in fibers prepared with IR only that the green signal grows significantly only after ~20 cm from the input end, as illustrated in Figure 3. This result does not mean that nothing happens in the first 20 cm of fiber. In fact, when the first part of the fiber is removed, there is virtually no green output for the same IR input [34].

Other experiments were performed where the side-light emission of fibers prepared by various seeding powers was studied [38]. It was found that distance from the input end where saturation is reached is reduced when higher SH signals are injected with the fundamental. It also was found that the sharp turn-on behavior illustrated in Figure 3 becomes less and less pronounced, and for relatively high SH input powers, the side-light curve grows quadratically with fiber length. These and the preceding results indicate that the case where the fiber is prepared with only IR input is a limiting

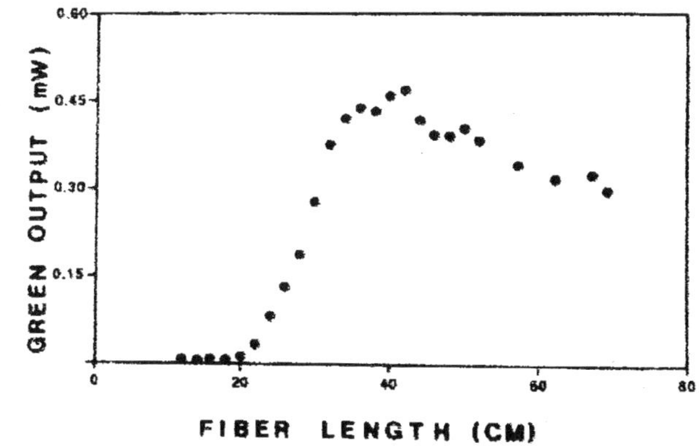

Figure 3. SH signal of a fiber prepared with IR only as a function of fiber length. The experimental points are obtained by cutting the fiber shorter and shorter starting from the output end.

case of seeded preparation with very low green input. Stolen and Tom suggested [29] that the initial SH light present in the fiber that makes it possible for preparation to start in the unseeded case comes from the weak interaction (magnetic dipole and electrical quadrupole, particularly at the core–cladding interface) that normally is neglected in the polarization expansion. The coherent interaction between the SH and fundamental light is then believed to play a major role in the preparation process.

Cut-back and side-light measurements have severe disadvantages. The latter technique is insensitive and can be used only when the fiber is already generating significant SH radiation. Therefore, it does not show how the preparation process evolves in time, particularly in the beginning. The former experimental procedure is destructive, which again makes it difficult to determine where the SH is generated at various stages of the preparation. Therefore, a technique was developed to look at the preparation process by scanning a small heating element along the fiber and measuring the total SH signal at the output of the fiber. The technique was named the heat scan [39]. The principle is the following: Whereas the harmonic generation is coherent, a phase shift introduced at any point affects the integrated signal. For example, if the phase shift is π, it should be possible to extinguish the total frequency-doubled signal at the output if the SH light created before the element exactly cancels the light generated after the heated point. In other words, it should be possible to determine the midpoint of the generation at various stages of the preparation for the same fiber. It turns out that the information gained by the heat scan is even more complete. It also is possible to determine the profile of the generating region, and this gives important clues to why the preparation saturates [40]. Figure 4 illustrates how the SH grows in time, and the extent and shape of the generating region [41]. The general trend is that frequency doubling occurs further and further into the fiber as the preparation evolves, but close to saturation the generating region reverses its movement and retreats toward the input end of the fiber.

We return to the physics of the saturation process in Section 2.17. The heat-scan technique was found to powerful

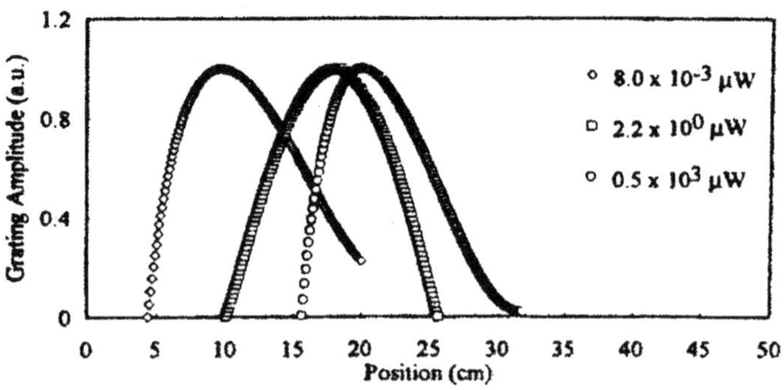

Figure 4. Time evolution of the $\chi^{(2)}$ grating profile. The curves are normalized. Their temporal order corresponds to the order of SH powers.

not only to characterize frequency-doubling fibers, but also fiber Bragg gratings [42], which are briefly discussed in Section 2.11.

2.5. A Grating for Phase Matching

As mentioned earlier, there are two major physical constraints for frequency doubling to be efficient: the existence of a $\chi^{(2)}$ in the fiber and of a phase-matching scheme. From the foregoing experiments and from the high levels of SH light generated, it seems that both requirements are met. Even though we lack knowledge of some of the most basic characteristics of the microscopic mechanisms that create a $\chi^{(2)}$ in glass, the way phase matching is obtained is now well established. It occurs through the creation of a periodic grating along the fiber that is the result of the coherent interaction of the second-harmonic light and the polarization wave produced in the material by the fundamental radiation of the laser. We come back later and discuss in more detail the original model where this scheme was suggested [29]. It is still not entirely clear if this electric field grating consists of free charges or polar defects, but there is overwhelming evidence that a grating does exists (see Fig. 5).

The grating is written by the interaction of the beams. The fact that SHG takes place over a limited fiber length suggests that the coherent interaction ceases after a relatively short distance. This happens because of three reasons. First of all, dispersion physically separates the green and IR pulses after ~1 m. Furthermore, self- and cross-phase modulation alter the relative phase of the waves, destroying phase matching. Finally, the finite bandwidth of the input results in different grating periods, which also limit the coherence length of the process. If the grating length is 10 cm and its period is 40 μm, the grating is composed of 2.5×10^3 individual elements. For all these to contribute constructively, the constraint is periodicity is one part in 10^4, indicating that spectral broadening and phase modulation must be avoided. The last reason is believed to be the one that sets an upper limit for the useful fiber length [43].

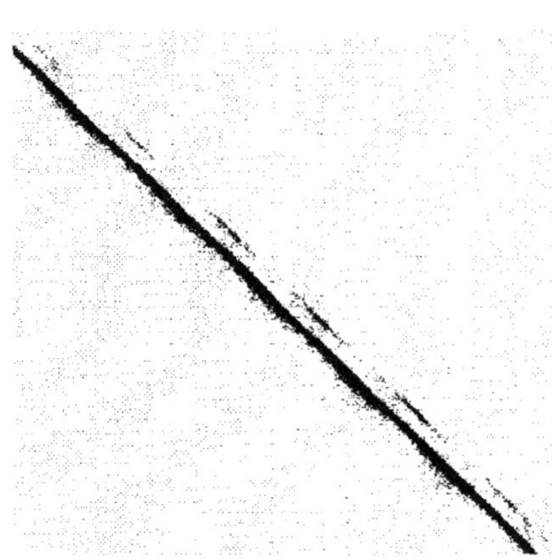

Figure 5. Image of an etched frequency-doubling fiber obtained in a light optic microscope.

The existence of a periodic grating in frequency-doubling fibers seems to be the only possible explanation of the high conversion efficiency found. However, direct evidence of a grating would still be more convincing and might also give further information about the mechanism of the preparation process. If the copropagating light waves of fundamental and second-harmonic frequency are able to modify the glass material permanently in such a way that the average inversion symmetry of the material breaks, we might expect that the same mechanism causes changes that are detectable by other means. For instance, the light scattering cross section might be influenced by these changes. Kamal et al. [44] used this idea. They launched laser light into a prepared frequency-doubling fiber and observed laterally scattered Raman light coming from the core of the fiber. The scattered light was observed though a microscope objective, and the whole light detection system was scanned along the fiber. The scattered light showed a periodic modulation with exactly the expected period of the $\chi^{(2)}$ grating. The average scattering intensity was found to be the same as that measured in an unprepared fiber. This was the first direct evidence of periodic structure in frequency-doubling fibers. To date, the results of Kamal et al. have in theoretical explanation. Such explanation might give important insight into the physics of the preparation process.

Later, an even more direct picture of the $\chi^{(2)}$ grating was obtained by etching frequency-doubling fibers with hydrofluoric acid [45]. The idea of this experiment was based on the hypothesis that the second-order nonlinearity was due to strong electric fields present in the glass. Initially the researchers intended to mark the high field regions with microscopic dust particles, which were expected to be attracted by the field gradient. However, this idea failed because the high field region is encapsulated in the almost "infinite" cladding layer. It was, therefore, necessary to remove this layer. However, if the glass material is removed chemically, it might be unnecessary to mark the high field region with dust particles. If the preparation process modifies the glass material, it could equally well modify the etching rate of the chemical process. Therefore, the etched fibers were submitted to visual inspection under a light optic microscope before exposing them to dust. In fact, the periodic structure showed up directly without any marker.

Figure 5 shows an image of the $\chi^{(2)}$ grating obtained this way. It was possible to relate the modification of etching rate to the electric field strength, and an analysis of the observed grating for different etching radii permitted the conclusion that the encoded high fields are most likely due to macroscopic charge migration and not to dipole orientation.

2.6. The Initial Stages—Modes

It is now generally believed that the coherent interaction of the second harmonic and the fundamental of the laser light writes a periodic grating along the fiber. In the unseeded case, it is believed that a weak SH signal is produced anyway by the fiber due to (normally neglected) electric quadrupole and magnetic dipole interactions, and this weak signal acts as a seed. It is natural to try to unify the description of the seeded and unseeded preparation, because so far all experiments suggest that one can be treated as the limiting case of the other. There is some disagreement in the literature as to the initial importance of this electric quadrupole interaction. It has been argued that there was cancellation of most terms, leading to a very weak net contribution [46], and also that far from the phase-matching condition, cancellation was not a big problem [47]. In either case, we are talking about total conversion efficiencies in the $<10^{-5}$ range.

To sort out this puzzle, an experiment was carried out with three fibers that had various Ge doping concentrations and Δn between core and cladding [48]. It was observed that the initial weak green signal did not depend on the Ge concentration. On the other hand, a larger value of $\Delta n(0.017)$ led to a 4 orders of magnitude higher initial SHG conversion efficiency than for a fiber with lower $\Delta n(0.008)$, suggesting that the refractive index change has strong influence on the amplitude of the SH signal generated initially. This can be interpreted as showing the importance of the role of the electric quadrupole interaction, which is very sensitive to the refractive index change across the interface due to the high gradient of the electric field [48]. However, a higher Δn is likely to be associated with a fiber that has much stress at the core–cladding interface after fabrication, so the role of stress relaxation under high power illumination should not be neglected.

Assuming that the electric quadrupole at the interface actually generates a weak SH signal that helps prepare the externally unseeded fiber, it is possible to understand observations that the green signal usually propagates in the LP_{11} mode for an IR input in the LP_{01} [34]. The gradient of the field across the center of the fiber goes through zero when the fundamental is in the LP_{01}, favoring odd symmetry modes for the SH. It is generally observed in seeded experiments that the mode pattern upon reading follows the mode of the seed. Therefore, this initial quadrupole green signal propagating in the LP_{11} should lead to generation of efficient SHG in the same mode. This, however, is not always the case. Occasionally fibers prepare (unseeded) in the LP_{01} mode, suggesting that the quadrupole odd symmetry requirement is not met.

Time evolution of the mode structure during preparation also was observed [48]. The same fiber with a large step index ($\Delta n = 0.017$) was used. For 6 KW peak power inputs (no external seed), the mode of the SH changed along the preparation process from LP_{11} to the LP_{01}. The time at which this change occurred was different for polarizations both parallel and perpendicular to the polarization of the laser input, and for higher preparation input powers, the mode evolution did not occur (i.e., the SH mode was LP_{11} until saturation). Although the mechanism behind these findings is not clear, this experiment gives clues to which parameters must be controlled to study initial stages of the preparation process, rendering less erratic observations of occasional mode change in fibers.

Because of the great loss of higher order modes, it is not surprising to see after a long preparation period that the green signal was produced in the fundamental mode. What was unexpected was an immediate switch to an LP_{11} mode when the fiber was released from the pencil. Could the stress during preparation have altered the phase-matching period of the grating? It seems obvious that many unanswered points remain as to the mechanisms involved in the initial stages of the preparation process.

2.7. Erasure by Light

Exposing frequency-doubling fibers to high levels of green light result in the strong reduction of the conversion efficiency [31, 49]. In other words, visible light can erase the grating created in the fiber. Two-photon excitation is believed to be responsible for this erasure, but here again it is not clear if previously trapped charges are released, if oriented defects lose the orientation, or if defects are created all over the fiber. The fact that two-photon excitation is involved implies that illuminating the prepared fiber with modest UV power, such that single-photon excitation would suffice, will result in similar erasure; this case was indeed verified [50]. Measuring the erasure rate for various UV optical powers inferred a single-photon process. It was suggested [31] that the growth of the SH signal to a high level in a fiber under preparation should lead to self-erasure, creating a mechanism for saturation. This prediction was quantified with the help of rate equations, and such behavior reproduced numerically [51]. An experiment also was carried out with highly Ge-doped fibers, where it was observed that for high power IR input, the grating was erased [52].

It is believed that optical erasure thus requires a 4ω quantum, arising from four IR photons, two 2ω photons, a single UV photon, or two IR and one 2ω photon. Access to the UV absorption band peaking at 244 nm seems to be required. It is interesting to note that the combination of quanta required for preparation (two IR photons and one SH photon) is also capable of destroying the grating. Experiments where the relative phase was varied between the preparing beams in a seeded experiment revealed that although it takes several minutes to reach the saturated SHG level, erasure takes only a few seconds. The efficient destruction of the grating in the presence of green and high intensity infrared radiation is believed to be the outcome of the saturation reached in the preparation regime with the Q-switched mode-locked laser generating high peak powers [40].

2.8. Erasure by Temperature

Defects in glass can be annealed by heating [53]. Whereas each defect is annealed at a particular temperature, it is natural to erase frequency-doubling gratings thermally in fibers so as to identify possible centers that are organized as a grating and to determine the characteristic activation energy.

The thermal dependence of the refractive index (11×10^{-6} K^{-1}) and, to a lesser extent, thermal expansion of silica (0.52×10^{-6} K^{-1}) alters the effective period of the grating. Consequently, the conversion efficiency is reduced dramatically when the temperature is changed, even if the defects are not erased. Therefore, the effect of annealing has to be probed when the fiber is at room temperature. Pieces of fiber were prepared inside a thin stainless steel tube, which was taken to an oven. After annealing the fiber conversion efficiency was probed. Even a few minutes of exposure to modest temperatures (~100 °C) was found to be [54, 55] sufficient to cause a 10 times reduction of the conversion efficiency in a fiber with Ge (only) doping. A similar effect was obtained in a fiber codoped with P at higher temperatures (~300 °C), suggesting that P introduces efficient traps to the glass matrix. These low temperature values imply that rather than annealing centers out of existence, the heating treatment randomizes the grating that is organized by light. These observations indicate that the preparation for SHG in fibers may have at least two stages, that of creation (or activation) and that of organization of centers (or free charges) that form a grating. Two values of the activation energy were obtained, suggesting that two types of species contribute to optical poling, one with lower value (0.28 for Ge-doped and 0.4 eV for Ge- and P-doped fibers) and one with higher value (~1 eV) [55–58].

2.9. Defects in Preforms and Fibers

Self-organization of a $\chi^{(2)}$ grating of color centers was proposed originally in [5]. Although lacking a mechanism that ensured phase matching, it originally was assumed that the grating was formed with the correct periodicity by the interaction of the IR and SH fields. Many studies of the defect centers that might be associated with SHG in fibers followed. The reader is referred to two excellent articles that deal with defects in glasses [59, 60].

It has been reported that neither pure Si core nor pure P-doped fibers can be prepared—Ge seems to be a necessary dopant for SHG induction [61, 62]. Even pure silica fibers can be prepared for SHG [63] if they are hydrogen loaded [64, 65]. Among the number of defects that might contribute to the frequency-doubling grating, the one that has attracted the most attention in the literature is the Ge E' center, which is created when a Ge atom is connected only to three oxygen atoms, leaving the fourth electron unbound. A similar defect in silica is the Si E' center. Electron spin resonance measurements indicate that fibers subjected to the preparation process have a much enhanced Ge E' defect concentration than before illumination and the increase of concentration correlates well with the conversion efficiency [66]. Note, however, that the increase in defect concentration extends over the whole fiber length, much beyond the length of the $\chi^{(2)}$ grating, suggesting again that orientation is a necessary stage of the preparation process.

The creation of Ge E' centers with an absorption peak at 215 nm usually is associated with the transformation of a germanium oxygen deficiency center (GODC) that has a characteristic absorption band centered around 244 nm and extends to close to 266 nm. The GODCs can be measured easily in a fiber preform. Figure 6 shows a spectrophotometer trace of a 0.6 mm slice of a fiber preform. The sample was apertured so that the transmission of either the core or the cladding was recorded. The trace of the absorption of the cladding is displaced for clarity, and it is apparent that the large peak is due to a Ge related center, because only the core shows this absorption peak. The large absorption at 242 nm extends all the way beyond 266 nm, the fourth harmonic of the 1.06 μm fundamental wavelength of the Nd:YAG laser.

There is enough evidence to suggest that the fourth harmonic of the Nd:YAG laser light plays an important role in the preparation process. First, (multiphoton) erasure with green and (single-photon) UV light, and the absorption peak in the 266 nm wavelength region in preforms involve the energy of UV photons. Also, electron spin resonance measurements of the creation of the Ge E' center under UV exposure find a parallel in the formation of Bragg grating in fibers. Early experiments where weak blue light was seen exiting fibers subjected to (unseeded) high IR powers led to the conclusion that the preparation process involved multiphoton mixing. The blue luminescence, which is associated with the presence of Ge in the core, can be efficiently excited by UV radiation and has a peak wavelength ~415 nm and

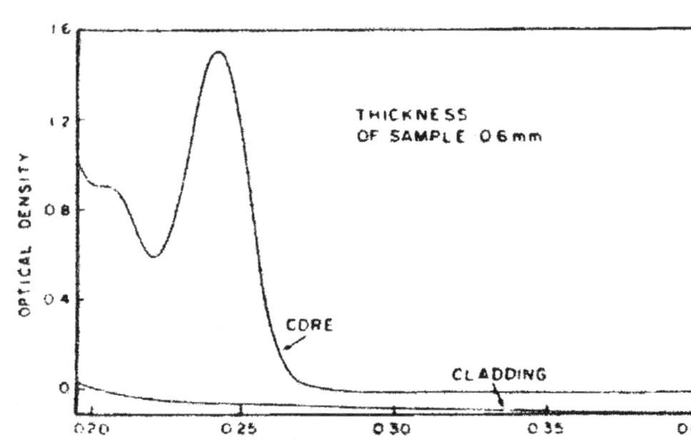

Figure 6. Absorption spectrum of core and cladding.

lifetime ~80 μs [67]. Under illumination with high power IR light and weak seed SH light, the allowed process via the term $\chi^{(3)}(4\omega = \omega + \omega + 2\omega)$ in fibers should be capable of creating UV light or at least a polarization wave at 4ω. Whether the 4ω wave actually propagates for a few micrometers in the fiber or whether a periodical UV signal exists has not been determined. Dianov and co-workers [68], however, have established that UV radiation is generated by the fiber during preparation.

2.10. Preparation Under UV Excitation

To examine the role of UV excitation, a Nd:YAG laser was frequency quadrupled to 266 nm with the aid of two crystals, so that radiation at ω, 2ω, and 4ω was available. Exposing fibers to blue or green light prior to the preparation process is known to lead to the creation of color centers and thus higher conversion efficiencies are obtained [69, 70]. Experiments were performed where fibers were exposed (1) to a preparation beam composed of the fundamental and the second harmonic of the laser (conventional preparation for reference purposes), (2) to the preparation beams after pre-exposure of the fibers to a high dose of UV for up to 1 h, and (3) to the three wavelengths simultaneously (1.06 μm, 0.53 μm, and 0.266 μm). In these experiments, UV radiation was coupled into the cladding modes of a short piece of fiber kept straight in a glass capillary tube, so that it bathed the core rather than propagated along it. The bright blue luminescence was used as a way to monitor and optimize alignment. The preparation rate increased by orders of magnitude when all the beams illuminated the fiber together, as illustrated in Figure 7 [71]. This result suggests that the UV indeed does play a role in the preparation of fibers, possibly providing the necessary energy for an electronic excitation that precedes the organization of centers. The lifetime of this excitation was not determined, but it was estimated to lie in the range 10 ns $\ll \tau \ll$ 1 ms. It was, therefore, speculated that the required excitation for the creation of the second-order nonlinearity was related to the luminescence associated with the emission of blue light at ~415 nm. It also was verified that under UV excitation, preparation required a minimum green input seed until the SH generated by the fiber was sufficient to replace the external seed.

Several questions remain to be answered, such as whether the saturation level is the same with or without UV light exposure. However, these results suggest that an incoherent high power excitation all along the fiber actually can contribute to the preparation process. This suggestion is apparently inconsistent [72] with preparation models based on selective absorption by mid-gap states that leads to charge separation. Enhancement of the preparation rate simultaneous with or following UV illumination also has been reported [73, 74].

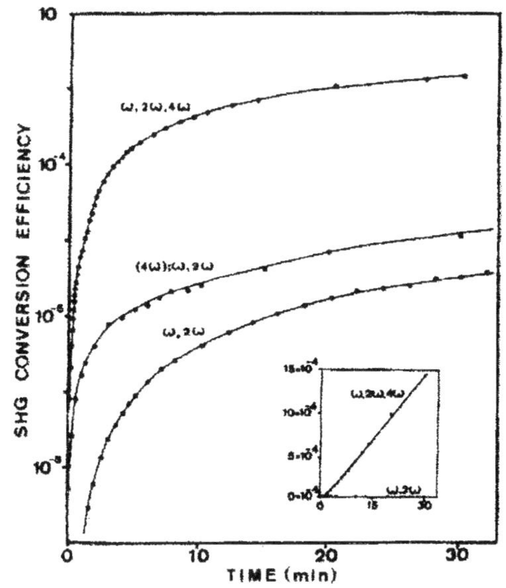

Figure 7. Enhancement of preparation speed by UV excitation. Previous exposure of the fiber to the fourth harmonic accelerates the preparation slightly: Curve (4ω); ω, 2ω. Simultaneous exposure to UV and the writing light accelerates the process by orders of magnitude: Curve ω, 2ω, 4ω.

2.11. Photoinduced Self-Organization in Fibers—Bragg Gratings

The slow growth of second-harmonic light generated by a fiber under IR excitation first reported in [14] is reminiscent of another photoinduced effect in fibers. Hill et al. [75, 76] observed that the transmission of a fiber illuminated by a continuous wave (CW) single-mode Ar^+ laser beam reduced slowly over time (minutes), and more and more of the light injected was reflected. Very good mechanical stability was necessary to provoke the effect, and it was concluded that a Bragg (Hill) grating had been created. The grating was formed by a periodical alteration of the refractive index along the fiber, and the period of the domains ($\lambda/2$) was such that there was destructive interference for the transmitted beam and constructive interference for the reflected signal. To establish the right periodicity, it was concluded that the back-reflected signal from the cleaved endface of the fiber interfered with the forward propagating light, producing a stationary periodic intensity distribution. Two-photon absorption of the green laser light led to the periodic excitation of defects, followed by a refractive index change. The linewidth of the mirrors thus created could be in the megahertz range.

The technique for producing Hill gratings in fibers is now well established [77, 78]. In addition to excellent mechanical stability, it is important to have a very well cleaved endface and high concentration of GODC in the fiber. A high reflectivity endface also can help the grating formation. The interest in Bragg gratings grew explosively after a technique was developed for writing the grating externally [79]. The light from an excimer laser (or a frequency-doubled Ar^+ laser) operating in the 250 nm wavelength region is divided into two components that interfere. The interference pattern is projected laterally onto the fiber and by adjusting the angle of the two beams, it is possible to control the spacing between fringes. The Bragg period, therefore, can be adjusted to form mirrors and filters for various wavelengths that are of obvious interest in fiber lasers, multiplexing systems, and so forth. A similar technique can be used to write polarization gratings, as long as the period equals half the beat length of the fiber.

It is now very common to make use of a phase mask that produces two diffracted beams at an angle, which leads to grating formation for the 1.5 μm region of the spectrum. The period of the grating can be varied along the fiber, so that a chirp is introduced. The quality of the fiber Bragg grating (FBG) can be improved by apodizing its profile (i.e., reducing the index profile at the beginning and at end), so that side bands in the reflected spectrum are suppressed. Photosensitivity can be enhanced greatly by diffusing hydrogen into the fiber all the way to the core. This can be accomplished by leaving the fiber in a hydrogen chamber for a couple of weeks at 200 atm room temperature. Long period gratings have been demonstrated, where light is coupled in the forward direction (generally into cladding modes). Such gratings work often as a loss element.

A huge variety of fiber components for optical communications presently are fabricated using the FBG technology, including narrow band filters for DWDM applications, DFB structures, gain equalizers in erbium-doped fiber amplifiers, and laser mirrors. Fiber Bragg gratings are an excellent example of a technology that developed in a few years from a scientific curiosity to a relatively mature methodology for making useful telecommunications components. The extensive subject of fiber Bragg gratings is covered by two books [80, 81].

The parallel between fiber Bragg gratings and optical poling is striking. Both photoinduced effects of SHG and Bragg reflectors in fibers have a grating written by light. In both cases, the germanium oxygen deficiency centers and the ~250–260 nm wavelength region are accessed, the writing process is slow (minutes), and the grating is stable at room temperature but can be annealed by light or by heat. Although the parallels are many, it remains to be determined to what extent the microscopic mechanisms behind these effects are related.

2.12. Polarization Measurements

One of the early physical characteristics of the process of frequency doubling in fibers that was studied was the polarization state of the light. Polarization also can give information about the mechanisms that make SHG possible. Österberg and Margulis [34] found in the simplest arrangement (preparation with only IR radiation) that the (linear) polarization of the output at the SH wavelength followed the (linear) polarization of the IR input used for probing. They also found that when the linearly polarized IR used for probing is rotated by 90° relative to the (linearly polarized) preparation IR beam, the polarization of the SH follows that of the probe beam, but the conversion efficiency drops significantly.

A more systematic and quantitative experiment was carried out [82] using single-mode fibers (at 1.06 μm and 0.53 μm) prepared by seeding with a CW mode-locked Nd:YAG laser. The objective was to verify the value or the ratio between the values of various tensor components involved in frequency doubling, and the findings were

$$\chi^{(2)}_{xxy} \sim \chi^{(2)}_{yyy} \sim \chi^{(2)}_{yxx} \sim 0 \quad \chi^{(2)}_{xyy} \sim \chi^{(2)}_{yyx} \quad \chi^{(2)}_{xxx}/\chi^{(2)}_{xyy} \sim 3 \qquad (33)$$

These results are consistent with a SHG process in which an internal dc electric field is established and acts upon the $\chi^{(3)}$ of the material, as discussed in more detail later. Mizrahi et al. also suggested that only through a fortuitous coincidence could the results be understood in terms of a SHG process that involves the creation of a macroscopic $\chi^{(2)}$ through the orientation of dipolar defects.

As mentioned earlier on, when the probing IR power in unseeded experiments was higher than ~5 KW, the power law that relates the green signal generated and the IR input deviated from a square law normally expected for SHG. Rather than a slope 2 on a log–log graph, a slope 3–4 is found. This effect was attributed [83, 84] to polarization- and intensity-dependent phase mismatch in prepared fibers. Qualitatively, if very high powers are injected into a fiber to write a grating, self- and cross-phase modulation introduce a modulation in its period. The intensity-dependent phase mismatch also depends on the relative polarizations of the fundamental and the SH light. When the grating is probed with low powers, only a fraction of the grating contributes for SHG, but the fraction increases as the probing power approaches the values present during the writing stage. The relative polarization of the signals determines the amount of cross-phase modulation in this process.

2.13. The Phase Problem

One of the most intriguing aspects of frequency doubling in fibers is the phase. It is reasonable to expect that the periodic grating created will peak at those points where the beating between the ω and 2ω waves that record the field goes through a maximum. One consequence of this situation is that the frequency-doubled signal generated by the fiber should be $\pi/2$ out of phase with the SH light used as the seed. The $\pi/2$ phase shift is a general result and is also found in the wave equation that describes frequency doubling in crystals and in other nonlinear optical effects.

The phase shift between the seed and the SH generated by the fiber can be measured. In one experiment [85], a fiber was prepared with a relatively strong seed and the process was interrupted at an early stage, so that the phase of the total green field was defined by the seed. For probing, the seed was attenuated without an intentional change in the relative phases of the waves. The resulting attenuated seed was then used as a reference wave. Interference between the second harmonic generated by the fiber and the seed with large modulation depth was observed. Because the phase of the SH light generated by the fiber was anchored to the phase of the IR input, it was possible to sweep the interference pattern by altering the phase of the IR input against the (seed) SH reference. Assuming that the relative phases had not been altered when the seed was attenuated, the initial point of the interference pattern contained the phase relationship between the seed and the SH signal generated by the fiber. The phase adjustment was accomplished by using the dispersion between the IR and the reference SH in a variable thickness water cell.

In the experiment described in [75], the attenuation of the SH was performed with a multilayer dichroic mirror that introduced a phase uncertainty, but several other studies revealed the existence of a $\pi/2$ phase shift that does not evolve significantly over time [86–88]. More recently, an interferometric preparation experiment performed with a twin-core fiber showed that the phase evolution of the SH during preparation is indeed small [63]. Although expected from the preceding considerations, this result is very puzzling. In unseeded experiments it is believed that the SH generated plays the role of the seed, so there can be no phase shift (e.g., $\pi/2$) in this case. In seeded experiments, as the SHG efficiency increases the phase of the total green signal in the fiber should alter over time, so the phase of the grating should change. This process leads to saturation when the SH generated by the fiber equals the seed but is π out of phase with it, that is, the total green light exiting the fiber should drop to zero. This never has been observed in practice. The phase of the second-harmonic signal does, in fact, evolve over time during the preparation process [89, 90]. Dianov et al. [91] demonstrated a dynamic phase shift of the grating due to inertial erasure. This phase shift provides the growth of the second-harmonic signal during preparation and attains a value of $3\pi/4$ at saturation. Temperature tuning of the phase shift to $\pi/2$ made it possible to increase the second-harmonic efficiency by a factor of 2 at saturation [91].

Other phase measurements have been carried out. When a second grating shifted in phase relative to the first grating was recorded in a fiber, the SHG efficiency dropped rapidly to almost zero and then recovered slowly to its initial value [92]. Phase fluctuations between the input IR and SH beams also were found to greatly limit the preparation process of seeded fibers [93]. Many other possibilities are opened by having control over the phase of the waves that write the gratings. It is possible, for instance, to write a series of frequency-doubling gratings with the necessary shift to enable Bragg reflection to occur. Obviously, much remains to be done and much remains to be understood with regard to the phase of the waves involved in SHG in fibers.

2.14. Electric Field Experiments

Several experiments have been performed to investigate the role of a dc electric field on fibers and glass preforms. Most of these were, in fact, designed to test specific models that have been put forward to explain SHG in fibers and are discussed in more detail in Section 2.16. The application of a strong external electric field was carried out first in preforms and then in fibers [94–96]. Although frequency doubling was observed in the bulk glass samples under the application of an external field due to the third-order nonlinearity, the results showed no permanent transformation of the fiber. When fibers were prepared for frequency doubling in the presence of a dc electric field, no significant difference was found from a set of control fibers prepared under the same laser conditions without the field. However, much more recently, fibers with internal electrodes were used to examine optical preparation under very strong electric fields that showed a significant enhancement of second-harmonic signal [97]. Fibers were prepared in the usual way by launching infrared and second-harmonic waves into the fiber. Then a strong electric field was applied across the fiber core that caused a 10 times higher second-harmonic signal. This observation can be explained by the coherent photovoltaic model [98].

Other dc field experiments, longitudinal or periodic along the fiber, also have been reported [99–102]. In the presence of visible or strong IR light, the poling field was found to induce a permanent change in the fiber and frequency doubling. Kashyap [102] suggested that two-photon absorption of green light provides an excitation and the orientation is provided by the dc field, much the same mechanism invoked to explain results of preparation under UV excitation. The possibility then exists to write the appropriate grating for the particular laser wavelength that is required to frequency double, with the simultaneous presence of an excitation and an orientation source. Such a scheme gained additional interest when UV poling was reported, where the two-photon process is replaced by a more efficient one-photon excitation, as described in Section 3 of this chapter. Also, the use of periodic structure electrodes under high voltages and heat is described in Section 4.6 as a means to record a quasi-phase-matched second-order nonlinearity in fibers for frequency conversion.

2.15. Other Experiments in Glass and in Fibers

Investigation on the generation of the second harmonic in different glass systems has been very promising. A $\chi^{(2)} =$ 0.8 pm/V in $Ge_{20}As_{25}S_{55}$ chalcogenide glass irradiated by an electron beam has been reported [103]. Experiments also were performed on $GeSe_4$ chalcogenide glass that was thermally poled to create a nonlinear susceptibility $\chi^{(2)}$ with a time scale of 5000 s [104]. Guillet et al. showed that lead silicate glass can be electrically poled and the necessary mobility of charge carriers is light induced with 800 nm radiation from a Ti:sapphire laser [105].

Several other experiments have been performed in fibers and other materials such as glasses, preforms, and semiconductor-doped glasses. The preparation of fibers for frequency doubling has been studied when fibers were subjected to radiation damage [106] and to exposure to an atmosphere of hydrogen [107], which was found to enhance SHG conversion efficiency. It also is possible to induce frequency doubling in multimode fibers [61] and even in glass preforms [28]. One interesting consequence of this is that it is possible to write holograms in glass and use them as optical memories (although at present conversion efficiencies are very low and high powers are needed). It was found that a minimum power exists above which the preparation process can take place, and the results also imply that SHG in fibers should not depend on gathering charges at the core–cladding interface, because in a preform the interface is very far indeed. Magnetic field effects in preforms also have been studied and the results indicate that the interaction of neighboring defects might be relevant [108].

Rare-earth-doped fibers show a large potential for frequency doubling [17–19], but have not been exploited extensively. The reason for their good performance could be the availability of excess electrons. Second-harmonic generation also has been studied in prepared semiconductor-doped glasses. These glasses have extremely high nonlinearities and, although each crystallite lacks inversion symmetry, the matrix is amorphous [109, 110].

2.16. Theoretical Models of Second-Harmonic Generation in Glass Optical Fibers

During the preparation process for frequency doubling fibers, the initial macroscopic inversion symmetry is broken. The resulting structure that is written into the fiber is remarkably stable at room temperature. Fibers that have been conditioned for second-harmonic generation do not forget their frequency doubling ability over decades. What kind of physical change occurs in the fiber during the light exposure?

Any attempt to understand this phenomenon has to explain how the symmetry of the glass material gets broken and how phase matching is obtained. The fundamental and second-harmonic waves definitely travel with different phase velocities and thus $\mathbf{k}_{2\omega} \neq 2\mathbf{k}_{\omega}$. However, an alternative exists to guarantee that the external force term in Eq. (14) oscillates with resonance spatial frequency $\mathbf{k}_{2\omega}$. Equation (14), rewritten for the second-harmonic signal, reads as

$$\begin{aligned} -\nabla \times (\nabla \times \mathbf{E}(\mathbf{r}, 2\omega)) + \mu_0 \varepsilon(2\omega) 4\omega^2 \mathbf{E}(\mathbf{r}, \omega) \\ = -\mu_0 \omega^2 8\chi^{(2)} \mathbf{A}\mathbf{A} \exp\{i(2\mathbf{k}_{\omega}) \cdot \mathbf{r}\} \\ + \text{other contributions} \end{aligned} \tag{34}$$

The right-hand side oscillates with the resonance spatial frequency if the susceptibility itself is not constant, but oscillates with spatial frequency $\mathbf{k}_{2\omega} - 2\mathbf{k}_{\omega}$. This means that the exposure of the fiber to the intense laser light not only breaks the macroscopic inversion symmetry of the material, but it does so in a smart way, by changing sign at the right places so as to guarantee phase matching. The conditioning process not only creates a second-order susceptibility, but a *periodic $\chi^{(2)}$ grating*.

The first successful theoretical model of efficient second-harmonic generation in optical fibers, which based phase matching on a periodic grating, was proposed by Stolen and Tom [29]. These authors realized that the third-order term in the polarization expansion gives rise to a spatially oscillating static electric polarization that has exactly the right wave vector $\mathbf{k}_{2\omega} - 2\mathbf{k}_{\omega}$ provided that already a second harmonic wave exists. Suppose that a second-harmonic wave exists in the fiber right from the beginning so that the total electric field is

$$\begin{aligned} \mathbf{E}(x, y, z, t) = \mathbf{E}_{\omega}(x, y) e^{ikz - i\omega t} \\ + \mathbf{E}_{2\omega}(x, y) e^{ik_{2\omega} z - i2\omega t} + \text{c.c.} \end{aligned} \tag{35}$$

Where $\mathbf{E}_{\omega}$ and $\mathbf{E}_{2\omega}$ are field amplitudes, and c.c. stands for complex conjugate of the previous expression. The existence of an initial, extremely weak second-harmonic field can be assumed because of the local core–cladding interface asymmetry. By substituting the total field into the third-order polarization (13), one obtain the static polarization.

$$\begin{aligned} \mathrm{P}^{(3)}(x, y, z, 0) = 3\varepsilon_0 \chi^{(3)}(\omega, \omega, -2\omega) \mathbf{E}_{\omega} \mathbf{E}_{\omega} \mathbf{E}_{2\omega} \\ \times \exp\{i(k_{\omega} - k_{2\omega})z\} + \text{c.c.} \end{aligned} \tag{36}$$

The interesting point is that this static polarization exhibits exactly the right spatial oscillatory behavior that the $\chi^{(2)}$ grating should have so as to guarantee phase matching. Therefore, Stolen and Tom assumed that the static electric field created by this polarization might orient electric dipoles. The oriented molecules provide the necessary breaking of symmetry that gives rise to the $\chi^{(2)}$ grating.

Stolen and Tom drew an important conclusion from their model: The presence of an initial second-harmonic wave is crucial for the formation of a $\chi^{(2)}$ grating. So what happens if we use the extremely faint signal produced by the quadrupole terms in conjenction with immediate launch of a fairly strong second-harmonic signal together with the pump light into the fiber? The glass material should learn much faster. In fact, by adding a seeding *second-harmonic* beam to the pump light, a fiber can be conditioned for second-harmonic generation in a few minutes. Not only that, it also was found that conditioning with the external seed could even prepare fibers for second-harmonic generation that were germanium doped and had no phosphorous codoping.

The Stolen-Tom model was an extremely important step in the history of second-harmonic generation in fibers. It stimulated many interesting experiments and opened the way for our understanding of the phenomenon. Nevertheless, it certainly cannot be taken as a realistic description of what is going on in a fiber during conditioning. Whereas the frequency-doubling ability of prepared fibers is stable at room temperature, we conclude that the possible orientations of the molecule, which are responsible for the $\chi^{(2)}$, should be separated by potential barriers that are much higher than

$kT \approx 4 \times 10^{-21}$ J $\approx 3 \times 10^{-2}$ eV. A rough estimate of the fields that can be generated from the polarization of Eq. (18) shows that these fields are far too weak to overcome such a potential barrier. The dc fields generated from the polarization (18) are typically on the order of 102 V/m^{-}. Supposing a dipole moment of one elementary charge times 1 Å yields a typical energy scale of 10^{-8} eV, which is 6 orders of magnitude too small to explain dipole orientation. Even on a purely macroscopic phenomenological level, the Stolen-Tom model does not explain all the experimental findings. The growth and saturation of the second-harmonic signal during the preparation process are not correctly described by the model.

Stolen and Tom's model is so elegant and simple that it is tempting to try to rescue it in spite of the difficulties. An excitation to a metastable state of defects has been suggested as a necessary step in the preparation process [86]. The excitation, with energy near $\eta 4\omega$, normally is achieved by absorption of two fundamental photons and one second-harmonic photon. During its lifetime, the excited center is then very free to be oriented even by a small electric field. However, the purely phenomenological problems of the Stolen-Tom model are essentially the same.

Several other authors suggested models based on oriented dipoles [5, 6, 111, 112] and ideas that go beyond dipole orientation [32, 113, 114]. In the Stolen-Tom model, the periodicity of the $\chi^{(2)}$ grating was included via the macroscopic third-order interaction. It was the participation of the very small third-order susceptibility that caused the main problem of the model.

All that we need to obtain an appropriate $\chi^{(2)}$ grating is a physical phenomenon that depends on the *relative harmonic phase* of the fundamental and second-harmonic wave. The relative phase of two waves, usually is defined as the difference of the phase, $\Phi_{\text{rel}} = \phi_2 - \phi_1$. The interference term of the total intensity depends on this relative phase and it is stationary only if the two waves have exactly the same frequency. Therefore, a hologram can be recorded in a photographic plate only if the object beam and the reference beam are generated by the same laser source. However, if the second wave oscillates with exactly twice the frequency of the first, we can define an alternative notion of relative phase that is stationary. Let us call it the relative harmonic phase: $\Phi = \phi_2 - 2\phi_1$. Any physical phenomenon that depends on this quantity is a possible candidate to form the $\chi^{(2)}$ grating. The $\chi^{(2)}$ grating can, in fact, be regarded as a sort of hologram, where the second-harmonic seeding light plays the role of the object wave and the fundamental beam is the reference wave [115].

Models cited in the literature [32, 113, 114] used the following type of phenomenon: Imagine some sort of polar center that appears in the glass material with random orientation. The fundamental wave induces a second-order response in each one of these centers. The second-order oscillation of the center then interacts with the external seeding oscillation. The effect caused by this interaction depends on the relative harmonic phase and the orientation of the polar center. In this way, it is relatively easy to imagine mechanisms that selectively eliminate or modify polar centers in a manner that depends on the local value of the relative harmonic phase.

Kamal et al. [32] used this kind of idea to hypothesize that the preparation process was based on selective ionization of certain centers. They measured the preparation speeds as a function of pump and seeding intensities with very careful experiments (minimizing self-phase modulation effects using only Q-switched laser pulses) and found an excellent agreement with their theoretical predictions. The preceding idea was used, assuming selective reorientation of certain centers [13, 14]. This scheme involved excited metastable states and the models made the following two predictions: (1) irradiating a prepared frequency-doubling fiber with the fourth-harmonic wave should erase the $\chi^{(2)}$ grating and (2) injecting the fourth-harmonic wave into the fiber together with the writing beams (ω and 2ω) should enhance the speed of preparation. Both predictions were confirmed experimentally [50, 71]. However, the fact that a model explains a couple of observed facts does not mean too much—cosmologists should bear this is mind. Dianov et al. [116] showed experimentally that the second-order susceptibility in prepared glass has a dominant component that should be attributed to macroscopic charge displacement and not to oriented dipoles.

There are still other arguments that indicate that the $\chi^{(2)}$ grating is not made up from oriented dipoles. First, it is uncertain whether the concentration of defects available could lead to the high conversion efficiencies observed, even if all centers were properly aligned [8, 82]. Second, the tensor components of the second-order susceptibility were measured [82] and the ratio found was consistent with an effective $\chi^{(2)}$ created by a strong electrostatic field actuating on $\chi^{(3)}$; see Eq. (29). The required fields have been observed at the fiber input end [32, 117] and explain the periodic variation of the etching rate of prepared fibers illustrated in Figure 5. Therefore, several models were developed to explain the creation of the strong electrostatic fields by light induced charge rearrangements.

Lawandy [118, 119] assumed that under intense nonresonant light field, deep localized electric states may become extended, leading to charge displacement. The delocalization of the charges from the centers is estimated to be on the order of 20–40 Å, and the model explains the sharp intensity threshold required to prepare fibers. However, the correct dependence on the relative harmonic phase to guarantee growth of the second-harmonic signal is unclear in that model. Another dc field model assumes that the charge centers are shallower, and after ionization and diffusion, the carriers would be trapped. Because of the nonuniform light intensities in the core, more carriers would be ionized from the center, leading to a radial distribution of charge [120]. This model is strongly dependent on mode structure, and has difficulty explaining the results of preparation of bulk glass (preforms) and even greater difficulty explaining the enhancement of the preparation rate under UV light excitation [71].

In 1990, Dianov et al. [58, 116] developed a model to explain the creation of the dc electric field based on a photovoltaic process. Other groups also explained the field by delocalization of electrons and Anderson et al. [121] developed models where the strong dc field was due to delocalization of electrons that were asymmetrically emitted and trapped. As we have seen, the combination of an oscillation of frequency ω with its second harmonic is a polar object (see Fig. 1). Thus, it is possible that a center subject to such polar oscillation emits electrons preferentially to one side. The direction of emission depends on the relative harmonic phase of the two oscillations. Thus the phenomenon is a good candidate to explain the formation of the $\chi^{(2)}$ grating. It is still a bit unclear how these models would explain the enhancement of the preparation rate under UV light excitation, but it might be that the emitting center has to get into a metastable excited state to emit. The correct description of the microscopic mechanism might be some combination of the models [32, 113, 114] and the asymmetric emission idea.

The experiment performed by the Dianov group to exclude oriented dipole models used a bulk silica sample instead of an optical fiber. These authors focused a writing beam (i.e., a mixture of a strong fundamental and weak second-harmonic beam) on a piece of germanium-doped silica sample. Both waves were linearly polarized in the same direction. During the preparation process the sample was moved slowly in the direction of the electric field vector. If the induced $\chi^{(2)}$ was due to oriented dipoles, we would expect to get an elongated area, more or less stadium shaped, full of dipoles facing in the direction of polarization; compare Figure 8. By focusing a reading beam (i.e., a beam with frequency ω), we expect to get a second-harmonic signal in the entire "stadium" with maximal efficiency when the polarization of the reading beam coincides with the writing polarization.

The asymmetric charge emission model predicts an entirely different configuration. The writing beam would push electrons to one side of the polarization direction. Let us suppose the relative harmonic phase is such that the electrons are pushed to the side to which the writing beam moves in the glass rest frame. This way the light spot would push a negatively charged cloud in front of it, leaving the whole stadium positively charged (compare Fig. 9).

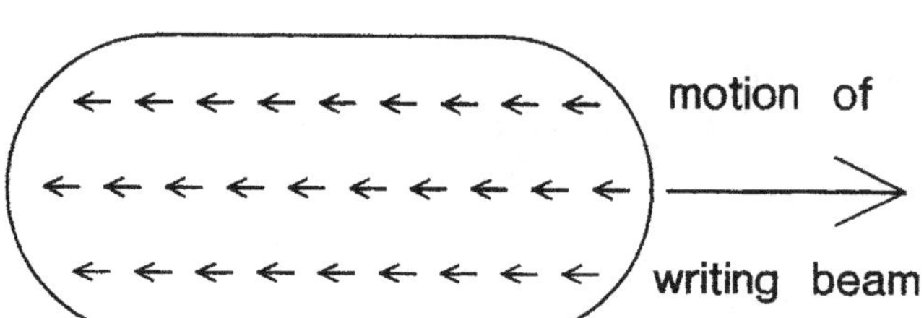

Figure 8. Prepared region according to a dipole model.

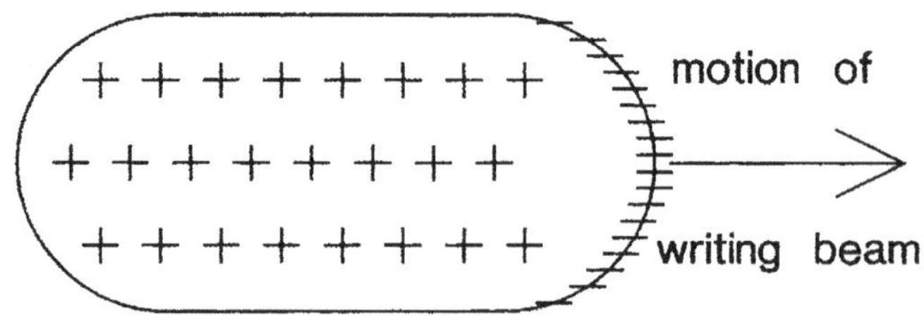

Figure 9. Charge distribution according to the asymmetric charge emission model.

Figure 10 shows the dc field configuration that was expect from this process. At the center of the stadium, point 1, we would not expect appreciable second-harmonic generation. At point 2, outside the stadium, there should be a nonzero second-harmonic signal that has maximum efficiency and polarization perpendicular to the writing signals. Finally, at point 3, between the negatively charged cloud and the positively charged stadium, we expect second-harmonic generation with maximum efficiency for the original polarization.

According to the photovoltaic model, a current density exists in the glass material during the preparation process that is the sum of an ohmic contribution and a current density $\boldsymbol{j}_{\text{ph}}$ due to asymmetric charge emission, $\boldsymbol{j}_{\text{total}} = \sigma \boldsymbol{E} + \boldsymbol{j}_{\text{ph}}$. In the stationary state, the total current is zero. This determines the written electric field $\boldsymbol{E} = -\boldsymbol{j}_{\text{ph}}/\sigma$, which in turn causes the effective second-order nonlinearity. The electrical conductivity σ of the glass material is itself a function of light intensities. The dependencies of $\boldsymbol{j}_{\text{ph}}$ and σ on the optical fields ultimately determines the saturation level of the preparation process. To maximize the saturation level, we can look for intensities such that $\boldsymbol{j}_{\text{ph}}$ increases without unduly increasing the photoconductivity. As previously mentioned, the asymmetric charge emission probably starts from some metastable excited state of some specific center. Therefore, $\boldsymbol{j}_{\text{ph}}$ may be enhanced and create such centers by specific treatments, such as previous UV irradiation, and by exciting the centers by UV pulses preceding the preparation pulses by a time shorter than the lifetime of the excited state. On the other hand, continuous UV exposure during preparation should increase the photoconductivity and, consequently, should decrease the saturation level. Such behavior indeed

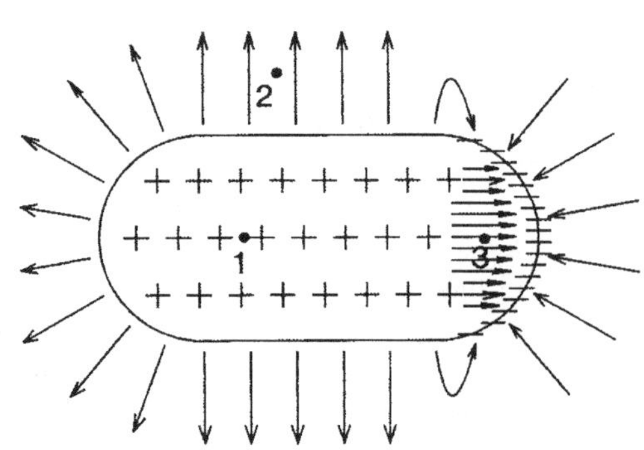

Figure 10. Field configuration according to the asymmetric charge emission model.

has been observed by Dianov et al. [73]. Many other theoretical models have been put forward to explain optical poling of fibers or part of the process [47, 51, 58, 72, 82, 86, 94–96, 100, 101, 122, 123].

2.17. Can Optical Poling Be More Efficient?

It is natural to expect that understanding the physical mechanisms that make preparation possible would lead to optimization of parameters and to higher conversion efficiencies at lower powers. One interesting application for a prepared fiber that can be envisioned is the obvious direct frequency doubling of a semiconductor laser or a fiber laser. For example, the intracavity flux of a fiber DFB laser should be sufficiently high even for CW operation to lead to efficient frequency doubling. In addition to the early demonstration with a Q-switched and mode-locked Nd:YAG laser that the conversion efficiencies of this forbidden process could be so high that it was possible to pump a dye laser [14], other applications have been demonstrated. These include the replacement of a SHG crystal in a picosecond pulse autocorrelator [27], the preparation of a GeO_2 silica ridge waveguide for frequency doubling [124], and the impressive synchronously pumping of a dye laser in the mode-locked (not Q-switched) regime with a rare-earth-doped frequency-doubling fiber [19].

Optical poling consists of a preparation process in which a dc electric field is recorded in the fiber by optical means. The strength of the electric field determines the efficiency of the nonlinearity induced: the higher the field, the better. In SHG in fibers, the frequency-doubling radiation helps in the preparation process, and the generated second harmonic increases with input intensity, also indicating the usefulness of a high input power for recording the dc field efficiently. Without additional fiber conditioning prior to the preparation process, a threshold is found in which a minimum optical power is required to induce a second-order nonlinearity. The higher the intensity, the easier it becomes to record the dc field. All these considerations point to the use of very high optical powers during the preparation process.

High optical intensities also have various counterproductive effects. The propagation of high power laser pulses in single-mode fibers leads to significant self-phase modulation after fiber lengths of a few centimeters that compromises the quality of the quasi-phase-matched grating recorded for SHG. Spectrally broad preparation beams lead to the formation of gratings of various periods, which compete and lead to destructive interference, as discussed in Section 2.5. High intensities also lead to erasure of the grating, as discussed in Section 2.7. High power fundamental, second-harmonic, and UV radiation, as well as a combination of such beams, can cause increased charge mobility and erasure of the recorded field.

Therefore, without preconditioning the fiber, we are trapped between the need for high optical powers to record strong electrical fields, and the need of a spectrally narrow and clean preparation beam that does not erase the dc field. Until the early 1990s, these two conflicting requirements could not be met simultaneously, and the conversion efficiencies for SHG stayed at the few percent level. In the last decade, with the development of externally written fiber Bragg gratings and the consequent huge expansion of the field of photosensitivity, much knowledge has been gained about how to affect an optical fiber prior to exposure to radiation. Treatments such as hydrogen loading were developed, where the sensitivity increases by orders of magnitude. Progress also was made in the dopants used, and the importance of stress-relaxation effects was identified. Somehow—and unfortunately—this knowledge was not incorporated into the field of optical poling.

Whereas the promise of higher conversion efficiencies did not materialize and whereas, in the early 1990s, the process of thermal poling was demonstrated where the field recorded in glasses can be $\sim 10^2$ times larger than that set by optical poling, most researchers abandoned optical preparation. Perhaps one day this technique for poling fibers will be revived, because it is a very simple and convenient way to modify the material and achieve phase matching, and it requires no more than laser radiation. The use of photosensitivity enhancement techniques now available could be very valuable then.

3. THERMAL POLING

3.1. Introduction

In 1991, Myers et al. [125] were poling a polymer film on a silica substrate by applying high voltages at a temperature of ~ 300 °C. After cooling to room temperature with the voltage applied, the poled sample was illuminated with a strong infrared laser beam. As expected, they found that the treatment frequency doubled the incoming radiation and some green light could be measured at the output. When a test was carried out to discard any contribution from the substrate to the frequency-doubled light, they discovered that it was the silica glass that was generating the green light and that they had induced a strong optical nonlinearity in silica by thermal poling [125]. Brueck and co-workers realized that the possibility existed to increase the recorded field by a factor 10^2–10^3 compared to what was achieved by optical poling. The prospect of true linear electro-optical modulation in optical fibers seemed just around the corner. High efficiency frequency doubling in glass, with applications in displays, UV and visible light generation, and optical and holographic memories also seemed closer than when only optical poling was available. Some of these exciting applications have been investigated, and are the subject of this section.

Thermal poling of dielectrics is a research field that goes back a couple of centuries [126] and is the basis of electrets [127]. The permanent or semipermanent polarization of dielectrics assisted by light or UV radiation also has been studied for many years [128] and is exploited in photocopying. Even thermal poling of silicate glass has been studied both theoretically and experimentally for decades, [129–132] and much is known from a materials point of view. Anodic bonding of metal to glass makes use of a procedure similar to poling, that is, subjecting the sample to ~300 °C and hundreds of volts. Ionic migration at temperatures reaching ~80 °C in glass isolation banks of high voltage transmission systems leads to dielectric weakening and demands costly maintenance. Therefore, it is obviously of great interest to understand the physical mechanisms behind poling, not only from a nonlinear optics perspective, but also from interest elsewhere in physics.

The first studies of poling in a nonlinear optics context naturally concentrated on the properties of bulk samples. Section 3.2 reviews some of the main reports in the literature. There, a collection of results obtained with various glass systems is presented, as well as a first approximation model that qualitatively describes the fundamentals of thermal poling. Some shortcomings of the model are introduced also. Various characterization techniques developed to study the nonlinearity are the subject of Section 3.3. Characterization of bulk poled samples has, in fact, attracted much of the work developed on thermal poling. The main reason for this is that the nonlinearity induced is generally (but not always) confined to a very thin region near the positive electrode. Whether the width of this layer is 1 or 10 μm can change the nonlinearity dramatically, so accurate characterization is quite crucial. Soon, applications started to emerge and poling of waveguides was demonstrated, as discussed in Section 3.4. The use of poled waveguides (fibers included) allows integration of the effect of the nonlinearity over a long length without allowing the probe beam to diverge spatially. Perhaps the most important devices that can be envision with poled fibers are those where the refractive index can be controlled with the application of voltage. This ability opens the doors to high-speed electro-optical switches and modulators, which are key components in wavelength division multiplexing (WDM), DWDM, and fiber-to-the-home systems, where wavelengths and ultimately packets can be routed and switched. Electro-optical fiber devices also have potential applications in add/drop multiplexers, electrically tuned attenuators, and Bragg reflectors. A number of nonlinear crystals have been developed that permit the construction of electro-optic modulators based on the Pockel's effect and that work with only a few volts. For example, $LiNbO_3$ modulators are available commercially to rates above 10 GHz. However, these devices have two drawbacks: (1) the crystals are too expensive to be used in a large scale, (e.g., individual homes; typical prices are in the range of $\$10^3$–$10^4$ and (a) the refractive index of these crystals is significantly different from the index of telecommunication fiber material. This index mismatch makes coupling to the optical system (i.e., fibers) difficult, and adds to the loss and price, because they require antireflection coatings. A silica-based fiber modulator or switch would have excellent transmission over a wide range of wavelengths, low dielectric constant (potentially high speeds), no photorefractive damage (such as $LiNbO_3$), and, above all, the promise of being inexpensive. Some of the work carried out on poled fibers for modulation and switching is discussed in Section 3.5. Section 3.6 examines some of the reports of periodic structures, and applications of thermal and UV poling as a means to tune electro-optically the reflectivity of fiber Bragg gratings and to convert quasi-phase-matched wavelength. These are some of the most promising devices to be made by poling, but the available literature is so far quite limited.

3.2. Thermal Poling of Bulk Samples and Films

In their original report, Myers et al. [125] demonstrated an induced second-order nonlinearity of ~1 pm/V, which on the verge of being sufficient for practical device realization. Their article then sparked a number of investigations that aimed to understand the effect and increase the nonlinearity even further. Myers et al. [125] found that, for the same poling voltage, temperature, and time, the nonlinearity obtained does not depend on the sample thickness. At first sight, it may be a surprise that the induced $\chi^{(2)}$ does not depend on the electric field, but on the poling voltage. However, it was determined that the region with the strong electric field was limited to the first few micrometers under the surface on the anode side. The fact that the nonlinearity is confined to a very thin region is a major feature of thermal poling. It has important consequences when the process is used to create a second-order nonlinearity in fibers. Various experiments show that all poling voltage drops in the vicinity of the surface. Consequently, the thickness of the bulk sample is expected to be of lesser importance. More recently, it was demonstrated that the thickness of the sample *does* play a role, and the poling time and even the induced nonlinearity are altered, depending whether the sample is 100 μm or 500 μm or 1 μm thick [133]. In thinner samples the field that moves the positive mobile ions is higher and the ions move with higher drift velocities. Therefore, the formation and progression of the ion depleted region is expected to be faster in thinner samples, which was, in fact, observed [133, 134]. On the other hand, the electric field in the depletion region is not a monotonically increasing function of time. After an initial growth, the intense field suffers a slow decay over time. Interestingly, this decay is also faster in thinner samples. It also was observed that the creation of the nonlinearity occurs only after a delay time that is dependent on the thickness of the sample.

The induced nonlinearity in thermally poled samples was inferred by Myers et al. [125] and Kazansky and Russel [135] to be due to the recorded high electric field created by ionic drift of metallic ions, such as Na^+. This inference is consistent with the findings of early experiments in the 1970s [129, 130, 132]. In the presence of the high voltage and temperature during poling, alkali ions are displaced from the anode side to the cathode side, leaving behind a thin ion depleted region under the anode (depletion region). The creation of such a depletion region was studied in detail in the 1970s, and various techniques, including Rutherford backscattering, were employed to measure ionic migration [130]. More recently, secondary ion mass spectrometry (SIMS) [136] and energy-dispersive spectrometry [137] measurements were performed to record the drift of cations from the anode side in thermally poled silicate glass. The electric field recorded in silica glass can reach $\sim 10^9$ V/m, which is close to the breakdown limit. In fact, such a value is believed to be determined solely by the material properties at high temperature. Myers et al. [125] also reported that the application of heat alone (>250 °C) could erase the nonlinearity. This effect can be attributed to the drift of cations rendered mobile by the heat.

Thermal poling of glass was extended with the demonstration that it was possible to use corona discharge rather than a contacting electrode on the anode side [138]. This technique is widely used in poling polymers like polyvinylidene Fluoride (PVDF), and has been used for poling glass films and waveguides [138, 139]. The decay rate of the induced nonlinearity was found to be too fast for practical applications [140]. Corona poling also has been used with non-silica-based systems such as telluride glass [141]. Electron [142] and proton [143] beams also have been used for poling. Kazansky et al. [144] found that it could be advantageous to use the beam to erase rather than to create the nonlinearity with good spatial resolution. In fibers, this can be exploited to create a periodic distribution of the nonlinearity, such that quasi-phase-matching takes place.

It was observed that poling of bulk silica and optical fibers showed better reproducibility when performed in vacuum (1.2×10^{-5} mbar), so that electric breakdown of air near the anode surface was avoided [145]. It was also demonstrated that the resolution could be improved to ~ 1 μm, which enables the fabrication of periodic structures for quasi-phase-matching, as discussed in more detail in Section 3.6.

The dynamics of the creation and destruction of the second-order nonlinearity in heated bulk silica glass samples was studied with the application, removal, and/or reversal of an external poling field [146]. The nonlinearity was assessed by real-time monitoring at the current flow and the growth and decay of the SH signal probed with a Nd:YAG laser beam. The initial creation of the nonlinearity was found to be very fast, but upon bias reversal, a long time period ellapsed until the original nonlinear layer was removed, an indication that at least two species of high and low mobility play a role in poling silica glass [147]. This conclusion is in qualitative agreement with early studies of poled soft glasses that showed the replacement of sodium from the depletion region by hydrogen [148, 149].

Measuring the SHG efficiency in terms of applied voltage, temperature, and poling time did not prove to be a straightforward task. Contradictory results about the expected quadratic voltage power dependency of the second-harmonic signal were found. Takebe et al. [150] observed second-harmonic signals proportional to the square of the applied voltage for samples poled both in air and in vacuum. On the other hand, Myers et al. [125] found a cubic dependence for voltages smaller than 3.5 kV and found only simple quadratic dependence when poling was carried out with indium titanium oxide transparent electrodes [151]. This difference may be due to the fact that both groups controlled the relevant parameters in different ways: Takebe et al. used the optimal poling time for any voltage, whereas Myers et al. used a fixed value for the poling time (15 min) for all voltages. In view of recent results [133], as the ion drift velocity depends on the applied field. Therefore, in experiments with a fixed and relatively short poling time of 15 min the poling process advanced further for higher applied voltages. This brings in an additional voltage dependence and may explain the cubic power law. For much longer poling times, the depletion region can become so large that phase mismatch also can alter the second-harmonic generation, leading to even more complicated dependencies. The maximum SHG efficiency reported to be at a temperature of ~ 280 °C for a bias of 3–5 kV [152] was later found also to depend on the range of applied voltages [153]. For a bias above 10 kV, a larger nonlinearity was induced at a higher temperature (400 °C). Although this is a very relevant observation, no similar result has been reported for fibers.

Often it has been observed that, given a poling voltage and temperature, the SHG efficiency can drop significantly for longer poling times [150, 154, 155]. This drop in $\chi^{(2)}$ conceivably results from the heated glass discharging, either through the migration of negative ions such as oxygen toward the anode or through an avalanche process that is often detected in the poling current and is attributed to field emission or local electrical breakdown [132, 148, 156]. In-diffusion of positive ions such as H^+ or H_3O^+ [147] under the anode to cancel the trapped negative charge also might play a role and is consistent with the dynamics studies that point to two types of ions contributing to the current flow.

To understand the parameters involved in the poling mechanism, a systematic study of the thickness evolution of the second-order nonlinear layer in thermally poled fused silica with the poling parameters (poling time, poling voltage, poling temperature) was undertaken by Qiu et al. [157]. They observed that the thickness of the second-order nonlinear layer increases linearly with the poling time and poling voltage, and depends on the poling temperature as a function of an Arrhenius plot.

Other types of glass systems in addition to SiO_2 have been thermally poled to induce large second-order nonlinearity. These glasses include, among others, Te:SiO_2 [141], Pb:SiO_2 [144], Ge:SiO_2 [144, 158, 159], Ti:SiO_2 [160], Ga:La:S [161], and As_2S_3 [162]. Some of these glasses are directly compatible with optical fiber manufacturing, so perhaps they could lead to practical devices in the future. Bulk soft glasses have not been used extensively [137, 163] because of the high conductivity of the glass, which leads to typical poling currents of several milliamperes. The samples then must endure high current flow during poling. Thermal runaway can be prevented by actively limiting the poling current. The long-term stability of the second-order nonlinearity induced in soft glasses is also very poor. A fast decay of $\chi^{(2)}$ has been reported in corona poled thin films [140], and in bulk soda lime and borosilicate glass, the SHG efficiency induced was found to decay both spontaneously and by a photostimulated process (optical erasure) [137, 163]. The initial nonlinearity induced by poling of soda lime and borosilicate samples is approximately the same as that of silica poled to the same bias voltage (sometimes marginally higher). This situation is consistent with the observation that it is possible to induce comparable second-order nonlinearity in silica samples that contain various amounts of alkali ions [125], although the alkali content here (~13% Na) is 4–5 orders of magnitude larger than in pure silica (~1 ppm Na). This result indicates that other types of glasses that are compatible with waveguide formation by ion exchange also can be poled, including semiconductor-doped glasses in a soda lime matrix.

At this stage, it is useful to have a simple, first-order approximation model to support the formation of a second-order nonlinearity by thermal poling. Under the application of an external electric field to a heated sample, mobile ions (in particular Li^+ and Na^+) migrate as a block toward the cathode. In soda lime, it is possible to physically see the excess sodium oxide on the cathode side after poling, but in pure silica, the amount of ions leaving the material is too small to be visible. As a consequence of this drift, the bulk of the glass is left unchanged, while the region under the anode becomes depleted of mobile ions, becomes negatively charged, and acquires high resistivity, as schematically illustrated in Figure 11. Gradually, more of the applied voltage drops across the depletion layer, until all bias appears across the few micrometers thick layer of high resistance under the anode. Interestingly, even modest voltages (<100 V) can lead to the creation of fields in excess of 10^8 V/m. The strong electric field is generally sufficient to account for the nonlinear optical effects measured through $\chi^{(2)}_{\mathrm{eff}} = 3\chi^{(3)}\mathbf{E}_{\mathrm{dc}}$. The growth of the electric field is limited by breakdown in the material. If the field exceeds this maximum value, the depletion layer becomes thicker. To a first approximation

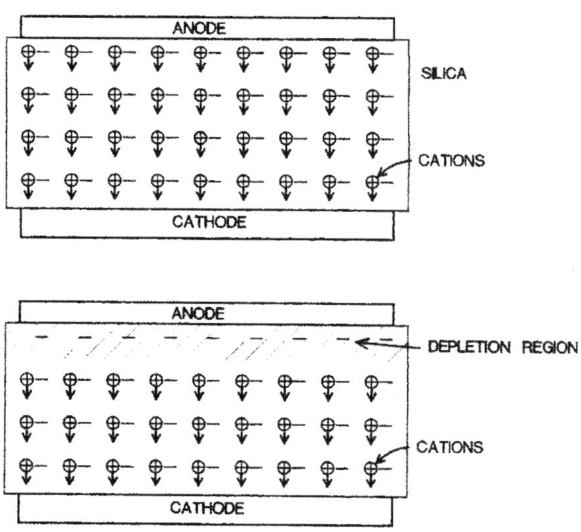

Figure 11. Formation of an ion depleted region by charge migration.

the width of the depletion region formed after saturation is expected to be equal to the breakdown field times the poling voltage. This estimate is independent of the thickness of the sample and in qualitative agreement with the original report of Myers et al. [125]. The movement of positive charge leaves the depleted region negatively charged. As the field increases, neutralization takes place by negative charge emission toward the anode, by positive ion diffusion into the depletion layer, or both. It is then expected that a neutral layer exists between the sample surface and a thin negatively charged layer at the innermost part of the depletion region.

It is possible to estimate how the thickness w of the depletion region increases in the regime where a single carrier dominates and no neutralization takes place. The drift velocity of the carriers of mobility, μ, away from the anode driven by the field $\mathbf{E}(t)$ determines the position of the edge of the region:

$$\frac{\partial w}{\partial t} = v_d = \mu \mathbf{E} \tag{37}$$

The electric field is initially given by the poling voltage V_{pol} divided by the sample thickness L, but with the establishment of the depletion layer, the driving field applied to the bulk of the sample drops. Assuming the field in the depletion region is equal to the breakdown field $\mathbf{E}_{\mathrm{lim}}$ allows us to write

$$\mathbf{E}(t) = \frac{V_{\mathrm{pol}} - \mathbf{E}_{\mathrm{lim}} w(t)}{L - w(t)} \approx \frac{V_{\mathrm{pol}} - \mathbf{E}_{\mathrm{lim}} w(t)}{L} \tag{38}$$

The solution of the differential equation for w is

$$w = \frac{V_{\mathrm{pol}}}{\mathbf{E}_{\mathrm{lim}}}\left(1 - \exp\left\{-\frac{\mu \mathbf{E}_{\mathrm{lim}}}{L} t\right\}\right) \tag{39}$$

This expression gives a feel for how the layer thickness is expected to evolve, growing rapidly initially. Unfortunately, the approximation of a single mobile ion breaks down even when the mobility of other ions that may contribute to the formation of the layer is much lower than that of the main carrier.

The foregoing simple model does not account for many of the features observed in thermal poling. For example, it is known that many species, both positive and negative, contribute to the ionic current, not only alkaline ions. When nuclear techniques such as Rutherford backscattering, nuclear reaction analysis, and resonant backscattering were used to study the ionic displacement, it was found that heavier ions such as Ca^{++} and even lighter species such as H^+ can accumulate at the inner end of the depletion region [148, 149]. Clearly, the ionic buildup and the contribution of all these species should be taken into account in a full description of the creation of the poled region, but this probably requires a numerical solution. Also, measurements showed that in sol–gel-derived SiO_2 glasses, the second-order nonlinear polarization induced by poling increased with the OH^- concentration and that the whole sample thickness was poled [164, 165]. Similarly, it was reported that the electric field region could be as thick as the sample in pre-annealed low water content fused silica [166]. In such cases the formation of the nonlinearity is better described by an orientational model. Side illumination and SHG by reflection in poled silica samples was used to conclude that a weak nonlinearity also is induced on the cathode side [167]. In high water content silica [166, 168], electron paramagnetic resonance data showed a correlation between the induced SH and the concentration of Ge E' defects in samples that were undoped (Ge concentration ~1 ppm). The results of these measurements are not accounted for in the simple model described.

Whereas a thin nonlinear region is attributed to the creation of a space–charge region formed by the depletion of mobile cations, the formation of a thick nonlinear layer is better described by the orientation of dipoles in the external field [135]. At present, no convincing encompassing explanation exists for the various reports that show a bulk polarization. Why and under which conditions this effect should dominate the creation of a depletion region is not known, but it is possible that orientation becomes more important when the contamination level of cations in the glass is very small, as is the case in optical fibers.

3.3. Characterization Techniques

Characterization is crucial to evaluate the efficiency of the poling procedure and to increase understanding of the process involved. To this end, we can attempt to measure the recorded electric field directly or indirectly through other important physical parameters, such as chemical composition, stress, and charge distribution. The easiest, and perhaps least ambiguous way to obtain information concerning the induced nonlinearity is to measure a second-harmonic signal generated from a poled sample. This method also can be performed in such a way as to obtain spatially resolved information. As explained earlier, because the nonlinear layer is very thin in most cases, high spatial resolution techniques are required to study the detailed carrier or dipole distribution. Etching, nonlinear conversion, and material surface analysis are some of the alternative characterization techniques used.

A thermally poled silica sample was etched in successive steps and after each step the residual SH signal was measured [125] with the probing beam traversing the sample. The results showed an exponential decay of the SHG as a function of etching time and that the field distribution becomes deeper in samples poled for longer periods. In an independent measurement, a 50 μm diameter laser beam was made to impinge on the side of a thermally poled silica sample that had been polished [142]. In this way, the SHG could be resolved spatially. The near-field pattern of SH light was analyzed with a microscope objective and a TV camera that revealed that the generation took place in a region that peaked, 12 μm under the anodic surface. This result contrast. With the results in [125].

A similar setup was used to characterize silica thermally poled in air with contacting electrodes at 3 kV and at 280 °C by probing samples from a single batch with a tightly focussed Nd:YAG laser beam. Figure 12 illustrates the experimental setup. The SHG was monitored through the polished side surfaces while the sample was scanned vertically. The reflected beam also was monitored, where the transition of the reflected light form minimum (0%) to maximum (4%) is a measure of the spot size of the probe beam and the half maximum value gives the location of the anodic surface of the poled sample. Data taken under the same conditions for two samples are illustrated in Figure 13. The spot sizes of the probe beams in these measurements were 2 and 5 μm and the anodic surfaces were located at 52 and 22 μm, respectively. While in the sample in Figure 13a the SHG peaked on the surface, whereas in the sample in Figure 13b, it was buried ~4 μm. Also, although the position of the peak was invariable in each sample, the shape of the curves obtained for the SH signal varied considerably for different points of the sample studied. This made it impossible to determine the field distribution. The profile measured in Figure 13a is in rough accordance with the etching measurements that showed that the SHG peaks on the surface [125], whereas the buried profile obtained in Figure 13b is in qualitative agreement with that of kazansky et al. [142]. Perhaps the main conclusion that can be drawn from these experiments is that very rigorous control is necessary in terms of sample fabrication, contamination, history, poling parameters, electrodes, and atmosphere to obtain reproducible and consistent results. Even under such ideal conditions, one should bear in mind that on a microscopic scale two glass samples are never identical because of the amorphous nature of the material. Similar measurements were used to probe SHG in a silica waveguide on silicon [169]. Frequency doubling peaked in the core of the waveguide.

Further high resolution experiments use the Maker fringe technique [125]. This method is based on measurements of intensity variations of a second-harmonic signal as a function of the angle of the sample and the reading beam. The idea of this method can be understood easily in using the driven harmonic oscillator analogy of the nonlinear wave equation (13) (Section 1.6). In the case without phase matching, the second-harmonic signal should behave in space like

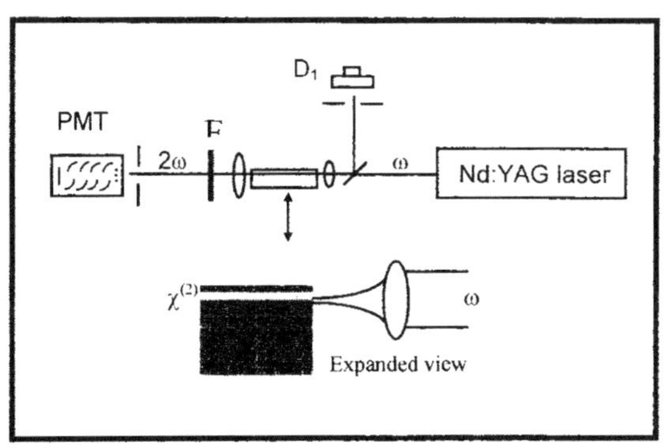

Figure 12. Lateral scanning of a thermally poled sample by means of second-harmonic signal—experimental setup.

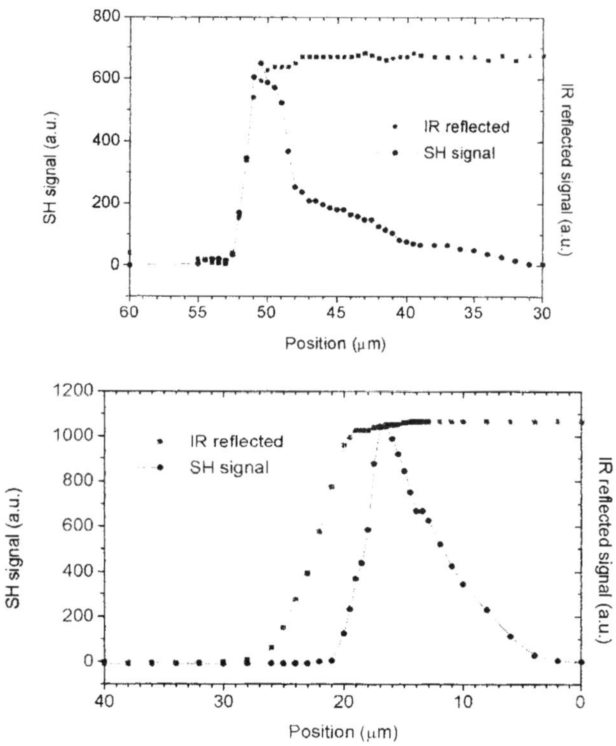

Figure 13. Lateral scanning of a thermally poled sample by means of second-harmonic signal. Sample (a) shows a nonlinear region near the surface and in sample (b) it is buried at $\sim$4 μm.

the temporal beating that results from the superposition of the proper frequency of an oscillator with the externally imposed frequency. That means the second-harmonic signal should vary periodically with the thickness of the nonlinear layer. In the case of Maker fringe measurements, the effective thickness is varied by tilting the sample with respect to the reading beam. The geometrical thickness can then be determined mathematically. Furthermore, using numerical solutions to the inverse problem, the profile of the nonlinearity in the depletion layer can be determined in large angle Maker fringe experiments [170]. The resolution of the Maker fringe technique was increased by the implementation of two noncolinear fundamental beams to generate the second harmonic [171]. This new configuration, was tested for poled silica glass and proved to be a powerful nondestructive technique to measure the depth of the nonlinear layer.

Although the transverse SHG ($\sim$60° incidence angle) makes it possible to probe the induced second order nonlinearity integrated along the depletion layer, it does not give direct information about the polarity and distribution of charges recorded in the poled glass.

The elegant technique of laser induced pressure pulse (LIPP) probing [149] was used with poled silica samples to address this question [172]. A short optical pulse illuminates an opaque film on the surface of the sample, creating a sound wave that gives rise to a current. The measurement of the current allows the sign and amplitude of the charges and their location to be inferred. Opposite results were obtained with two groups of samples. In the first group a positive surface charge and a 5.5 μm deep negative charge were inferred, whereas in the other group of samples, a positive space–charge was measured between two negative layers. Furthermore, the charge distribution was found to evolve over time, and after a few days the measurements showed a reverse behavior. One problem with the technique is that the spatial resolution is limited at best to a few micronmeters and some of the features can be much closer to the surface, leading to ambiguity in some measurements. The conflicting results reported [172] are difficult to understand and perhaps for this reason, they have been somewhat neglected in the literature.

An alternative approach to the characterization of poled glasses is to selectively etch the surface. It has been shown that the etching rate of glasses by HF can be affected by the existence of a depletion region in silicate glass [132, 173] and by the presence of a periodic SHG grating in a fiber [45], as convincingly illustrated in Figure 5. Whereas etching can be carried out slowly and the etched thickness of the sample can be determined very accurately, the creation of a depletion region due to ion migration can be measured with sub micrometer accuracy. An in situ interferometric technique was then developed that provided real-time submicrometer resolution measurements of the etching rate [174]. A similar approach has been used for real-time characterization of diamond film growth [175]. The poled glass sample under test was mounted horizontally on an adjustable positioner. A He-Ne beam (633 nm) crossed the sample nearly vertically, creating two reflections from top and bottom surfaces that overlapped spatially. The resulting interference pattern was detected. Only the anodic (bottom) surface was immersed in HF so that the optical path of the beam reflected off this surface changed in time. A periodic harmonic interference pattern was expected from an unpoled sample, indicative of a constant etching rate with a period corresponding to half the wavelength of the beam in the material ($\lambda/2n = 0.217$ μm). When the glass was poled and the acid attacked a region subjected to an electric field or charged by ion depletion, the expected etching rate was different. Therefore, information about the charge–electric field distribution was gained by modulates the frequency of the interference pattern of a poled sample, as compared to a harmonic pattern. The variation of etching conditions over time and from sample to sample were taken into account by simultaneously monitoring the etching of the poled and unpoled regions of the sample with two He-Ne laser beams.

In practice, the etching rate of unpoled silica glass (reference) was constant, as expected. In contrast, the etching rate of the anode side of the poled region showed significant frequency modulation. Expect for the first quarter of a cycle, the traces initially had a period $\sim$2 times larger than the reference signal that is, much slower etching rate than unpoled glass. The results differed considerably over the first $\sim$0.1 μm from the anodic surface, where in some cases a high frequency was detected. A few micrometers into the samples the etching rate becomes equal to that of the reference. Figure 14 shows a plot of the etching rate as a function of depth, for one of the silica samples studied. The abrupt change in etching rate that occurs at ~6 μm can be associated with the edge of the depletion region. This feature was observed in the other samples etched. The abrupt transition follows a region of relatively constant etching rate (in Fig. 14, between 2 and 5 μm).

Interpretation of the results described in the foregoing step requires further knowledge about the etching process of silica. The question that should be addressed is what causes the increase or reduction in etching rate. Various mechanisms are known to affect the rate of chemical attack of glass. For instance, when an interferometer was used to monitor the etching rate of ion exchanged soda lime glass, it clearly revealed the region of the in-diffused ion in the unpoled, electrically neutral glass [176]. Mechanical stress, hydrostatic pressure, compaction, trapped charges, and even UV radiation lead to an alteration in how fast the glass is attacked; therefore, experiments are required to quantify each contribution.

Precise interferometric measurements of the etching rate of samples etched in the presence of applied electric fields indicate that the etching rate is also influenced by electrostatic fields [177]. Quantitative analysis of the field dependence shows that the observed alternation of etching speed in poled glasses is predominantly caused by this field dependence. Therefore, the interferometric etching technique provides a convenient way to observe the electric field with high spatial resolution. Indeed, a detail that occupies 0.1 of a period of the harmonic pattern can be measured, and that corresponds to only 22 nm. The constant etching rate in the depletion region has been associated with the similarity of the layer to a charged capacitor. For a constant sample thickness (1 mm) and poling time (90 min), it was inferred that the recorded field is constant (~4×10^8 V/m), not only along the depletion region, but also from sample to sample, for a range of poling voltages [178]. The amplitude of the recorded field in the samples also was found to grow until ~15 min poling time, and then to decrease slowly [155].

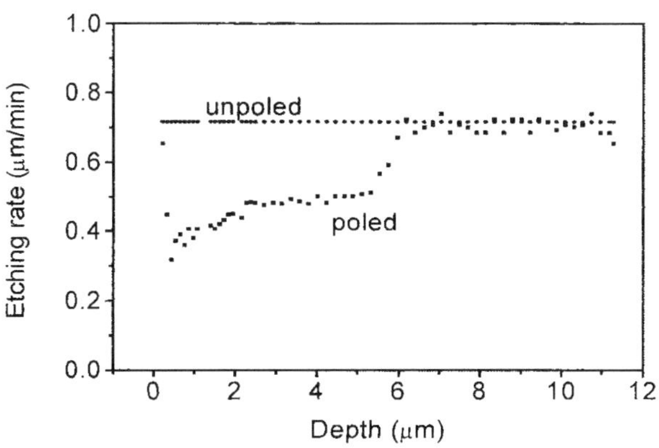

Figure 14. Interferometric measurement of chemical etching rate.

By using atomic force microscopy (AFM) after etching, a simple and extremely sensitive determination of the resulting profile of the sample can be obtained (but the measurement does not accompany the etching rate in real time and integrates the rate of chemical attack). For example, samples of silica glass were thermally poled, cleaved, and etched, so that the recorded field was parallel to the etched surface. An AFM was employed and submicrometer resolution was again obtained [179]. Some of the features observed with the optical interferometer also were seen in the AFM studies, including the edge of the depletion region.

The combination of HF etching with AFM measurements in poled optical fibers has revealed the presence of frozen in fields. [180, 181]. A twin-hole fiber with internal wire electrodes was poled, and after, removal of the electrodes, the fiber was cleaved and etched, revealing a ring around the anode. The ring was associated to the electric field present in the depletion layer. The fact that a circular structure is formed may be a consequence of the fact that in thermal poling it is the voltage and not the electric field that ultimately determines the creation of the depletion region.

The fast dynamics of the creation of the second-order nonlinearity [132] was studied by real-time monitoring of the growth and decay of the SH signal upon switching the bias on or the off. The second harmonic and the current were measured when the sample was subjected to a rapid bias sign reversal. The measurement showed that a single carrier model for the ionic transport in thermal poling is incomplete; at least two species of high and low mobility are required [147]. Thermally polled samples also were analyzed by SIMS which show that Li and Na ions form the depletion region, and that K and Na are injected from the anode surface into the depletion region [136]. These conclusions are in qualitative agreement with nuclear studies of poled soft glasses that showed the replacement of Na from the depletion region by H [148, 149]. The in-diffusion of H^+ or H_3O^+ to the depletion region can cancel to a large extent, the negative charge left by thermal poling following the migration of Na^+ and other alkali ions. Information about the poling mechanism also can be inferred from measurements of the current that appears when a poled sample is heated again, destroying the polarization in a short circuit configuration [181]. The results obtained in this work also indicate that the poling is predominantly due to charge migration and not to dipole orientation.

A nondestructive method for characterization of the second harmonic in bulk silica and of the charge distribution was proposed by Guillet de Chatelus et al. [182] Their method is based on a noncollinear frequency conversion of two femtosecond pulses and the results are in agreement with the model of a built-in electric field associated with the diffusion of two ionic species during the poling process.

A powerful characterization technique was presented by Markmann et al. [183] who used a Bragg grating written on a silica waveguide that was subsequently poled thermally.

The grating wavelength shift was then used to measure the third-order nonlinearity, the induced second-order nonlinear coefficient, and the dc electric field.

It is not obvious that the second-order nonlinear susceptibility induced for frequency doubling, which only mixes optical waves, is the same susceptibility induced for the linear electro-optic effect when both optical and low frequency fields are combined. The third-order nonlinear susceptibility that is the basis for both effects may have dispersion, as discussed elsewhere [184]. Therefore, although frequency doubling is an easy way to monitor the effect of poling, it is often necessary to characterize directly the linear electro-optic effect (LEO) induced.

The growth of the LEO coefficient in a poled fiber with internal electrodes was described by Wong et al. [185]. The fibers had holes with inserted wires that were connected to a high voltage supply, and the whole device was positioned in a small heater. This device was then placed in a Mach–Zehnder interferometer (MZI) and the induced phase shift was measured as a function of the applied voltage Xu et al. [186] improved the MZI by making in situ measurements of the LEO coefficient while poling the fibers. A dc poling voltage was applied to the fiber device placed in one arm of the MZI that was added to an ac testing voltage of small peak-to-peak amplitude. Therefore, the sum of dc and ac voltages was applied to the fiber device. In the reference arm of the MZI, a linear ramp phase modulator was introduced to shift the measurement away from dc by a few hertz and reduce thermal perturbations, thereby increasing the signal-to-noise ratio. With these measurements, positive and negative thermal poling of fibers was studied.

3.4. Poling of Waveguides

It is important to make use of waveguide structures to exploit efficiently the nonlinearity created by thermal poling in the thin region under the anode. In addition to fibers (discussed in Section 3.5), planar waveguides have been poled [159, 169, 187–192]. One important motivation to study this configuration is that complex functions can be performed with integrated optics components. As mentioned earlier, the replacement of $LiNbO_3$ and groups III–V semiconductors by glass as a substrate material that has a significant second-order optical nonlinearity would be of great interest for switching and actively controlling splitting, with the added bonus of low coupling loss.

A film of radiofrequency sputtered Corning 7059 glass on a Pyrex glass substrate was corona poled [138]. Waveguide dispersion was used to compensate for the material dispersion. To find the particular thickness for which the phase matching condition was satisfied, the film was deposited in a tapered geometry. The second-order nonlinear coefficient d_{33} induced was estimated to be ~0.5 pm/V. Thermal poling was also demonstrated [187] in a waveguide written on silica by a 30 keV electron beam. The channel waveguide was not erased at the poling temperature (270 °C). A modulator was constructed with a useful interaction length of 4.8 mm, and a maximum phase shift of 32 mrad was measured at $\lambda = 0.633$ μm at an applied field of 7.3 V/μm. Full switching was demonstrated with a 2×2 silica-on-silicon integrated Mach–Zehnder interferometer, constructed with a GeO_2-doped channel waveguide fabricated by flame-hydrolysis deposition and reactive ion etching [159]. The interferometer was poled at 300 °C and 4 kV. The device had a 36 cm long phase shifter, and the distance between the poling anode and the waveguide was 9 μm. A V_π voltage of 1700 V, an extinction ratio of 20 dB, and a risetime of ~100 ns were demonstrated.

Although the waveguide geometry especially attracts application-oriented research, fundamental questions also can be investigated with advantages in poled optical waveguides [169, 189–192]. The activation energy of decay of stable electro-optic effects was measured in a waveguide structure [190]. The activation energy was found to be (0.43 ± 0.05) eV, which is surprisingly close the value of the activation energy of UV-poled glass, although nonlinear effects in UV-poled glasses are fairly unstable at room temperature in contrast to thermally poled glasses. Arentoft et al. [169] addressed the old question of whether the induced nonlinearity in poled glasses is due to charge displacement or to oriented dipoles. The ratio of nonlinear coefficients $\chi^{(2)}_{33}/\chi^{(2)}_{31}$, which should be 3 if the second-order effect are due entirely to charge displacement, was found to be significantly different from that number in a poled waveguide. The ratio value was close to 1 if electro-optic measurement was performed and ~2 when observed by SHG.

In Arentoft et al. [189] a larger second-harmonic signal in a poled waveguide and a second-order nonlinear susceptibility in the range of 25 pm/V was reported, despite the fact that the linear electro-optic effect was only small. Diffusion of positive silver ions from the anode material into the waveguide was measured that introduces a resonance in the value of the third-order nonlinear susceptibility associated with the optical frequencies, but not in that of low frequencies. The very large amplitude of the second-harmonic signal detected confirms the useful consequence of a resonance in the third-order susceptibility. So far, this is the only work where such a resonance as exploited for frequency conversion in poled glasses.

Finally, the increase in $\chi^{(3)}$ reported [185] for optical fibers and discussed in the next section also was observed in planar waveguides [191, 192]. It was found that poling at voltages as low as 1 kV always enhances the third-order nonlinear susceptibility by a factor of 1.9 ± 0.4. Puzzling enough, no enhancement was observed in other studies with planar waveguides [183]. Theoretical understanding of this alteration may furnish important tools for fabrication of optoelectronic devices.

3.5. Poling of Fibers—Switches and Modulators

A good review article by Kazansky et al. [193] the use of poling to create the linear electro-optical effect in fibers. Such an effect allows the potential of poled fibers be exploited to the limit to modulate each bit in a telecommunications system. An equally important application, but with more modest speed requirements, is the use of electro-optical fibers for switching. If the device speed is restricted to 100 ns or slower, the aim is no longer to affect each individual bit, but rather to affect streams of data (packages or bursts). Even here, it would be extremely useful to have cheap spliceable fiber devices that operate at a few volts bias signals. In the following discussion, we describe both applications (modulation and switching) indiscriminately, although a travelling wave arrangement would be necessary in modulation at a few gigabits per second, thereby adding further requirements to the devices.

In the 1980s, before the technological explosion of optical communications and WDM, Li et al. [194] reported an impressive fiber modulator based on the Kerr effect. Superimposing high voltage on an ac signal and using a 30 m long piece of fiber with internal liquid electrodes they showed a $V_\pi = 40\ V_{rms}$. In addition to a performance that only few groups are capable of equaling even many years later, they used an excellent high pressure technique to make long electrode devices that were nearly forgotten over the years. In the 1980s, poling was generally optical, and poled fiber modulators had a very low nonlinearity (0.005 pm/V) and poor performance [100]. Therefore, the creation in silica of a nonlinear coefficient $\sim 10^3$ times greater than that induced by optical poling reported by Brueck and co-workers was an exciting prospect. Soon after the initial paper on thermal poling of bulk silica, various groups embarked on a research program to pole fibers for electro-optical modulation.

Whereas in most cases the nonlinearity from thermal poling is induced only in the region closest to the anode electrode, it is not possible to apply the electric field externally to a conventional optical fiber, because the nonlinearity would be induced mainly in the $\sim 60\ \mu m$ thick cladding. As a consequence, light travelling in the core never experiences the second-order nonlinearity recently acquired by the fiber. Therefore, a good deal of engineering is required to maximize the overlap of the poled region with the core of the fiber (ideally the overlap integral should be unitary) and thus allow for the fabrication of an efficient fiber-based electro-optical device.

It is not straightforward to ensure that the difference of potential available from the high voltage power supply during poling and from the lower power supply after poling is applied only to the fiber core. The voltage developed in the cladding is regarded as an unwanted waste of electrical power that should be minimized. This condition is made more difficult by the fact that the electrical conductivity of the Ge-doped core of the fiber is higher than that of the pure-silica cladding. From a simplistic voltage-division point of view, if identical thicknesses of Ge-doped glass and pure-silica cladding are placed in series in a biased arrangement, most of the voltage develops across the cladding. The round cross section of the core and charge trapping at the core–cladding interface add to the difficulty to model the field distribution in the fiber adequately during and after poling.

The greatest problem in realizing a fully operational fiber modulator or switch is undoubtedly the relatively low nonlinearity achieved even after poling. Development of a way to induce a permanent nonlinear coefficient in silica glass comparable to the one in $LiNbO_3$ controlled way in a reproducible and would be enormously important. The most natural way to achieve this is to study glass properties and to exploit such knowledge. Special dopants can be used to enhance the nonlinearity via resonances, but so far this has been done only to a very limited extent. For example, the work in waveguides showed an anomalous $\chi^{(2)}$ following in-diffusion of silver ions, as discussed in the previous section. Likewise, it could be advantageous to make use of fibers that are not particularly "clean". It has been shown that although various types of fused quartz glasses (Homosil™, Optosil™, and Herosil™ that contain parts per million traces of Na) give the same SH level after poling, synthetic processed fused silica (such as Suprasil™), which has a purity comparable to that of fibers (Na content ≤ 0.1 ppm), gives a tenth of that level [125]. The extreme purity of fibers developed to minimize transmission loss over propagation lengths of kilometers is a disadvantage here, because the density of cations available is perhaps insufficient to build up a significant field after displacement by poling.

We leave the issue of enhancement of the nonlinearity as a nearly unexplored avenue, and concentrate on what has been achieved and reported in the literature. From the difficulties already discussed (particularly the issues of overlap integral and field distribution), it should come as no surprise that the optical nonlinearity induced in optical fibers ($\leq$0.2 pm/V) [195, 196] is generally lower than in bulk glass ($\sim$1 pm/V) [158].

Application of an electric field to a poled fiber affects the effective refractive index experienced by the propagating light. If the fiber is placed between crossed polarizers, this can lead to amplitude modulation and switching, because the system is anisotropic. Another way to accomplish amplitude modulation is to place the poled fiber in one arm of an interferometer. The effective change in index can be translated into a change of optical intensity, going constructive to destructive interference and vice versa. The most common configuration employed involves the use of a (MZI), as schematically represented in Figure 15. The difference in optical path between the two branches determines whether the output state is high (constructive interference) or low (destructive). When the poled fiber is biased, its effective index changes and the interference state can be switched or the output signal can be modulated. Such a scheme has been demonstrated by various groups [185, 197–199]. The major drawback of the fiber MZI is that although the two arms are independent to enable controlling the output, external perturbations also lead to a change in the output signal. Such perturbations can be of mechanical, acoustic, and thermal nature, for instance. Assuming the branches to be 1 m long and given $dn/dt = 11 \times 10^{-6}\ K^{-1}$ even a 0.1 degree difference in temperature between the two arms leads to a change

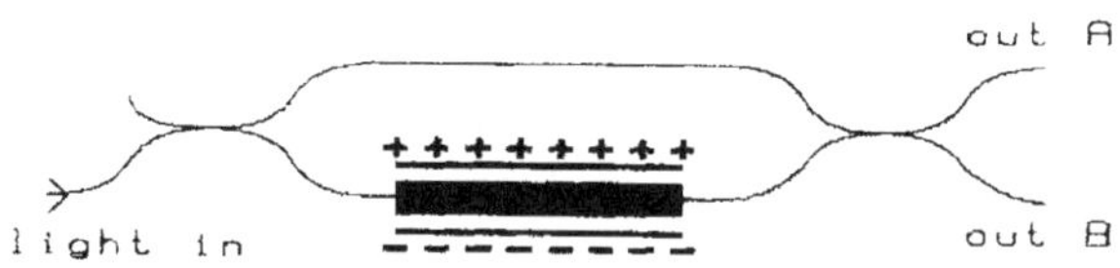

Figure 15. Scheme of an electro-optic modulator.

in optical path comparable to the optical wavelength. In practice, without careful packing, the output of the interferometer wanders with a time constant of ~1 s due to temperature effects alone, and noise and vibrations only make matters worse.

To tackle the thermal and mechanical instability problem, the two arms of the interferometer must be kept in close contact. This possibly can be accomplished by packaging two independent fibers together (e.g., by fiber ribbon techniques) or, alternatively, by making use of the two cores of a twin-core fiber as the interferometer [200]. In the latter case, there is no need for two fiber couplers to close the MZI. However, the liberty of having two independent output ports is lost, because the twin cores are too close to each other to allow for splicing two output fibers.

Various approaches have been tested to bring the anode to close contact with the core. These include D fibers [154, 197], standard fibers etched to the core and then mounted for poling [201], and fibers with holes on either side of the core to accommodate internal electrodes [202, 203] similar to those employed in [100].

D fibers can be pulled from an appropriate preform. Although the resulting fiber is uniform, splicing and handling are very difficult. The fiber tends to curl over itself, and it is neither straightforward to cleave it nor to splice it to a standard cylindrical fiber. Alternatively, standard fibers can be side-polished over a short length so that the polished section gains the D profile [204]. In this case, cleaving and splicing become straightforward, but polished fibers have microcracks and become very brittle. Furthermore, the core–air distance is nonuniform along the polished region, leading to poorer reproducibility. One additional problem with D fibers is the ~60 μm thick cladding underneath the core. During fabrication it is possible to ensure that the core is only approximately 1 μm away from the flat face of the fiber where the poling anode is placed. This ensures that the depletion region overlaps the core of the fiber. As explained before, however, after poling a large part of the voltage bias for the fiber modulator develops in the cladding region, making it necessary to etch or to further polish the fiber, which is a delicate and expensive operations. Nevertheless, two groups independently reported thermal poling of D-shaped fibers in 1994 [195, 196], and modulators were constructed and optimized [154, 197].

Long et al. [195] members of the research group at the University of New Mexico in Albuquerque, poled a 6 cm long piece of polarization-maintaining D fiber with core diameter ~1.25 μm and 9 μm core–air thickness. The thickness of the cladding was reduced to 6 μm by etching with HF. This piece of fiber was then assembled on electrodes and poled at 275 °C and 5 kV for 13 h. The poled fiber was placed in a Mach–Zehnder interferometer and the linear electro-optic coefficient was measured to be $r = 0.05$ pm/V ($\chi^{(2)} = 0.1$ pm/V). An improved arrangement was later developed, where the fiber was surrounded in polyimide for mounting, electrical isolation, and mechanical rigidity [154]. After a second polish to a final thickness of 34 μm, a nonlinear coefficient of 0.3 pm/V was inferred, even assuming perfect overlap between the optical and electrical fields. The fiber modulator thus constructed had a V_π as low as 75 V for a 12 cm long device, and the induced nonlinearity decayed only 10% over a 3 month period at 90 °C [205].

In 1994, the research group at the Optical Research Center in Southampton demonstrated a second-order nonlinearity as high as $\chi^{(2)} \sim 0.2$ pm/V (up to 0.5 pm/V, depending on the assumed value of the overlap integral) [196]. Conventional fibers were wheel polished and placed on the top of a silica substrate. The device was then assembled between two electrodes with the anode on the polished fiber surface. Thermal poling was performed at 4.3 kV and 280 °C for 15 min. Experiments also were carried out with electron beam poling by irradiating a side-polished fiber with a 40 kV beam at 0.3 nA current, and the estimated induced $\chi^{(2)}$ was 0.1 pm/V. To improve the poor reproducibility of the induced nonlinearity in thermal poling, this group carried out poling experiments in vacuum (1.2×10^{-5} mbar). Kazansky et al. [145] then found that the reproducibility and quality of the induced second-order susceptibility was improved significantly.

An alternative approach to making a fiber modulators–switch involves using pieces of standard telecommunication fiber, the central part of which is etched to the core ad subsequently poled [201]. After etching, the fibers have a diameter of ≤ 10 μm and therefore are referred to as microfibers. Microfibers are surprisingly strong for their diameter and are very uniform. Also, they can be fabricated from any conventional (cylindrical) fiber, the two ends that are not etched are easily spliceable, and the applied voltage during poling and in later use of the modulator develops mainly in the core, ensuring a large overlap integral. The difficulty with this approach is preventing electrical breakdown during poling with high voltages and developing handling procedures so that reproducible performance is achieved. Also, so far the schemes reported with microfibers (and etching in general) involve a large amount of handicraft, and devices poled in this way are inherently expensive. Poling was carried out at a typical temperature of 200–250 °C. A thin layer of a silicon rubber helped with both the mechanical stability of the system and the electrical insulation during and after poling. A charging resistor that consisted of a 3 mm thick piece of soda lime glass was used to circumvent the breakdown problem. After poling, the charging resistor was removed from the electrical circuit. The anode and cathode were pieces of silicon, and the poled microfiber was 5 cm long. From the SH signal it was estimated that the induced second-order nonlinear coefficient was ~0.2 pm/V. The fiber was modulated by 50 V square pulses to a frequency of 5 MHz, and the risetime of the detected signals was 200 ns. No resonances were encountered, suggesting that the modulation observed was indeed induced electro-optically.

The third and perhaps most promising approach demonstrated for making a fiber electro-optical modulator utilizes fibers with internal electrodes. To fabricate internal electrode fibers, holes can be drilled in the preforms. As demonstrated by the research group from Sydney University in Australia, by polishing the fibers from the sides, access can be gained to the holes so that electrical contact can be established with the electrodes while still enabling conventional splicing with the fiber ends. The electric fields that can be applied in internal electrode fibers are extremely high (~10^8 V/m [160]) because the electrodes are buried in glass. As mentioned in 1986, Li et al. reported a 30 m long fiber Kerr modulator that had liquid metal electrodes. Over the years, the technique for injecting liquid electrodes was replaced by using very thin wires that were manually inserted into the holes. This fabrication has severe disadvantages in terms of reproducibility (the field is not uniform along the device during and after poling, and devices made in this way are unequal) and the delicate operation of inserting the wires prevents the manufacture of devices that are long and cheap. More recently, the molten metal technique was reinvented, and low melting temperature alloys (solid at room temperature) were used as internal electrodes [200]. By using pressure and melted metals, handling is minimized, which paves the way for cheap component manufacture. Also, because the entire hole can be filled, reproducible performance is ensured.

Various groups have fabricated devices using internal electrode fibers, above all in an attempt to reproduce the exciting reports by Fujiwara et al. as described in detail in Section 4. Fujiwara [202] described thermal poling of the same fiber used in the original UV poling experiment. A 3.3 cm long fiber modulator was reported that had a switching voltage of 68 V for the transverse electric (TE) mode and 151 V for the transverse magnetic (TM) mode at 633 nm. This switching voltage corresponds to a V_π of 224 and 498 V/cm respectively, so that a TTL compatible 5 V switch would need a length of 45 cm. The fiber was poled at high temperature (250 °C) and large field, but at relatively low bias voltage (800 V). Even assuming perfect overlap between the region of recorded field and the optical field in the fiber, the nonlinear optical coefficient calculated was 0.84 pm/V for the TE propagation mode, which is a very large value. These results are to be compared with a previously reported switching voltage of 210 V/cm after poling another piece of this fiber for 60 min at room temperature [198].

A similar fiber was poled by the group at Lille University [203] using either thermal or UV excitation (Section 4). The nonlinear coefficients induced there were orders of magnitude lower, and the sign of the nonlinearity induced by UV poling was opposite to that of thermal poling. It can be speculated that both the sign reversal and the increase in current after poling result from imperfect contact of the thin wire inside the fiber with the walls of the holes.

A large number of experiments were carried out on fibers with internal electrodes and poling by Wong, Xu, Blazkievich, and Fleming. They studied the recording of the linear electro-optic effect in real time and described the process as being dominated by a competition of three fields. Following application of the external electric field E_{ext}, charges migrate and a shielding field develops in the opposite direction to E_{ext}. As the depletion region is formed, charge emission leads to the creation of an ionizing field. Although the dynamics of the processes observed with these three fields can be explained, so far he shielding field and the ionizing field model has not been used extensively by other groups. Xu et al. [206] compared negative with positive thermal poling by using a twin-hole fiber with the core placed closer to the negative or to the positive electrode, respectively. Their results corroborates the space–charge electric field model. The Australian group are, in fact, the only ones who identified (by etching) unambiguously the creation of depletion region in a thermally poled fiber. They also used the knowledge gained from their experiments to design new fibers with potentially improved performance for poling [181]. For example, a ring structure of B-doped silica was introduced as an electron trap, while Er doping was exploited and studied as a source of electrons. Er-doping was found to enhance the poling current. The B-doping ring increased the lifetime of poled devices to a few years rather than hundreds of days.

As mentioned in Section 3.4, a highly relevant observation in poled fibers was the reported increase in the value of $\chi^{(3)}$ [185].

Another issue that should be examined is the time response. The (weak) nonlinearity of silica glass is very fast, and various examples exist of the material responding to electromagnetic perturbations within 50 fs. Therefore, poled fibers should be able to respond intrinsically to electric fields within a fraction of a picosecond. Of course, other considerations come in when a real device is being contemplated, such as resistance, capacitance, inductance, and loss caused by radiation and the skin effect. As already discussed, many useful functions potentially can be performed by poled fibers, ranging from a variable attenuator to an analog multigigahertz amplitude modulator. To exploit such applications at high speeds, a travelling wave bias arrangement has to be used, because the length of the poled fibers envisioned is several centimeters and the propagation time becomes comparable with the response times desired. A travelling wave arrangement has been discussed [200]. The problems that are faced moles dispersion between the optical and electrical signals (O–E dispersion), dispersion of the electrical pulses, and loss of electrical power and bandwidth due to the contacts and the electrodes. It was found that by placing an internal electrode fiber in proximity to a substrate of appropriate dielectric constant, it was possible to minimize the O–E dispersion to a few picoseconds. It also was found that the dispersion of electrical pulses could be tolerable for pulses as short as 100 ps. The most serious problem for the transmission line arrangement is loss. For typical cross sections (~40 μm diameter) the electrodes present a series resistance of several hundreds of ohms per meter of fiber. If the impedance of the source is 50 ω, the resistance of the wires is intolerably high. For poled fiber devices that operate at lower speeds (~100 ns or slower) for switching, for instance, travelling wave arrangements may not be required, and resistive loss becomes less of a problem.

Electrical loss in long lengths of fiber with internal electrodes is accompanied by optical loss. Although this issue has not been discussed in the literature so far, the proximity of the electrodes to the core of the fiber is bound to attenuate the optical signal. The nonsymmetrical geometry of the fiber and the effect of poling also are bound to lead to unwanted polarization-dependent loss.

All these difficulties with poled fiber devices are to be added to both short and long term stability considerations that have to be addressed. Packaging is a main issue that so far has not been discussed in the literature. Summing up, the commercialization of poled fiber devices, and, in particular, of fiber switches and modulators still has several hindrances to overcome.

3.6. Poling of Fibers—Wavelength Conversion and Electro-optically Tuned Gratings

We saw in Section 2.5 that, in the case of optical poling, efficient second-harmonic generation in fibers is obtained by a quasi-phase-matching scheme, where the necessary periodic grating naturally builds up during the preparation process. It is instinctive to try to imitate nature and write periodic gratings by hand using the much more efficient process of thermal poling. In addition to the much higher values of nonlinear coefficients, appropriate gratings can be built to convert any desired laser wavelength. Several researchers have succeeded in creating efficient frequency-doubling devices in this way [97, 101, 102, 207–210].

The application of a periodic transverse field to fibers with the intent of poling in conjunction with Ar^+ laser light at 514 nm was reported by Kashyap [102]. His experiment led to the permanent preparation of a fiber that could be probed at a wavelength determined by the electrode spacing. Fine tuning was done by rotating the interdigital arrangement in relation to the fiber, so that the apparent electrode spacing was altered. Two-photon absorption at 514 nm provided an excitation, and the orientation was provided by the dc field. Kashyap et al. [207] also periodically poled 2 mm of fused silica to generate a phase-matched second harmonic at 1.064 μm. Poling was performed at 250 °C at 5 kV for 2 h using periodic electrodes prepared by lithography. With this setup, a second-order nonlinearity equivalent to $d_{11}/200$ of quartz was induced.

By using D-shaped fibers with photolithographically engraved electrodes on the flat surface, Kazansky et al. [211] thermally poled optical fibers in vacuum to generate blue light by CW quasi-phase-matched frequency doubling. A CW Ti:sapphire laser was used as a reading beam for the frequency doubling. A pitch of ~28 μm was introduced by a 20 μm periodic anode with a 45° tilt to guarantee quasi-phase-matching. The measured bandwidth of SH generation in the poled silica fiber was 0.78 nm. The maximum blue radiation detected was 20 pW for ~200 mW IR pump radiation in the core. The low conversion efficiency of this early experiment attests to the fact that the devices were not optimized. Indeed, the Southampton group later was able to demonstrate conversion efficiencies in excess of 20% in periodically poled fibers [212] with high power probing pulses. D-shaped germanosilicate fibers were used with periodic electrodes defined by standard photolithography. The gratings were up to 75 mm long and were fabricated for quasi-phase-matched frequency doubling of 1532 nm nanosecond pulses from a high-power erbium-doped fiber amplifier. Average second-harmonic power as high as 6.8 mW and peak power greater than 1.2 kW at 766 nm were generated, with average and peak conversion efficiencies as high as 21 and 30%, respectively

One interesting application of poled fibers was demonstrated by Srinivasan and Jain [213], who exploited the second-order nonlinearity induced by poling in the fiber to create an electro-optical tunable fiber Bragg grating filter. Hydrogen-loaded D-shaped fibers were used to fabricate the Bragg grating using 193 nm radiation from an ArF excimer laser. The fiber was subsequently thermally poled. A small spectral shift of 10 pm, measured in the reflection peak of the grating with the application of external voltage to the device, was attributed to the electro-optical effect. Thus, some controlled frequency doubling can be achieved in thermally poled fibers. Note that for typical values achieved in thermal poling of fibers ($\chi^{(2)} \leq 0.1$ pm/V, $\mathbf{E}_{ext} \leq 10^7$ V/m), the refractive index change induced externally is only $\leq 10^{-6}$, and for a device operating at 1550 nm, the expected shift in the peak wavelength of the grating is only a few picometers. Perhaps this is too little to lead to practical applications, except when extremely narrow linewidths are used (as in DFB fiber lasers). On the other hand, very sharp cutoff gratings can be fabricated, and even a small wavelength shift can lead to very large changes in the transmission of the grating near the spectral edge. The combination of poling (by UV excitation) and fiber gratings already has been discussed [214], and large transmission changes of about 0.01 nm near the edges of the gratings can be achieved. UV poling was performed by using a pulsed ArF excimer laser (193 nm), as described in detail in the subsequent text. A phase mask was used to write a grating at 1.53 μm while a voltage was applied across the internal electrodes. The transmission spectra for various applied voltages across the electrodes showed a Bragg wavelength shift proportional to the applied electric field [214].

4. UV POLING

In 1994, Fujiwara et al. [198] employed a fiber with two holes on either side of the core (similar to that used in early optical poling experiments [100]) and introduced the technique of UV-assisted poling. The replacement of blue or green light by UV radiation for the creation of defects in glass fibers previously has been found to be extremely successful. It greatly accelerated the creation of a second-order nonlinearity in optical poling [71] and opened the possibility to make Bragg gratings in fibers by side exposure [79, 215]. An ArF laser was used as a source of 193 nm wavelength radiation. Here again, very high electric fields ($>8 \times 10^7$ V/m) could be applied to the core, with the insertion of 25 μm wires into the holes for biasing. After 10 min irradiation at a density of 36 mJ/cm^2 at 10 Hz, the nonlinearity reached the saturated value. For polarization parallel to the direction of the applied field, the electro-optical coefficient induced was 5.8 pm/V (and 1.7 pm/V for the orthogonal polarization). This figure is much higher than achieved by thermal poling and is comparable to the r_{22} nonlinear coefficient of $LiNbO_3$. As a consequence of the high nonlinear coefficient induced, Fujiwara et al. [198] reported a 1.8 cm long fiber modulator that needed only 18 V to induce a π phase shift, so that a 5 V switch would required a length of only 6.5 cm, which is perfectly feasible.

The product $V_\pi \times L = 32.4$ V/cm at 5 MHz was measured to be nearly the same as that found for dc (34 V/cm). The relevance of the comparison of the performance of the device at dc and at 5 MHz is that acoustic resonances can increase the sensitivity of fiber modulators significantly at particular frequencies via the electrostrictive effect. In electrostriction, the electric field induces a stress in the material, leading to a deformation and a change in refractive index via the photoelastic effect. Such an effect was quantified with care by Liu et al. [184]. Their findings indicated that the electrostrictive contribution to the nonlinearity of poled silica near a resonance can be as large as 28 pm/V. The resonance frequency expected for a 200 μm diameter fiber as used by Fujiwara et al., assuming the sound velocity to be 5.96×10^3 m/s in silica, is close to 30 MHz, and this frequency is far from the 5 MHz at which the modulator was tested. The presence of 40 μm air-filled holes, however, should introduce a resonance at 8.5 MHz, a frequency lying close to the 5 MHz where the measurements were performed. The comparison with the response of the modulator at dc is therefore useful.

Observe that Fujiwara et al. poled the fiber with a very high electric field but relatively low voltage (800 V). As previously remarked, in thermal poling, the voltage is more important than the field, and in bulk silica glass, the nonlinear coefficients induced at <1 kV are generally insufficient for practical use. This is per se a surprising observation. These investigators also reported that applying just 800 V dc reduced the product $V_\pi \times L$ to only 210 V/cm after 60 min of poling at *room temperature*. This is a remarkably low voltage: a 10 cm long device would require only 21 V for switching without ever being exposed to heat or UV. Also, it is unlikely that glass can change permanently with the application of dc at room temperature, and such an effect may have been caused by a gradual displacement of the wires in the holes of the fiber. Anyway, the mechanisms here seem to differ from those in thermal poling, but what makes this fiber so special is unknown. Attempts by the same group to manufacture identical fibers after 1995 failed. Could modal interference be playing a role, because the fiber was tested with He-Ne laser light and the fiber was not single mode at 633 nm? Could contamination of the preform when the holes were drilled have affected the resulting fiber? Could microcracks and unusually high stress of the preform have resulted in an exceptionally highly photosensitive fiber? The bulk experiments, summarized in the following text, seem to confirm the original reports of very high effective $\chi^{(2)}$ induced by UV poling. Perhaps just as happened with the discovery of photosensitivity by Hill et al. [75], it is not always possible to repeat the exact conditions under which the original experiments were performed, because the critical parameters are not always identified. Renewed confirmation of these results would be very welcome.

Poling heavily Ge-doped silica fiber preforms under UV excitation also has been described. The technique used consisted of preparing slabs 1 cm × 1 cm × 1 mm from VAD fiber preforms. The glass studied had 15.7 mol% GeO_2 and 84.3 mol% SiO_2. Grooves were cut into the samples so that they could be subjected to a transverse poling field as high as 30 kV/mm. An ArF laser (193 nm) was used as the excitation source and 10^4 shots of energy density 100 mJ/cm^2 and duration ~20 ns were fired at 10 Hz. Probing the SH signal was carried out with a Nd:YAG laser at 1.064 μm [216–219]. Although SH was measured to quantify the second-order nonlinearity induced, third-harmonic generation was used as a means to determine an eventual change in the $\chi^{(3)}$ of the samples exposed.

The incidence angle was adjusted in Maker fringe pattern measurements, where extremely well behaved data were obtained [216]. The nonlinear coefficients d_{33} and d_{31} for orthogonal polarizations (s and p, respectively, for the fundamental; s for the SH) were 3.4 and 1.1 pm/V, and the exact relationship 3:1 allows assignment of the nonlinearity to the dc electric field through $\chi^{(2)}_{\mathrm{eff}} = 3\chi^{(3)}\mathbf{E}_{\mathrm{dc}}$. The large value of the nonlinearity induced in the bulk sample (that exceeds d_{22} in $LiNbO_3$) is of great practical interest. It also was found that although the ordinary refractive index for the fundamental radiation (p polarization) increased with poling field, the extraordinary index (s polarization) was found to decrease, evidencing the fact that poling breaks the symmetry of the material. The penetration depth of the UV in these experiments is ~0.22 mm, that is, much greater than the coherence length (~30 μm). The perfect fit of the theory to the data indicates that poling occurs in the whole exposed thickness, not only in a thin depletion region under the anode.

Studies of the centers involved in UV-excited poling were carried out. The difference in optical absorption spectra measured for various poling fields revealed that the concentration of defects Ge(1), Ge(2), NOMV (neutral oxygen monovacancy center), and Ge E' depend on the applied field. The conclusion was drawn that Ge E' centers are converted from the electron trap Ge(1) and (2) centers, and that the conversion rate is independent of applied electric field [217]. In studies carried out at temperatures ranging from 300 to 600 °C, a thermal equilibrium reaction was found between NOMV centers and the Ge E' center [218]. The decay of the second-order nonlinearity induced by UV excitation was studied [219], where each poled sample was subjected to an elevated temperature (in the range 200–500 °C) for a period of time sufficient for the SHG efficiency at normal incidence to drop at least 2 orders of magnitude. Unlike the defect concentration of samples exposed to UV but no high voltage, in UV-poled samples, all decays measured followed an exact single time constant exponential, one for each temperature. An Arrhenius plot of the decay time constants gave activation energy 0.41 ± 0.05 eV, and the time constant measured for a temperature 15 °C was (only) 287 days [219]. This energy correlates well with that associated to the Ge E' center. Clearly, the rapid decay of UV poling would be a problem in commercial devices. Further studies were carried out by the same research group, where doping was used as a means to trap electrons and stabilize the induced second-order optical nonlinearity. It was reported that various dopants, and in particular H_2 worked well as electron scavanges. With such doping, the predicted lifetime of poled devices increased to approximately 7 years [220].

One of the most remarkable results reported in the context of UV poling is crystallization. Under the presence of high electric fields and UV exposure, some thin film samples of GeO_2-SiO_2 composition showed the appearance of a crystalline phase. The microcrystals had a typical size of 10–50 μm. The crystalline phase was not unambiguously determined, but the crystals had enhanced Ge concentration as compared to the rest of the sample. Studies by third-harmonic generation showed that $\chi^{(3)}$ in the neighborhood of the crystals was up to 15 times larger than that of unpoled glass [221]. Whether this increase was due to matrix stress around the crystals or stress in the crystals themselves was not reported. We note, however, that the transmission of light through the crystals is very limited due to the reflection on its facets, and it is unlikely that any third-harmonic radiation generated inside the crystals would ever reach the detector. Although this consideration was never put forward by the authors of the work, we suggest that this indicates that the enhancement of $\chi^{(3)}$ comes from the stressed glass matrix. Another remarkable fact is that the crystallites as well as the enhanced value of $\chi^{(3)}$ remain stable even after the quick decay of second-order nonlinearity due to a thermal treatment that erases the dc field [222]. This indicated that the effective $\chi^{(2)}$ is not due to some inherent second-order nonlinearity of the crystallites.

To sum up, the high nonlinearity induced in Ge-doped glass by UV poling was attributed to the recording of a dc field and to a large increase in the $\chi^{(3)}$ of the glass via $3\chi^{(3)}\mathbf{E}_{dc} \equiv \chi^{(2)}_{eff}$. Although the dc field was built by the displacement and organization of charge associated with the creation of Ge E' defect centers, the large $\chi^{(3)}$ was associated to the creation of microcrystals and stress in the glass matrix. The crucial problem with UV poling has been the poor reproducibility of the results. An extension of this work to include UV poling of H-loaded thin films improved the decay of SHG [223].

5. POLING OF POLYMER WAVEGUIDES

The idea to create second-order nonlinearity artificially by some kind of poling process is not restricted to silica glass materials. Thermal poling of polymers has a large range of applications, and poled polymer waveguides are widely used in integrated optics. In fact, both kinds of processes—thermal poling and optical poling—have been used to make nonlinear optical polymers. Polymeric nonlinear optical materials offer the potential for low switching voltages. Also, large bandwidth electro-optic devices have been built that exploit the relatively small velocity mismatch between radiofrequency and optical signal in these materials with traveling wave configurations [224, 225]. For example, the refractive index at $\lambda = 1.3$ μm of methacrylate polymer with an N-methylaminonitrostilbene side chain is $n = 1.65$ and the dielectric constant at radiofrequency is $\varepsilon = 2.6$, which gives a mismatch of $\Delta n = \sqrt{\varepsilon} - n = 0.038$. Modulators with 40 GHz of 3 dB electrical bandwidth with very low distortion have been demonstrated with switching voltages of only a few volts [224]. Electro-optic coefficients of ~100 pm/V have been obtained [226]. During the last two decades, considerable research work has been devoted to understanding poling of polymer waveguides [224–233].

The polymer materials suitable for thermal poling consist of optically nonlinear groups chemically attached to a polymer backbone. The nonlinear part usually contains a chemical group that has exceptionally large molecular second-order polarizability and a permanent dipole moment. The optical waveguides are fabricated by spin coating layers that form the cladding and guiding regions, and lithography. The optical nonlinearity generally is obtained by corona poling by applying high (10–200 V/μm) electric fields to the sample at the glass transition temperature of the polymer. The high temperature is then switched off and the electric field is maintained until the sample has cooled down. Thus, from the operational point of view, thermal poling of polymers is quite similar to silica glass poling. However, the physics is different, because the nonlinearity obtained in silica is due to mainly static frozen fields due to ion migration, where the nonlinearity of polymers is due mainly to the orientation of polar molecules. In principle, the physics behind thermal poling of polymers is simpler than for silica, because the large hyperpolarizability of the molecules guarantees a large effective $\chi^{(2)}$ after poling. Thermally poled polymers have been used in a large number of demonstrations of electro-optic devices such as modulators, grating assisted codirectional coupler filters [230], and high bandwidth Mach–Zehnder modulators [224, 225]. The performance of more recently reported devices is indeed impressive (e.g., 100 GHz response at <2 V switching). However, problems exist with polymer waveguides, in addition to the intrinsic disadvantages in comparison with silica, such as splicing limitations, lower damage threshold, and instability. Usually the poling of such materials enhances the optical losses. Teng et al. [229] showed that these losses are due to light scattering at inhomogeneities that increases with the poling process. The greatest problem that afflicts poled polymer components is the limited lifetime of devices based on such materials. The large molecules and low processing temperatures generally lead to a decay in performance over the years, in a field where component specifications often need to be met for a 20 year span. The lifetime issue has hindered the industry form embracing this technology to any large extent.

Optical poling of nonlinear polymer material also has been demonstrated [231–233]. Again the operational aspect is similar to the case of optical poling of germanium-doped silica glass, but the physical process is different. The poling is obtained by simultaneous exposure of the material to a fundamental and a copropagating second-harmonic beam. The polar oscillation of the electric field is capable of switching molecules between different configurations or orientation-dependent modifications (orientational hole burning). Interestingly, the dipole orientation models of second-harmonic generation on glass optical fibers can be applied here with success. The second-order nonlinearity induced is generally short lived even at room temperature.

6. FUTURE WORK AND CONCLUSIONS

It has been known for more than a decade that it is possible to record a permanent second-order nonlinearity in glass systems. Optical poling has been complemented by thermal and by UV poling, where heat and short wavelength radiation are used as excitation sources, and much higher nonlinearity is achieved. Poled devices such as fiber switches and wavelength converters (frequency doublers) have been demonstrated in the laboratory, but so far the technology has not reached the telecommunications and sensors markets.

All the literature available on second-order nonlinearities in optical fibers and other glass systems is written from a physics perspective or, at most, from a device demonstration point of view. So far, no articles have been published on the systems aspects of poled fibers or on the impact that fibers with a second-order nonlinearity could make on the optical network. Cheap, spliceable fiber modulators, switches, and wavelength converters with nanosecond response times could profoundly affect the optical layer. Scientific work on how such devices could help bring about optical packet switching and how poled fibers could be used as logical building blocks in the management of the network would be valuable to guide research, which so far has concentrated solely on more fundamental aspects. The literature is also incomplete when it comes to poled fiber device fabrication. Few articles, if any, address practical but crucial aspects of manufacture, such as packaging, reliability, mass production, and expected lifetime (aging of components).

Research is also needed on polarization issues and loss. Although fibers intrinsically can have low polarization-dependent loss, it is not obvious that the same applies to fibers with internal electrodes or to any fibers that are poled so that the transverse symmetry is broken. Likewise, fibers with a second-order nonlinearity are supposed to be the solution to the problem of coupling loss in semiconductor devices and ferroelectric crystals. Indeed, spliceability is preserved even in internal electrode fibers. However, splicing fibers with large holes to standard telecommunications fibers is an issue. Perhaps even more important is the question of the optical loss introduced by the electrodes. As explained in Section 3, the nonlinearity of (thermally) poled fibers is generally confined to a thin region neighboring the anode. To guarantee a strong interaction of the optical field with the depletion region, the fiber often is designed so that the core lies only a few micrometers away from the holes that will accommodate the anode electrode. It is obvious that the evanescent optical field reaches the electrode and causes loss. Therefore, a compromise needs to be found between the nonlinearity experienced by the light propagating in the device and the loss that can be tolerated, particularly for long devices. None of the articles that presented various demonstrations of fiber modulators and switches discussed the loss introduced.

A clear link between optical, thermal, and UV poling has not been established, and a multitude of explanations seem to be required to describe the process of recording a strong electric field in glass. Focusing on thermal poling, it is important to understand why poling sometimes takes place over a thickness that exceeds the depletion region, indicating that orientation of defects dominates over space–charge effects. This central question remains unanswered. Studies also are needed to clarify the role of various carriers (positive and negative), as well as of defects. The mechanisms of charge neutralization, the postpoling evolution of the field distribution, and even more straightforward optimization conditions (applied voltage, poling time, atmosphere, and temperature) also deserve more study.

From an applications perspective, the essential point is how large the induced nonlinearity can be. In the original 1991 article by Myers et al. on thermal poling, a nonlinearity of ~1 pm/V was reported. Since then, the highest nonlinearity level achieved is nearly unchanged, in spite of the effort by various research groups to increase the nonlinear coefficient. This limitation could be fundamental, depending on the breakdown field in silica and on the characteristic value of $\chi^{(3)}$ for the material. It is difficult to increase the breakdown limit of silica, but charge cancellation should be prevented so as to ensure that the field recorded by poling stays at that limit.

Apparently, the only way to significantly increase the second-order nonlinearity is to increase $\chi^{(3)}$. A number of articles have reported an increase in the third-order nonlinearity by thermal and UV poling, but at this stage there are no good explanations for why such an effect takes place. Perhaps an increase in the third-order nonlinearity by a factor of 2 or 3, as reported by the group from Sydney [206], should not be too surprising, because glass is no longer the original material with inversion symmetry after poling. The nearly integer factors measured could arise from some degeneracy being broken. However, the reported increase in the intrinsic third-order nonlinearity of the glass by a factor of 15 following UV poling is quite unexpected [221]. If confirmed, it must be associated with a dramatic change in the material properties, such as those caused by the phase change following crystallization. There are other ways to achieve enhanced third-order nonlinearity, as mentioned in Section 3. It is known that near a resonance, the material not only exhibits an absorption peak, but also an anomalous behavior of the refractive index. This has been exploited only to a limited extent with the in-diffusion of silver ions into a planar waveguide [189]. Systems doped with semiconductors, noble metals, and rare earths present other possible candidates to be examined. The contribution of material science experts here is greatly needed.

While researchers are struggling to increase the nonlinear coefficient of poled silica to above 1 pm/V, those working with polymers have poled materials with effective $\chi^{(2)}$ that is 2 orders of magnitude larger. Devices have been demonstrated that have a couple of centimeters that have $V_\pi <$ 2 V and frequency response close to 100 GHz. The question can then be raised as to whether poling silica is justified. From a nonlinear performance point of view, perhaps not. However, the advantages of silica glass over organic materials are striking. In terms of transparency, reproducibility, splicing possibility, and, above all, stability, silica-based fiber devices are much superior to polymer-based materials. The hard silica cladding protects the waveguide against the environment in a way that polymer waveguides cannot, and in addition to thermal effects, chemical degradation, radiation, and UV exposure all contribute to a short lifetime for poled polymer components. Perhaps we should not stop trying to improve the lifetime of polymer components, but it is likely that if the nonlinearity of silica components is shown to suffice, then silica fibers will play a major role in telecommunications applications.

It is likely that a true packaged device for use in telecommunications or for sensing will be achieved in the near future, but several problems have to be solved. From the long list of work that remains, both from a fundamental and applications points of view, it is clear that second-order nonlinear effects in fibers still represent a lively research field. The specialized literature on fibers with second-order nonlinearities has a void on system aspects, severe limitations on device considerations, and, even from a physics point of view, the process behind the creation of second-order nonlinearity in fibers still hides many secrets.

IT = Information Technology
KDP = Potassium dihydrogen phosphate
KTP = Potassium Titanyl Phosphate
BBO = β Barium Borate
DWDM = Dense Wavelength Division Multiplexing
DFB = Distributed Feedback
VAD = Vapor-phase axial deposition

ACKNOWLEDGMENTS

The elaboration of this chapter was made possible only by the experience acquired by the authors through collaborative work with a large number of colleagues in Brazil and in Sweden. The authors express their gratitude to all collaborators involved in the research work, and in particular to U. Österberg, P. M. P. Gouvea, M. M. Lacerda, E. N. Hering, F. C. Garcia, L. C. Guedes Valente, C. Mattos, D.C.Reigada, C. M. B. Cordeiro, Michele F. Moreira, Hélio R. Carvalho, Adriana L. C. Triques, and F. Laurell. The authors also thank the technical and secretarial staff at the Physics Department of PUC-Rio for their supporting effort. We gratefully acknowledge support for this work from Ericsson Telecomunicações S.A. Brazil, from the Brazilian agencies Conselho Nacional de Desenvolvimento Científico e Tecnológico (CNPq), Coordenação de Aperfeiçoamento de Pessoal de Nível Superior (CAPES), Programa de Apoio ao Desenvolvimento Científico e Tecnológico (PADCT), and Financiadora de Estudos e Projetos (FINEP), Fundação de Amparo à Pesquisa do Estado do Rio de Janeiro (FAPERJ), from NEDO International Joint Research Program (Japan), and from IRECO (Sweden). Partial financial support from the European Project GLAMOROUS also is gratefully acknowledged. We thank I. S. Bott, M. S. Malheiros, R. R. Avillez, and M. F. S. Lopez for the use of facilities.

REFERENCES

1. N. Bloembergen, "Non-linear Optics," Addison–Wesley, Reading, MA, 1992.
2. G. P. Agrawal, "Nonlinear Fiber Optics," Academic Press, San Diego, 1995.
3. H. S. Nalwa and S. Miyata, "Nonlinear Optics of Organic Molecules and Polymers," CRC Press, Boca Raton, FL, 1997.
4. S. S. Viana, *Opt. Commun.* 116, 67 (1995).
5. M. C. Farries, P. St. J. Russel, M. E. Fermann, and D. N. Payne, *Electron. Lett.* 23, 322 (1987).

6. E. M. Dianov, A. M. Prokhorov, V. O. Solokov, and V. B. Sulimov, *JETP Lett.* 50, 13 (1989).
7. R. H. Stolen, "Second Harmonic Generation in Optical Fibers, Nonlinear Waves in Solid State Physics," Plenum Press, New York, 1990.
8. V. Mizrahi and J. E. Sipe, *Opt. Photon.* 2, 16 (1991).
9. E. M. Dianov, P. G. Kazansky, and D. Yu. Stepanov, *Proc. SPIE* 1516, 81 (1991).
10. E. M. Dianov and D. S. Starodubov, *Opt. Fiber Technol.* 1, 3 (1994).
11. Y. Fujii, B. S. Kawasaki, K. O. Hill, and D. C. Johnson, *Opt. Lett.* 5, 48 (1980).
12. Y. Ohmori and Y. Sasaki, *Appl. Phys. Lett.* 39, 466 (1981).
13. Y. Ohmori and Y. Sasaki, *IEEE J. Quantum Electron.* 18, 758 (1982).
14. U. Österberg and W. Margulis, "Proceedings of the Optical Society of America Conference on Lasers and Electro-Optics," San Francisco, 1986, p. 102.
15. U. Österberg and W. Margulis, *Opt. Lett.* 11, 517 (1986).
16. M. C. Farries, *Laser Focus World* 10, 12 (1988).
17. A. S. L. Gomes, private communication.
18. D. M. Kroll and J. R. Simpson, "Proceedings of the Optical Society of America Quantum Electronics Laser Science Conference," Baltimore, 1991, post-deadline paper.
19. D. M. Kroll and J. R. Simpson, *Opt. Lett.* 16, 1650 (1991).
20. N. M. Lawandy, J. P. Bernardin, G. Demouchy, and R. L. MacDonald, *Electron. Lett.* 27, 1264 (1991).
21. T. J. Driscoll, N. M. Lawandy, A. Killian, L. Rienhart, and T. F. Morse, *Electron. Lett.* 27, 1729 (1991).
22. I. B. Couceiro, I. C. S. Carvalho, W. Margulis, T. Carlnäss, and A. Lidgard, *Opt. Mater.* 3, 217 (1994).
23. E. M. Dianov, L. S. Kornienko, A. O. Rybaltolvsky, P. V. Chernov, and Yu P. Yatsenko, *Opt. Lett.* 19, 439 (1994).
24. W. Margulis and U. Österberg, *Proc. SPIE* 864, 180 (1987).
25. B. Valk, E. M. Kim, and M. M. Salour, *Appl. Phys. Lett.* 51, 722 (1987).
26. M. C. Farries and M. E. Fermann, *Electronic Lett.* 24, 294 (1987).
27. U. Österberg and W. Margulis, *IEEE J. Quantum Electron.* 24, 2127 (1988).
28. W. Margulis and U. Österberg, *J. Opt. Soc. Am. B* 5, 312 (1988).
29. R. H. Stolen and H. W. K. Tom, *Opt. Lett.* 12, 585 (1987).
30. E. M. Dianov, P. G. Kazansky, and D. S. Starodubov, *Sov. Lightwave Commun.* 3, 247 (1993).
31. A. Krotkus and W. Margulis, *Appl. Phys. Lett.* 52, 1942 (1988).
32. A. Kamal, R. W. Terhune, and D. A. Weinberger, *Proc. SPIE* 1516, 137 (1991).
33. V. Dominic and J. Feinberg, *Opt. Lett.* **17**, 1761 (1992).
34. U. Österberg and W. Margulis, *Opt. Lett.* 12, 57 (1987).
35. M. D. Selker and N. M. Lawandy, *Opt. Commun.* 77, 339 (1990).
36. M. D. Selker and N. M. Lawandy, *Electron. Lett.* 25, 1440 (1989).
37. Y. Hibino, V. Mizrahi, and G. I. Stegeman, *Electron. Lett.* 26, 1578 (1990).
38. B. Batdorf, C. Krautschik, U. Österberg, G. Stegeman, J. W. Leitch, and T. F. Morse, *Opt. Commun.* 73, 393 (1989).
39. W. Margulis, I. C. S. Carvalho, P. M. P. Gouvea, and B. Lesche, *Opt. Lett.* 18, 1016 (1993).
40. M. M. Lacerda, I. C. S. Carvalho, W. Margulis, and B. Lesche, *Electron. Lett.* 30, 732 (1994).
41. M. M. Lacerda, I. C. S. Carvalho, W. Margulis, and B. Lesche, *Opt. Lett.* 19, 948 (1994).
42. S. Sandgren, B. Sahlgren, A. Asseh, W. Margulis, F. Laurell, R. Stubbe, and A. Lidgard, *Electron. Lett.* 31, 665 (1995).
43. H. W. K. Tom, R. H. Stolen, G. D. Aumiller, and W. Pleibel, *Opt. Lett.* 13, 512 (1988).
44. A. Kamal, D. A. Weinberger, and W. H. Weber, *Opt. Lett.* 15, 613 (1990).
45. W. Margulis, F. Laurell, and B. Lesche, *Nature* 378, 699 (1995).
46. R. W. Terhune and D. A. Weinberger, *J. Opt. Soci. Am. B* 4, 661 (1987).
47. F. Payne, *Electrons Lett.* 23, 1215 (1987).
48. U. Österberg, R. I. Lawconnell, F. J. Seiler, L. A. Brambani, C. G. Askins, and E. J. Friebele, *Opt. Lett.* 16, 132 (1991).
49. F. Ouellette, K. O. Hill, and D. C. Johnson, *Opt. Lett.* 13, 515 (1988).
50. I. C. S. Carvalho, W. Margulis, and B. Lesche, *Electrons Lett.* 27, 1497 (1991).
51. A. Kamal, R. W. Terhune, and D. A. Weinberger, *Proc. SPIE* 1516, 137 (1992).
52. Y. Hibino, V. Mizrahi, G. I. Stegeman, and S. Sudo, *Appl. Phys. Lett.* 57, 656 (1988).
53. D. L. Griscom, E. J. Friebele, K. J. Long, and J. W. Fleming, *J. Appl. Phys.* 54, 3743 (1983).
54. I. C. S. Carvalho, P. M. P. Gouvea, W. Margulis, J. P. von der Weid, and B. Lesche, "Proceedings of the Optical Society of America Quantum Electronics Laser Science Conferences, Baltimore, 1991, Vol. 11, p. 38.
55. P. M. P. Gouvêa and W. Margulis, *J. Opt. Soc. Am. B* 11, 1515 (1994).
56. T. J. Driscoll and N. M. Lawandy, *J. Opt. Soc. Am. B* 11, 355 (1994).
57. T. J. Driscoll and N. M. Lawandy, OSA "Technical Digest, Annual Meeting," Optical Society of America Albuquerque, New Mexico, 1992, Vol. 23, p. 50.
58. E. M. Dianov, P. G. Kazansky, and D. Yu. Stepanov, *Sov. J. Quantum Electron.* 19, 575 (1989).
59. T. E. Tsai and D. L. Griscom, *Proc. SPIE* 1516, 14 (1991).
60. P. S. Russel, L. J. Poyntz-Wright, and D. P. Hand, M. J. Digonnet, Ed. *Proc. SPIE* 1373, 126 (1991).
61. M. A. Saifi and M. J. Andrejco, *Opt. Lett.* 13, 773 (1988).
62. J. R. Taylor, "Workshop on SHG in Optical Fibers," Imperial College London, 1987 (on published).
63. M. Fokine and W. Margulis, "Proceedings of the Optical Society of America Conference of Bragg Gratings, Photosensitivity, and Poling in Glass Waveguides," Stuart, Florida, 2000, p. 385.
64. P. J. Lemaire, R. M. Atkins, V. Mizrahi, and W. A. Reed, *Electron. Lett.* 29, 1191 (1993).
65. M. Fokine and W. Margulis, *Opt. Lett.* 25, 302 (2000).
66. T. E. Tsai, M. A. Saifi, E. J. Friebele, D. L. Griscom, and U. Österberg, *Opt. Lett.* 14, 1023 (1989).
67. L. N. Skuja, A. N. Trukhin, and A. E. Plaudis, *Phys. Status Solidi a* 84, k153 (1984).
68. I. A. Bufetov, M. V. Grekov, K. M. Golant, E. M. Dianov, and R. R. Khapko, *Opt. Lett.* 22, 1394 (1997).
69. L. J. Poyntz-Wright, M. E. Ferman, and P. St. J. Russel, *Opt. Lett.* 13, 1023 (1988).
70. K. D. Simmons, S. LaRochelle, V. Mizrahi, G. I. Stegeman, and D. L. Griscom, *Opt. Lett.* 16, 141 (1991).
71. I. C. S. Carvalho, W. Margulis, and B. Lesche, *Opt. Lett.* 16, 1487 (1991).
72. N. M. Lawandy, T. J. Driscoll, and C. L. Adler, *IEE Proc. J.* 139, 133 (1992).
73. E. M. Dianov, P. G. Kazansky, A. M. Prokhorov, D. S. Starodubov, and D. Y. Stepanov, *Sov. Lightwave Commun.* 2, 147 (1992).
74. T. J. Driscoll and N. M. Lawandy, *Opt. Lett.* 17, 571 (1992).
75. K. O. Hill, Y. Fujii, D. C. Johnson, and B. S. Kawasaki, *Appl. Phys. Lett.* 32, 647 (1978).
76. K. O. Hill, Y. Fujii, D. C. Johnson, and B. S. Kawasaki, *Opt. Lett.* 3, 66 (1978).
77. D. K. W. Lam and B. K. Garside, *Appl. Opt.* 20, 440 (1981).
78. R. J. Campbell and R. Kashyap, *Opt. Lett.* 16, 898 (1991).
79. G. Meltz, W. W. Morey, and W. H. Glenn, *Opt. Lett.* 14, 823 (1989).
80. R. Kashyap, "Fiber Bragg gratings," Academic Press, San Diego, 1999.
81. A. Othonos and K. Kalli, "Fiber Bragg Gratings: Fundamentals and Applications in Telecommunications and Sensing," Artech House, Boston, 1999.
82. V. Mizrahi, Y. Hibino, and G. Stegeman, *Opt. Commun.* 78, 283 (1990).
83. F. Ouellette, *Opt. Lett.* 14, 89 (1989).
84. F. Ouellette, *Electron. Lett.* 26, 740 (1990).

85. W. Margulis, I. C. S. Carvalho, and J. P. von der Weid, *Opt. Lett.* 14, 700 (1989).
86. W. Margulis, I. C. S. Carvalho, and B. Lesche, *Proc. SPIE* 1516, 60 (1991).
87. K. Koch and G. T. Moore, *Opt. Lett.* 16, 1436 (1991).
88. Yu. E. Kapitzky and B. Ya. Zel'dovich, *Opt. Commun.* 78, 227 (1990).
89. P. Chmela, J. Petrácek, and K. Gniadek, *Opt. Quantum Electron.* 30, 271 (1998).
90. E. M. Dianov and D. S Starodubov, *Opt. Fiber Technol.* 1, 3 (1994).
91. E. M. Dianov, P. G. Kasansky, D. S. Starodubov, and D. Yu Stepanov, *Sov. Lightwave Commun.* 1, 395 (1991).
92. B. Ehrlich-Holl, D. M. Krol, R. H. Stolen, and H. W. K. Tom, *Opt. Lett.* 17, 396 (1992).
93. D. M. Krol, M. M. Broer, K. T. Nelson, R. H. Stolen, H. W. K. Tom, and W. Pleibel, *Opt. Lett.* 16, 211 (1991).
94. V. Mizrahi, U. Österberg, J. E. Sipe, and G. I. Stegeman, *Opt. Lett.* 13, 279 (1988).
95. V. Mizrahi, U. Österberg, C. Krautschik, G. I. Stegeman, J. E. Sipe, and T. F. Morse, *Appl. Phys. Lett.* 53, 557 (1988).
96. V. Mizrahi, U. Österberg, C. Krautschik, G. I. Stegeman, J. E. Sipe, and T. F. Morse, *Appl. Phys. Lett.* 54, 2277 (1989).
97. P. G. Kazansky and V. Pruneri, *J. Opt. Soc. Am. B* 14, 3170 (1997).
98. P. G. Kazansky and V. Pruneri, *Phys. Rev. Lett.* 78, 2956 (1997).
99. M. E. Fermann, L. Li, M. C. Farries, and D. N. Payne, *IEE Conf. Proc.* 292, 135 (1988).
100. M. V. Bergot, M. C. Farries, M. E. Fermann, L. Li, L. J. Poyntz-Wright, P. St. J. Russel, and A. Smithson, *Opt. Lett.* 13, 592 (1988).
101. R. Kashyap, *J. Opt. Soc. Am. B* 6, 313 (1989).
102. R. Kashyap, *Appl. Phys. Lett.* 58, 1233 (1991).
103. Q. Liu, F. Gan, X. Zhao, K. Tanaka, A. Narazaki, and K. Hirao, *Opt. Lett.* 26, 1347 (2001).
104. H. Zeghlachè, G. Martinelli, Y. Quiquempois, P. Bernage, J. Troles, and F. Smektala, "Proceedings of the Optical Society of America Conference on Bragg Gratings, Photosensitivity, and Poling in Glass Waveguides," Stresa, Italy, 2001, p. BThC37-1.
105. H. G. Chatellus, S. Montant, and F. Freysz, "Proceedings of the Optical Society of America Conference on Bragg Gratings, Photosensitivity, and Poling in Glass Waveguides," Stresa, Italy, 2001, p. BFA2-1.
106. E. V. Anoikin, E. M. Dianov, P. G. Kazansky, and D. Yu. Stepanov, *Opt. Lett.* 15, 834 (1990).
107. F. Ouellette, K. O. Hill, and D. C. Johnson, *Appl. Phys. Lett.* 54, 1086 (1989).
108. N. M. Lawandy and R. L. MacDonald, *J. Non-Cryst. Solids* 141, 47 (1992).
109. R. L. MacDonald and N. M. Lawandy, *J. Opt. Soc. Am. B* 8, 1307 (1991).
110. R. L. MacDonald, T. J. Driscoll, and N. M. Lawandy, *Electron. Lett.* 27, 1717 (1991).
111. P. Chmela, *J. Mod. Opt.* 37, 327 (1990).
112. P. Chmela, *Opt. Lett.* 16, 443 (1991).
113. B. Lesche, *J. Opt. Soc. Am. B* 7, 53 (1990).
114. B. Lesche, *Proc. SPIE* 1516, 125 (1991).
115. N. B. Baranova and B. Ya. Zel'dovich, *JETP Lett.* 45, 717 (1987).
116. M. Dianov, P. G. Kazansky, A. M. Prokhorov, and D. Yu. Stepanov, *Mol. Cryst. Liquid Sci. Technol. B* 3, 329 (1992).
117. A. Kamal, M. L. Stock, A. Szpak, C. H. Thomas, D. A. Weinberger, M. Fankel, J. Ness, K. Ozaki, and J. Valdmanis, "Optical Society of America Annual Meeting," Boston, 1990, Postdeadline paper PDP25.
118. N. M. Lawandy, *Opt. Commun.* 74, 180 (1989).
119. N. M. Lawandy, *Phys. Rev. Lett.* 65, 1745 (1990).
120. D. Z. Anderson, V. Mizrahi, and J. E. Sipe, *Proc. SPIE* 1516, 154 (1991).
121. D. Z. Anderson, V. Mizrahi, and J. E. Sipe, *Opt. Lett.* 16, 796 (1991).
122. P. Chmela, *Opt. Lett.* 13, 669 (1988).
123. Y. Chen, *Appl. Phys. Lett.* 54, 1195 (1989).
124. R. Kashyap, B. J. Ainslie, and G. D. Maxwell, *Electron. Lett.* 25, 206 (1989).
125. R. A. Myers, N. Mukherjee, and S. R. J. Brueck, *Opt. Lett.* 16, 1732 (1991).
126. S. Gray, *Philos. Trans. Roy. Soc. London, SEPE A* 37, 285 (1732).
127. G. M. Sessler, "Topics in Applied Physics 33: Electrets," Springer-Verlag, Berlin, 1980.
128. J. R. Freeman, H. P. Kallmann, and M. Silver, *Rev. Mod. Phys.* 33, 553 (1961).
129. D. E. Carlson, K. W. Hang, and G. F. Stockdale, *J. Am. Ceram. Soc.* 55, 337 (1972).
130. D. E. Carlson, *J. Am. Ceram. Soc.* 57, 291 (1974).
131. D. E. Carlson, *J. Am. Ceram. Soc.* 57, 461 (1974).
132. D. E. Carlson, K. W. Hang, and G. F. Stockdale, *J. Am. Ceram. Soc.* 57, 295 (1974).
133. D. Faccio, V. Pruneri, and P. G. Kazansky, "Proceedings of the Optical Society of America Conference on Bragg Gratings, Photosensitivity, and Poling in Glass Waveguides," Stresa, Italy, 2001, p. BFA3-1.
134. D. Faccio, V. Pruneri, and P. G. Kazansky, *Appl. Phys. Lett.* 79, 2687 (2001).
135. P. G. Kazansky and P. St. J. Russel, *Opt. Commun.* 110, 611 (1994).
136. T. Alley and S. R. J. Brueck, *J. Appl. Phys.* 86, 6634 (1999).
137. F. C. Garcia, I. C. S. Carvalho, E. N. Hering, W. Margulis, and B. Lesche, *Appl. Phys. Lett.* 72, 3252 (1998).
138. A. Okada, K. Ishii, K. Mito, and K. Sasaki, *Appl. Phys. Lett.* 60, 2852 (1992).
139. S. Horinouchi, H. Imai, G. J. Zhang, K. Mito, and K. Sasaki, *Appl. Phys. Lett.* 68, 3552 (1996).
140. H. Imai, S. Horinouchi, Y. Uchida, H. Yamasaki, K. Fukao, G. Zhang, T. Kinoshita, K. Mito, H. Hirashima, and K. Sasaki, *J. Non-Cryst. Solids* 196, 63 (1996).
141. K. Tanaka, K. Kashima, K. Hirao, N. Soga, A. Mito, and H. Nasu, *Jpn. J. Appl. Phys.* 32, L843 (1993).
142. P. G. Kazansky, A. Kamal, and P. St. J. Russel, *Opt. Lett.* 18, 1141 (1993).
143. L. J. Henry, B. V. McGrath, T. G. Alley, and J. J. Kester, *J. Opt. Soc. Am. B* 13, 827 (1994).
144. P. G. Kazansky, A. Kamal, and P. St. J. Russel, *Opt. Lett.* 18, 693 (1993).
145. P. G. Kazansky, L. Dong, and P. St. J. Russel, *Electron. Lett.* 30, 1345 (1994).
146. N. Mukherjee, R. A. Myers, and S. R. J. Brueck, *J. Opt. Soc. Am. B* 11, 665 (1994).
147. T. G. Alley, S. R. J. Brueck, and R. A. Myers, *J. Non-Cryst. Solids* 242, 165 (1998).
148. U. K. Krieger and W. A. Lanford, *J. Non-Cryst. Soilds* 102, 50 (1988).
149. F. L. Freire, Jr., C. M. Lepiensky, and C. A Achete, *Nucl. Instrum. Methods Phys. Res. B* 68, 227 (1992).
150. H. Takebe, P. G. Kasansky, P. St. J. Russel, and K. Morinaga, *Opt. Lett.* 21, 468 (1996).
151. S. R. J. Brueck, "Optical Society of America Topical Meeting on Photosensitivity and Quadratic Nonlinearity in Glass Waveguides," Portland CR, 1995.
152. R. A. Myers, N. Mukherjee, and S. R. Brueck, *Proc. SPIE* 2044, 2 (1993).
153. A. C. Liu, D. Pureur, M. J. F. Digonnet, G. S. Kino, and E. J. Knystautas, "Proceedings of the Optical Society of America Conference on Bragg Gratings, Photosensitivity and Poling in Glass Fibers and Waveguides: Fundamentals and Applications," Williamsburg, VH, 1997, p. 302.
154. X. C. Long, R. A. Myers, and S. R. J. Brueck, *J. Lightwave Technol.* 8, 277 (1996).
155. A. L. C. Triques, M. F. Moreira, H. R. Carvalho, I. C. S. Carvalho, B. Lesche, and W. Margulis, "Proceedings of the Optical Society of America Conference on Bragg Gratings, Photosensitivity, and Poling in Glass Waveguides," Stresa, Italy, 2001, p. BThC34-1.
156. C. J. S. Matos, I. C. S. Carvalho, E. F. da Silveira, W. Margulis, R. R. Pinho, and B. Lesche, *J. Non-Cryst. Solids* 273, 25 (2000).

157. M. Qui, S. Egawa, K. Horimoto, and T. Mizunami, *Opt. Commun.* 189, 161 (2001).
158. X. C. Long, R. A. Myers, and S. R. J. Brueck, *Opt. Lett.* 19, 1819 (1994).
159. M. Abe, T. Kitagawa, K. Hattori, A. Himeno, and Y. Ohmori, *Electron. Lett.* 32, 893 (1996).
160. K. Tanaka, K. Kashima, K. Hirao, N. Soga, S. Yamagata, A. Mito, and H. Nasu, *Jpn. J. Appl. Phys.* 34, 175 (1995).
161. V. Pruneri, P. G. Kazansky, D. Hewak, J. Wang, H. Takebe, and D. N. Payne, *Appl. Phys. Lett.* 70, 155 (1997).
162. Y. Quiquempois, A. Villeneuve, D. Dam, K. Turcotte, J. Maier, G. Stegeman, and S. Lacroix, *Electron. Lett.* 36, 733 (2000).
163. M. Qui, F. Pi, G. Orriols, and M. Bibiche, *J. Opt. Soc. Am. B* 15, 1362 (1998).
164. H. Nasu, H. Okamoto, A. Mito, J. Matsuoka, and K. Kamyia, *Jpn. J. Appl. Phys.* 32, L406 (1993).
165. H. Nasu, H. Okamoto, K. Kurashi, J. Matsuoka, K. Kamyia, A. Mito, and H. Hosono, *J. Opt. Soc. Am. B* 12, 644 (1995).
166. L. J. Henry, A. D. DeVilbiss, and T. E. Tsai, *J. Opt. Soc. Am.* 12, 2037 (1995).
167. A. Le Calvez, E. Freysz, and A. Ducasse, *Opt. Lett.* 22, 1547 (1997).
168. L. J. Henry, *Opt. Lett.* 20, 1592 (1995).
169. J. Arentoft, K. Pedersen, S. I. Bozhevolnyi, M. Kristensen, and P. Yu, *Trends Opt. Photon.* 33, 397 (1999).
170. D. Pureur, A. C. Liu, M. J. F. Digonnet, and G. S. Kino, *Opt. Lett.* 23, 588 (1998).
171. D. Faccio, V. Pruneri, and P. G. Kazansky, *Opt. Lett.* 25, 1376 (2000).
172. P. G. Kazansky, A. R. Smith, P. St. J. Russel, G. M. Yang, and G. M. Sessler, *Appl. Phys. Lett.* 68, 269 (1996).
173. P. Regnier, Y. Serruys, and A. Zemskoff, *Phys. Chem. Glasses* 27, 185 (1986).
174. W. Margulis and F. Laurell, *Opt. Lett.* 21, 1786 (1996).
175. C. D. Zuiker, D. M. Gruen, and A. R. Krauss, *Mater. Res. Soc. Bull.* 20, 29 (1995).
176. C. M. B. Cordeiro, C. Borges, L. C. G. Valente, I. C. S. Carvalho, B. Lesche, and W. Margulis, *J. Non-Cryst. Solids* 247, 183 (1999).
177. B. Lesche, F. C. Garcia, E. N. Hering, W. Margulis, I. C. S. Carvalho, and F. Laurell, *Phys. Rev. Lett.* 78, 2172 (1997).
178. A. L. C. Triques, C. M. B. Cordeiro, V. Balestrieri, B. Lesche, W. Margulis, and I. C. S. Carvalho, *Appl. Phys. Lett.* 76, 2496 (2000).
179. T. G. Alley and S. R. J. Brueck, *Opt. Lett.* 23, 1170 (1998).
180. D. Wong, W. Xu, and S. Fleming, *Trends Opt. Photon.* 29, 278 (1998).
181. P. Blazkiewicz, W. Xu, D. Wong, S. Fleming, and T. Ryan, *J. Lightwave Technol.* 19, 1149 (2001).
182. H. Guillet de Chatelus, S. Montant, and E. Freysz, *Opt. Lett.* 25, 1723 (2000).
183. C. J. Markmann, G. Genty, Y. Ren, J. Arentoft, and M. Kristensen, "Proceedings of the Optical Society of America Conference on Bragg Gratings, Photosensitivity, and Poling in Glass Waveguides," Stresa, Italy, 2001 p. BFC3-1.
184. A. C. Liu, M. J. F. Digonnet, and G. S. Kino, *J. Opt. Soc. Am. B* 18, 187 (2001).
185. D. Wong, W. Xu, S. Fleming, M. Janos, and K.-M. Lo, *Opt. Fiber Technol.* 5, 235 (1999).
186. W. Xu, J. Arentoft, D. Wong, and S. Fleming, *IEEE Photon. Technol. Lett.* 11, 1265 (1999).
187. A. C. Liu, M. J. F. Digonnet, and G. S. Kino, *Opt. Lett.* 19, 466 (1994).
188. W. Margulis and F. Laurell, *Appl. Phys. Lett.* 71, 2418 (1997).
189. J. Arentoft, M. Kristensen, K. Pedersen, S. I. Bozhevolnyi, and P. Shi, *Electron. Lett.* 36, 1635 (2000).
190. J. Arentoft and M. Kirstensen, *Trends Opt. Photon.* 33, 388 (1999).
191. F. C. Garcia Polarização Eletrotérmica de Vidros, Fibras ópticas e Guias de Onda Planares, PhD Thesis, Pontifícia Universidade Católica do Rio de Janeiro, 2000.
192. R. Kashyap, F. C. Garcia, and L. Volgelaar, on published manuscript.
193. P. G. Kazansky, P. St. J. Russel, and H. Takebe, *J. Lightwave Technol.* 15, 1484 (1997).
194. L. Li, R. D. Birch, and D. N. Payne, "IEE Colloquium on Advanced Fibre Waveguide Devices," England, 1986, p. 10.
195. X. C. Long, R. A. Myers, and S. R. J. Brueck, *Electron. Lett.* 30, 2162 (1994).
196. P. G. Kazansky, L. Dong, and P. St. J. Russel, *Opt. Lett.* 19, 701 (1994).
197. P. G. Kasansky, P. St. J. Russel, L. Dong, and C. N. Pannell, *Electrton. Lett.* 31, 62 (1995).
198. T. Fujiwara, D. Wong, Y. Zhao, S. Fleming, S. Poole, and M. Sceats, *Electron. Lett.* 31, 573 (1995).
199. X. C. Long and S. R. J. Brueck, *IEEE Photon. Technol. Lett.* 9, 767 (1997).
200. M. Fokine, L. E. Nilsson, Å. Claesson, D. Berlemont, L. Kjellberg, L. Krummenacher, and W. Margulis, "Proceedings of the Optical Society of America Conference on Bragg Gratings, Photosensitivity, and Poling in Glass Fibers and Waveguides: Fundamentals and Applications," Stresa, Italy, 2001, postdeadline paper PD4.
201. I. C. S. Carvalho, C. M. B. Cordeiro, L. C. Guedes Valente, E. N. Hering, W. Margulis, F. Laurell, and R. Stubbe, "Proceedings of the Optical Society of America Conference on Bragg Gratings, Photosensitivity, and Poling in Glass Fibers and Waveguides: Applications and Fundamentals," Williamsburg, VA, 1997, OSA Technical Digest Series, Vol. 17, p. 313.
202. T. Fujiwara, *IEEE Photon. Technol. Lett.* 7, 1177 (1995).
203. Y. Quiquempois, G. Martinelli, P. Benage, M. Douay, P. Niay, E. Delevaque, H. Poignant, B. Loisel, and J. F. Bayon, "Proceedings of the Optical Society of America Conference on Bragg Gratings, Photosensitivity, and Poling in Glass Fibers and Waveguides: Applications and Fundamentals," Williamsburg, VA, 1997, OSA Technical digest Series, Vol. 17, p. 207.
204. C. D. Hussey and J. D. Minelly, *Electron. Lett.* 24, 805 (1988).
205. S. R. J. Brueck, T. G. Alley, R. A. Myers, and X. C. Long, "Proceedings of the Optical Society of America Conference on Bragg Gratings, Photosensitivity, and Poling in Glass Fibers and Waveguides: Applications and Fundamentals," Williamsburg, VA, 1997, OSA Technical Digest Series, Vol. 17, p. 84.
206. W. Xu, D. Wong, and S. Fleming, "Proceedings of the Optical Society of America Conference on Bragg Gratings, Photosensitivity, and Poling in Glass Fibers and Waveguides," Stuart, FL, 1999, p. 115.
207. R. Kashyap, G. J. Veldhuis, D. C. Rogers, and P. F. Mckee, *Appl. Phys. Lett.* 64, 1332 (1994).
208. V. Pruneri and P. G. Kazansky, *Electron. Lett.* 33, 318 (1997).
209. V. Pruneri, G. Bonfrate, P. G. Kazansky, C. Simonneau, P. Vidakovic, and J. A. Levenson, *Appl. Phys. Lett.* 72, 1007 (1998).
210. V. Pruneri and P. G. Kazansky, *IEEE Photon. Technol. Lett.* 9, 185 (1997).
211. P. G. Kazansky, V. Pruneri, and P. St. J. Russel, *Opt. Lett.* 20, 843 (1995).
212. V. Pruneri, G. Bonfrate, P. G. Kazansky, D. J. Richardson, N. G. Broderick, J. P. de Sandro, C. Simonneau, P. Vidakovic, and J. A. Levenson, *Opt. Lett.* 24, 208 (1999).
213. R. Srinivasan and R. K. Jain, *IEEE Photon. Technol. Lett.* 12, 170 (2000).
214. T. Fujiwara, D. Wong, Y. Zhao, S. Fleming, V. Grishina, and S. Poole, "Proceedings of OFC'95," 1995, postdeadline paper PD6-1.
215. J. Albert, *Mater. Res. Soc. Bull.* 23, 36 (1998).
216. T. Fujiwara, M. Takahashi, and A. J. Ikushima, *Appl. Phys. Lett.* 71, 1032 (1997).
217. M. Takahashi, T. Fujiwara, T. Kawachi, and A. J. Ikushima, *Appl. Phys. Lett.* 71, 993 (1997).
218. M. Takahashi, T. Fujiwara, T. Kawachi, and A. J. Ikushima,*Appl. Phys. Lett.* 72, 1287 (1998).
219. T. Fujiwara, M. Takahashi, and A. J. Ikushima, *Electron. Lett.* 33, 982 (1997).

220. J. Khaled, T. Fujiwara, and A. J. Ikushima, *Opt. Mater.* 17, 275 (2001).
221. T. Fujiwara, S. Matsumoto, M. Ohama, and A. J. Ikushima, *J. Non-Cryst. Solids* 273, 203 (2000).
222. S. Matsumoto, T. Fujiwara, Y. Seno, Y. Hirose, M. Ohama, and A. J. Ikushima, *J. Appl. Phys.* 88, 6993 (2000).
223. J. Khaled, T. Fujiwara, M. Ohama, and J. Ikushima, *J. Appl. Phys.* 87, 2137 (2000).
224. C. C. Teng, *Appl. Phys. Lett.* 60, 1538 (1992).
225. D. G. Girton, S. L. Kwiatkowski, G. L. Lipscomb, and R. S. Lytel, *Appl. Phys. Lett.* 58, 1730 (1991).
226. W. H. Steier, M.-C. Oh, H. Zhang, A. Szep, L. R. Dalton, C. Zhang, H. R. Fetterman, D. H. Chang, E. Erlig, and B. Tsap, "Optical Fiber Conference," Anaheim CA, 2001, paper MJ1-1.
227. M. A. Mortazavi, A. Knoesen, S. T. Kowel, R. A. Henry, J. M. Hoover, R. F. Kubin, and G. A. Linsay, *Appl. Phys. B* 53, 287 (1991).
228. K. D. Singer, M. G. Kuzyk, W. R. Holland, J. E. Sohn, S. L. Lalama, R. B. Comozzoli, H. E. Katz, and M. L. Schilling, *Appl. Phys. Lett.* 53, 1800 (1988).
229. C. C. Teng, M. A. Mortazavi, and G. K. Boudoughian, *Appl. Phys. Lett.* 66, 667 (1995).
230. S. W. Ahn and S. Y. Shin, *Opt. Commun.* 197, 289 (2001).
231. F. Charra, F. Kajzar, J. M. Nunzi, P. Raimond, and E. Idiart, *Opt. Lett.* 18, 941 (1993).
232. C. Fiorini, F. Charra, J. M. Nunzi, and P. Raimond, *Nonlinear Opt.* 9, 339 (1995).
233. X. Yu, X. Zhong, Qu Li, S. Luo, Y. Chen, Yu Sui, and J. Yin, *Opt. Lett.* 26, 220 (2001).

Index